Re-Examining the Hist
Econo

Jeffrey K. Hass
Editor

Re-Examining the History of the Russian Economy

A New Analytic Tool from Field Theory

palgrave
macmillan

Editor
Jeffrey K. Hass
Department of Sociology & Anthropology
University of Richmond
Richmond, VA, USA

Faculty of Economics, Department of Economic Theory
St. Petersburg State University
St. Petersburg, Russia

ISBN 978-3-030-09225-2 ISBN 978-3-319-75414-7 (eBook)
https://doi.org/10.1007/978-3-319-75414-7

Cover illustration: VIEW Pictures Ltd / Alamy Stock Photo

Printed on acid-free paper

This Palgrave Macmillan imprint is published by the registered company Springer International Publishing AG part of Springer Nature.
The registered company address is: Gewerbestrasse 11, 6330 Cham, Switzerland

To intellectual community across borders, and in the hopes of continuing Russian-American fellowship—intellectual, social, and personal

Acknowledgments

As editor, I would like to thank my colleagues at St. Petersburg State University for supporting the idea of this edited volume, for their patience in the long process of working out the themes and structure of the book as they evolved over time, and for the efforts they put into writing and editing their contributions, all the while putting up with various queries. I would also like to thank them for their collegiality in the years that they have let me be part of their department and for making me feel welcome in that home-away-from-home. I would like to thank the University of Richmond for the various forms of support over the years for research and conference travel to Russia. I would also like to thank St. Petersburg State University, especially Elena Chernova (Senior Vice-Rector for Economics), for giving me the opportunity to join the faculty. It has been my honor and pleasure to work alongside a diverse and talented group of academics. Further, our work together has been an inspiration on two fronts: that sociologists and economics can work side by side and have far more in common than sometimes seems the case (at least in the United States) and, more importantly, that in this strained day and age, Russians and Americans remain bound together by shared interests and values, which are not only academic. I would also like to thank Nikita Lomagin and Maksim Storchevoi for helping me become better acquainted with Russian academe and helping me navigate the process that has brought me there. I am grateful to the staff of Crossroads, Chainikoff, Sideriia, and Rare Olde Times for providing venues for productive writing. Finally, I thank my family for enduring yet another book.

Members of our authors' collective would like to express our thanks to St. Petersburg State University for the various forms of support that made not only this volume but also our general academic work possible. Additionally, Danila Raskov would also like to thank the Russian Foundation for Basic Research for the financial support that made possible his broader project on Nikolai Sieber's life and work.

Finally, we all would also like to express our heartfelt gratitude to Laura Pacey and Clara Heathcock, our editors at Palgrave Macmillan, for having faith in the project and moving it along smoothly from its initial inception to the finished product.

Contents

Notes on Contributors

Aleksandr Bartenev is an Associate Professor in the Department of Economic Theory in the Faculty of Economics at St. Petersburg State University. His research interests lie in the area of socioeconomic convergence, goal-setting processes, and financialization (theory and practical implementation).

Anton Leonidovich Dmitriev is a candidate in economic sciences and an Associate Professor in the Department of General Economic Theory and History of Economic Thought and in the Department of Economic Cybernetics in the Faculty of Economics at St. Petersburg State University. His interests include the history of economic thought and statistics in Russia, the history of methods and modeling in economics, and microeconomic analyses. He is the author of more than 180 works on these subjects.

Jeffrey K. Hass is an Associate Professor at the Department of Sociology and Anthropology at the University of Richmond (USA) and a part-time Professor at the Department of Economic Theory in the Faculty of Economics at St. Petersburg State University. In the course of visiting and working in St. Petersburg since 1991, he has written on post-socialism and economic sociology, including various articles and three books: *Economic Sociology* (2007); *Power, Culture, and Economic Change in Russia* (2011); and *Rethinking the Post-Soviet Experience* (2012). He continues to work on post-socialist economics and politics. He is also working on a major project about political economy, survival, and fields of power and practice in the Blockade of Leningrad, which has resulted in many publications and three books (in progress).

Maxim Markov is an Associate Professor in the Department of Economic Theory in the Faculty of Economics at St. Petersburg State University. His areas of interest are the economics of regional markets, labor markets, economies and power, and the history of economic thought. His chapter in this volume is part of a wider project on the 1905 Revolution and economics at St. Petersburg State University. He has published on these and related topics in a wide variety of journals.

Denis Melnik is an Associate Professor in the Department of Theoretical Economics at the National Research University Higher School of Economics (Moscow). His current research focuses on the history of Russian and Soviet economic thought and on theories of economic development. During recent years, he was a visiting scholar at the New School for Social Research (New York, USA) and Kanagawa University (Yokohama, Japan). He has published on this topic a range of work, including in such prestigious journals as *Slavic Review*.

Ainur Musaeva is a candidate in economics and senior lecturer in the Department of Economic Theory in the Faculty of Economics at St. Petersburg State University. She is the author of 14 scholarly papers and works in the areas of institutional economics, theory of property rights, contract theory, and intellectual property. In 2017 she was made a laureate of St. Petersburg State University "for educational and methodological work."

Aleksandr Protasov graduated from the Higher Military School of the Soviet Ministry of Defense in 1984, and in 1996 he graduated with honors from the Faculty of Economics at St. Petersburg State University, Department of Economic Theory. In 2000 he defended his candidate dissertation and since 2001 has been working at the Department of Economic Theory in the Faculty of Economics at St. Petersburg State University, where he is an Associate Professor and deputy chair of the department. He has published over 50 scholarly papers on inflation in Russia. His main areas of interest are methods of economic research, macroeconomic problems of Russia's development, cyclical dynamics of inflationary processes, and economic policy. He has been honored by the Ministry of Education and Science of the Russian Federation.

Danila Raskov is an Associate Professor at the Department of Economic Theory and Head of the Center for the Study of Economic Culture at St. Petersburg State University. His research interests cover the history of Russian economic thought (populism, Marxism), economic history and institutions of heterodox religious communities, and economic methodology. In 2012 he published *Economic Institutions of Old Believers*. He is currently working on two projects: Nikolai Sieber's heritage in classical political economy and rhetoric of institutional economics.

Mikhail Rumiantsev is a Doctor of Economics and Professor in the Department of Economic Theory in the Faculty of Economics at St. Petersburg State University, as well as a member of the Academy of the Philosophy of Economics and of the philosophical and economic Scholarly Assembly of Moscow State University. He is a deputy editor-in-chief of the journal *Problems of Modern Economics* and author of more than 150 papers. His scholarly interests include economic philosophy, economy and religion, economic and social change, institutional economics, and innovative economies.

Svetlana Rumyantseva is a candidate in economics and an Associate Professor in the Department of Economic Theory in the Faculty of Economics at St. Petersburg State University. She is the author of more than 100 papers and a specialist in the theory of economic cycles (especially Kondratieff cycles), innovation and innovation politics, evolutionary economics, and methodology. She is a First Prize Laureate of St. Petersburg State University (2005) and a winner of the N. D. Kondratieff bronze medal (2017).

Viktor Ryazanov is a Doctor of Economics and a Professor in the Department of Economic Theory at St. Petersburg State University. From 1989 to 1994, he was the Dean of the Faculty of Economics at St. Petersburg State University, and since 1995 he has been the chair of the Department of Economic Theory. He is the author of over 250 scholarly papers, including eight monographs, and much of his work has been translated into English, German, Chinese, and Polish. His major work includes: *Ekonomicheskoe razvitie Rossii. Reformy i rossiiskoe khoziaistvo v XIX–XX vv* (1998); *Khoziaistvennyi stroi Rossii na puti k drugoi ekonomiki* (2009); *(Ne)realnyi kapitalizm. Politicheskii krizis i ego posledstviia dlia mirovogo khoziaistva i Rossii* (2016). He has been honored for his professional work in the Russian Federation.

Aleksandr Shevelev has been affiliated with the Faculty of Economics at St. Petersburg State University since 1990, as student and now as faculty in the Department of Economic Theory. His interests include institutions and institutional economics, institutional analyses of social processes, and contemporary political economy. He has written more than 50 works in these areas published in such journals as *Problemy sovremennoi ekonomiki* and *Filosofiia khoziaistvo*. Lately he has turned his attention to developing field theory.

Leonid Shirokorad is a Doctor of Economic Sciences and a Professor in the Department of Economic Theory in the Faculty of Economics at St. Petersburg State University. His interests include the history of Russian economic thought, Russian economic history, and methodological issues in economic theory. He is the author of more than 100 works, including two books on the political economy of socialism and many articles on political economy and economic history, some of which were published in English, German, and Chinese journals. He has also received several awards for his service to the discipline.

List of Figures

List of Tables

1

Fields in Russian Economic History

Jeffrey K. Hass

This volume resulted from the collective endeavor of an American sociologist and Russian economists. Despite being in different disciplines, we all share more than one might expect. In a departure from all-too-often interdisciplinary struggles and sniping, we all know the same literature from both disciplines and share a healthy appreciation, and skepticism, for the social sciences. In particular, two shared interests made this volume possible. The first is Russia's economy—past, present, and future—which drives our research and teaching. Each chapter reflects one facet of our myriad interests in how Russia's economy has operated, changed, evolved, or broken down. The general, abstract theory of economics and sociology comes alive, warts and all, when we engage that theory with Russia's reality across the centuries. And we all share the conviction that

J. K. Hass (✉)
Department of Sociology & Anthropology, University of Richmond,
Richmond, VA, USA

Faculty of Economics, Department of Economic Theory,
St. Petersburg State University, St. Petersburg, Russia

J. K. Hass (ed.), *Re-Examining the History of the Russian Economy*,
https://doi.org/10.1007/978-3-319-75414-7_1

Russia's stories have something important to add to human knowledge about economics, just as economics and sociology have something to say about Russia's experiences.

This brings us to a second shared interest: possible insights and potential of field theory. Hass had been working with variations of field theory since his days in Princeton's graduate program in sociology in the 1990s, where he learned from one of field theory's founders, Paul DiMaggio. His dissertation and first project involved fields, *habitus*, and economic practices in Russia's post-socialist experience. Members of the Department of Economic Theory of the Faculty of Economics at St. Petersburg State University also knew the work of Pierre Bourdieu, and they had been applying various facets of that framework in their work on religion and economy, Old Believers, the historical roots of Russian economic theory, inflation, and so on. We all found it odd that Russia was absent from the general corpus of field scholarship (but we mention a few exceptions later). In discussions over a few years, we concluded that it was natural to combine our interests, knowledge, and efforts to address this oversight, both to expand the horizons of field theory and to open up Russia's rich economic history to a new analytic approach—one that retained insights of political economy and appreciated of culture in a non-ad hoc manner. This serendipitous congruence of interests and cross-disciplinary knowledge bred this project.

One of our goals is to bring Russia and fields into closer proximity to see how theory and reality can inform each other. Some of us have written explicitly on fields and Russia, and others' scholarship has come close enough to these two topics that it was natural for them to take the next step of engaging Russia and fields. We have also not been content with the existing state of affairs in much field theory. Neoinstitutionalist field theory, usually employed to organizational analyses, has worked well for cases of stable capitalist economies (especially the United States), but we suspect that this has also needlessly narrowed the possible territory a field framework could cover. For example, the usual field theory tends to focus on organizational and institutional fields, privileging one particular level of analysis (the "meso"). While we will also engage the meso level here, some authors also point to multiple levels of fields intersecting: for example, fields of interpersonal networks nested inside organizations and institutions still might have their own dynamics as a community of actors oriented to particular rules of engagement and goals of strategies and practices. And inasmuch as institutions can create or shape networks by bringing actors into proximity, network-based fields can in turn affect higher-level institutional fields if those actors have positions or luck to propagate particular rules or ideologies.

Expanding Field Theory, with Help from Russia's Economic Reality

Field frameworks were embraced in the natural sciences, but in the social sciences and humanities, the pace of development has been slower. Psychologists in the Gestalt tradition grounded their theory in fields of perception, and Max Weber's sociology has a field logic running through it (Martin 2011). Much social science continues to focus on correlations between actors as bundles of traits essential to that actor (years of education, "gender," employment) when trying to explain tastes or consumption, careers, wages, and so forth. Studies of policies and development are little different, except actors are not individuals but institutions and the explicandum is economic structure, productivity, or growth. Much scholarly progress was made in this logic of analysis, yet there were limits, such as the persistence of seemingly irrational behavior or the failure of regression to norms (whether economic policies and structures, or everyday practices).

Enter field theory, which made strides after the appearance primarily of Paul DiMaggio and Walter Powell's (1983) famous article in *American Sociological Review*, Pierre Bourdieu's (1984) study of tastes (*Distinction*), and Neil Fligstein's (1990) analysis of corporate strategies and structures. In the 1990s, this spawned follow-up studies and helped generate work in political sociology (cf. Fligstein and McAdam 2012). As field theory seemed to reach a lull, Fligstein and McAdam (2012) and John Martin (2011), drawing on different theoretical traditions, tried to expand the foundations and scope of field theory. Martin has proposed less of a framework than notes toward a framework, and in this work, we draw primarily on neoinstitutionalism and Bourdieu. In the neoinstitutional framework (DiMaggio and Powell 1983, 1991; Fligstein 1990, 2001), which draws in part on Bourdieu and Max Weber, Anthony Giddens'

(1984) structuration theory, and on other strands of structuralist work (in particular, ideas of structural equivalence Harrison White), neoinstitutionalists posit that organizations operate in fields of other organizations, all of whom share an accepted affinity and similar structural locations. Usual economic theory claims that organizations orient primarily to consumers (who make up markets that firms need to service to survive), with competitors and regulatory states shaping the context in which firms address consumers' wants and needs. While neoinstitutionalists do not deny the importance of consumers and markets, they do not elevate them above states and communities of firms. Rather, organizational elites pay close attention to others in their particular field for ideas to common problems and for legitimacy, and to state and other powerful players (such as financial organizations) that wield legal or dependency power (Roy 1997). Organizational elites then adopt strategies and structures that conform to those of leaders in their field so as to retain legitimacy and a sense that they are playing by accepted rules. These three forms of influence are *isomorphic mechanisms* (DiMaggio and Powell 1983): mimetic, normative, and coercive.

Neoinstitutionalist field theory is primarily structural and meso-level: actors themselves are buffeted by isomorphic forces within fields. Bourdieu, however, adds actors to his framework to make sense of how collective practices and structures are continuously reproduced in the first place, especially when not everyone in a field is a winner. An important facet of a field framework is how actors are conceptualized. Rather than being a bundle of "preferences" for consumption or gain that a rational agent seeks to maximize, we have actors that are intersections of various relations, which in turn shape that actor's broader dispositions and knowledge. The usual instrumental approach of microeconomics and much political science, unfortunately, misses two important facets of economic practice: the source of actors' own preferences and "tool kits" of perceptions and strategic responses, and emergent properties of institutional systems and actors. So, Bourdieu's schema begins with *habitus*: crudely put, an individual's structured knowledge and how to use that knowledge. This constrains and enables how one interprets and responds to the world. A second leg of Bourdieu's framework is *capital*, existing resources actors deploy: social (e.g. networks and reputation), economic (money or shares), cultural (tastes and behavioral skills), and symbolic (status symbols, such as credentials). (One could add institutional capital, i.e. formal access to formal rules and organizations.) Bourdieu suggests that how actors use capital depends, first, on *habitus*—do they know how to gain and use capital—and on the rules of the concrete institutional context that govern the status and use of said capital. This brings in the third leg of Bourdieu's framework, *fields*, which are arrangements of actors and rules. (Bourdieu uses two metaphors to illustrate what he means: a magnetic field orienting actors in a particular way, e.g. categories and strategies of action, and a field of battle with actors arrayed in alliances and confrontations.) Behavior in the field is governed by *doxa*, taken-for-granted rules of entry into and engagement within the field.

These three entities interact in Bourdieu's framework. Field location shapes an actor's *habitus* and capital, the first from experience and the second from rules of resource access. Actors internalize field rules, and resulting *habitus* influences how they judge and respond to contexts (opportunities, threats, etc.), although *habitus* does not overwhelm individual agency. This suggests that post-socialist economic change has not been only competing elites and interests. Rather, it has been competing assumptions and knowledge of how a "normal" economy operates, and conflict over ritualizing and normalizing these assumptions—and primacy of particular knowledge, *habitus*, and capital—in organized fields of property and governance. This suggests that economic organization is not merely the evolutionary emergence of efficient means for producing, trading, and making profit. Rather, economic organization is the institutionalization of norms and logics of what constitutes a "normal" economy. Actors compete and struggle to defend and enforce what they consider to be the ultimate meaning of economic action, which acts as a measuring rod for the status and legitimacy of economic tactics and relations. Victors in such struggles impose their versions of normality via laws, organizational structures and procedures, and arrangements of property ownership. In this regard, post-socialist economic change has been no different than the emergence of capitalism or state socialism. To better understand the post-socialist process, we must broaden our vision, beyond usual political economy of immediate interests of state and business elites, to logics of economic action—logics inculcated in their biographies and manifest in *habitus* of knowledge, strategies, and practices.

In its neoinstitutional version and to an extent in its Bourdieusian version, fields do not replace institutions or structures. Rather, they add a dimension of *real practice*. In new institutional economics and much political economy, "institutions" are formal rules and procedures that shape costs and benefits of action; in the sociological understanding, known institutions are also categories and schemas of action, position, and identity. However, these notions are too broad; nearly anything routine, it seems, can be an "institution." Separating the rule facet from other components would make it easier to pin down what institutions are and allow us to make better sense of how they operate. (The same goes for "structures.") Here, fields can do much of the heavy lifting. Structures and institutions—say, corporate schemas and personal networks, and rules and relations of labor markets and professions—position individuals in particular experiences, shaping *habitus*. However, actors don't follow rules and relations mechanistically. They calculate—but this involves *how others treat and employ* institutions and structures.[1] Actors monitor each other—but in orienting to each other and to rules and structures, they have created a field of organized striving and practice.[2] Note that Bourdieu allowed for politics of fields, as actors maneuver to use rules for gain, defense, and so on.

Field theory does not suggest humans are automatons programmed by positions or *habitus*. Fields can compel action into a particular form and orientation, and limits of *habitus* (knowledge and dispositions) can channel what we think we are capable of or desire to do. However, nothing here means we can or should dismiss politics and contingencies. Field rules and *habitus* also provide tools as well as constraints. Agency is variable, depending on knowledge, capital, and field position. As Fligstein (2001) noted, institutional entrepreneurs can use capital and social skills to act strategically within institutional fields and attempt to alter field rules of status and practice. This means that field theory can build on insights of usual political economy: power, calculations, and contingencies matter. However, they are embedded not in single, isolated institutions (or in sets of institutions), but rather in fields of multiple and (imperfectly) interconnected institutions with emergent properties (the field). Thus, field theory promises to add dimensions and correct potential oversights of political economy (e.g. assuming the rational actor, or taking microfoundations of actors' motives and perceptions into consideration at all).

For all the exciting insights field theory can provide, the framework still requires more development and refinement. Certainly, Russia's post-Soviet history provides much food for thought. Russia's post-socialist *doxa* unraveled with radical reforms and emergence of new agents developing new claims and acting on new interests. Owners and oligarchs wanted economic capital to be triumphant. Managers preferred social capital and technical knowledge to be hegemonic. State officials preferred that institutional capital (the state) dominate other forms of capital. And naturally, all had different forms and degrees of social capital, playing off networks with local elites, different Kremlin insiders and "clan" representatives, and alliances with parties and other groups in an attempt to advance their claims and conceptions of the normal *doxa*. Further, fields are contexts for strategic and collective action, and their reconstruction should be a matter of contention, and Russia's economic history certainly bears this out. Even more contentious is contention within and over the "master field" or "field of power"—the specific field that, through a constellation of material symbolic resources, has the greatest potential force to shape general principles of practice and structure for other (e.g. economic) fields. The battle over *doxa*, especially in the master field, was linked to property and principles of control, and it pitted managers, property owners, and state officials against each other.

Yet the post-socialist era is not the only wellspring for possible insights Russia can provide for field theory, and vice versa. Arguably, Russia's entire history is one of contentious fields and multiple struggles: between a state expanding its scope and trying to define identities and rules versus emerging professions and elites trying to situate themselves and articulate rules of economic and political normality; between different elites over status vis-à-vis the state and over boundaries of authority; and between Russia's state and elites and their counterparts in Europe, to which

Russians looked with a combination of awe and envy. And even the history of Soviet political economy has its field dynamics: much politics over how to run that massive command economy were not only about power and gain, but also about defining the rules of that economic gain, and just what a "normal" Soviet economy should look like in the first place. Rather than tell a story of how existing fields function and shape economies at one or several points in time, we have the opportunity to begin to sketch a dynamic picture of the politics of creating various fields: for example, older and rising elites after the collapse of Soviet socialism, or the economics profession trying to figure out its place, status, mission, and understanding of "economics."

This is not the first volume to bring field theory to Russia. Yoshiko Herrera (2005) used a field framework to make sense of dynamics of change and variation across Russia's regions. In her analysis, interests and institutions alone could not adequately explain. Regional interests were not "objective," but rather constructed. Constructions of a "region," its boundaries and position in a broader polity, the nature of actors, and their interests all take place in a context of discourses unevenly weighted by institutional support or linkage to other important categories, identities, and discourses. That is, "regional" actors constructed "regionalism" in a context of structured and aligned relations and practices. Russian economic sociologist Vadim Radaev (2003) explored the construction of Russian markets (e.g. real estate) not simply as a function of laws, actors' capital, and actors' interests, but also as a community of actors sharing and trading ideas of what a "normal" market should look like in the first place. And the editor of this book has applied field theory to make sense of the dynamics and trajectory of post-Soviet economic and organizational change (Hass 1999, 2011a, b). The politics of enterprise restructuring and privatization were not simply stories of competing material interests (property and profit). Rather, there was a serious battle over what a "normal" Russian economy was. Even enterprises less affected by struggles over property ended up struggling to make sense of "production," "sales," and the like because important actors were embedded in multiple fields: Soviet-era networks and organizational communities reproducing Soviet-era logics, and wider fields that included foreign actors, which imported and introduced new logics of business practice.

Beyond Russia, Stark and Bruszt (1998) used a field framework to explore trajectories and variation in post-socialist economic reform policies in East Europe. While they did not eschew interests, they noted that elites were embedded in fields of network relations. This is an important facet of structure: Stark and Bruszt did not examine "networks" simply as particular relations between individual actors, but as a broader "community" of relations actors sensed and acknowledged, even if they did not have direct relations with every other actor in a community or clan.

From these few examples, we see the potential field frameworks bring to making sense of Russian and Soviet economic practices, structures, and histories. Can Russia and the USSR return the favor? Pierre Bourdieu's pathbreaking work on fields focused on contemporary France, from class and tastes to class and education (Bourdieu 1984, 1998). The bulk of Neil Fligstein's work (e.g. Fligstein 1990, 2001) has focused on American corporate history, although he has applied his field framework to the rise of the European Union (Fligstein and Mara-Drita 1996). Yet what is striking about these empirical cases is what they share: institutionalized bureaucratic politics, especially stable states, parties, and rule of law; relatively stable economic elites and sectors, such that even new dynamic sectors (e.g. information technology) are embedded in a context with more continuity than change; and a stable conception of state–society relations. What happens outside cases of developed, stable, and institutionalized polities and economies? Perhaps in contexts with less institutionalization of organizations and roles, then fields will not have soil from which to grow. We disagree with this proposition. First, we follow John Martin's (2009) insights about the nature of networks. Network structures come first, and institutions and fields follow. Hierarchies of cliques and clans—the basic foundation of European monarchies and post-socialist Russia—are the blueprint for relations of order and authority. However, over time such structures of *personal* relations of loyalty and obedience are abstracted, turned into more generic principles of organization—and from this, *institutions* are born as recipes and templates for organizing any set of actors. Once institutions are in place, then fields arise from communities of institutions and organizations, whose elites construct an affinity of traits and practices.

It might seem Martin agrees that fields matter less when institutions are weak or underdeveloped—but this need not be the case, as Martin hints in his later work (Martin 2011). A field is a community of actors engaged in "organized striving," that is, a set of actors shares a basic understanding of appropriate rules and goals for their context. In this approach, fields can exist whenever and wherever multiple actors align their orientations to goals and rules.[3] We can speak of organizational and network fields simultaneously: organizational elites are embedded in a community of organizations (replete with formal procedures and structures, output, and status) *and* in a community of actors who are not all in the organizational field, but who all have some sense of shared context and aligned interests, identities, and relations.[4] In fact, we will see different fields crosscut, augment, and contradict, and that relations of these different fields help drive economics and policy in Russia, for example, in the creation of "Russian economists" as a formal profession (organizational) and as a faction of a wider international community of scholars who shared a serious concern with economic issues but who took different approaches (network field).

But Why Fields and Field Theory?

We should not multiply theories needlessly—between all the social science disciplines, there is more than enough chaff obscuring the wheat. Field theory holds promise, perhaps less as a "theory" proper and more as a framework or even paradigm. Neoinstitutional field theory originally emerged in response to institutional economics and organizational ecology. Paul DiMaggio and Walter Powell (1983) asked a simple question: Why did it seem that so many organizations, in business and cultural production, had more in common in their strategies and structures than one would think of stakeholders and managers thought in terms of their own interests or profit maximization? William Roy (1997) expanded on this in his history of the modern corporation, where he proved Alfred Chandler wrong. The modern corporate form did not emerge out of efficiency considerations. In fact, patterns of incorporation often were the reverse of what Chandler would have expected and even claimed. Roy confirmed what Fligstein had earlier discovered: that business strategies and structures were governed by far more than mere profit maximization. Certainty and stability, politics, and legitimacy were quite important, as were power relations between corporate actors and elites, the state, and other gatekeepers (such as financial actors). Equally important, commonality and variation seemed to follow a pattern of organizational fields—patterns that could not be explained by reference to transaction or agency costs, for example.

Left here, we have a potentially powerful framework that has pointed out what the average participant in business might admit: there is more to the economic world than clients, competition, and profit. Status and power matter mightily, sometimes more than profit (and sometimes less). But there is more field theory could offer: appreciation of greater connectivity between actors or institutions. Current economic theory presumes utility-maximizing, atomized actors. (I leave out behavioral economics, which has not—yet—had much effect on the mainstream paradigm.) To the extent context matters, it is markets, that is, the distribution of resources—capital, labor, inputs, and so on—and competition over them. (Technology is not "context," but rather one form of input or capital.) Institutional economics adds to context "institutions" as rules that shape costs and benefits of decisions, in addition to markets. This narrow, but relatively precise, conceptualization of "institutions" has much to recommend it. However, it is as if we expand atomization to one level of social organization higher. In addition to atomized actors, we have atomized institutions—separate, with their own spheres and consequences.

This picture—admittedly an oversimplification to get to the core of economic theory—seems to mirror crude introductory-level physics. Yet physicists understand that particles or bodies can influence or interact with each other even if they do not touch or somehow directly react with each other. Bodies in space attract even though they do not act upon others' bodies—we call this force "gravity," and the arrangement of such forces a "gravitational field." Fields came to play an important role in physics: they described organized movement by celestial bodies, metal shavings, and other objects. The field logic came to complement and even replace mechanistic explanations that required one object "touching" another to exert some influence (such as one billiard ball colliding with another and transferring energy and altering its inertia).

DiMaggio and Powell noticed that organizations with seeming affinity came to resemble one another, and Roy later noticed that business organizations came to adopt the corporate form, even in cases where it brought little, if any, economic benefit. Such stable and consistent organizational behavior could not be due to profit maximization alone. Nor could it be due to institutions alone, given that economic institutions are not so well integrated and aligned. Further, actors in different institutional domains—such as professors of art, art critics, and artists—might be at home in different institutional settings, and yet, as Pierre Bourdieu noted, they tend to orient to similar norms, goals, and conceptions of what imbues what with status. These scholars noticed that markets and institutions by themselves were not enough to explain important aspects of everyday economic (or political, or artistic) life. Some other force connected these different actors.[5]

I hesitate to say that "fields" are a causal force that aligns these interrelated actors.[6] Rather, fields are an effect of forces that entice alignment: institutions, power, structured signals that compel behavior (Zucker 1977), and isomorphic mechanisms (DiMaggio and Powell 1983). We see a field as a community of institutions and structures with presumed affinity and symbolic boundaries, along with rules of participation and practice that actors understand, even if such rules are not formally spelled out (as is the case for institutions). Actors navigate these institutional and structural communities, monitoring each other for signals of legitimate norms and goals and opportunities to act in pursuit of some gain or to defend themselves against potential threats. That is, a field is in part a heuristic that forces us to look beyond calculation of market prices and institutional costs or benefits, and to dynamic (and political) interactions of a set community of actors. We would look not only for actors' measurements and considerations of gain and loss, but also perceptions and politics of entry into that community, status within that community, and legitimate practices and strategies. In sum, a field framework forces us to look beyond individual decision-making actors (individuals or firms) and to take a holistic view of actors, practices, and contexts as some bigger whole that might have emergent properties.

Our Exploratory Use of Fields

A key word in this edited volume is "exploring." Each contributor has engaged the concept of fields in different ways and in relation to different economic phenomena and intellectual histories. I have my own approach to fields, and other contributors have their own conceptions of how fields fit into broader explanatory frameworks. This volume embraces *pluralism* so as to probe different directions and possibilities for employing a field framework. Because we do not take a collective, unified approach to what fields are and how they operate—no such unified model was imposed at any time—one might argue that there is the risk of cacophony (although often enough this is in the eye, or ear, of the beholder). However, the upside of this intellectual pluralism offsets any such risk, in that we are casting a broader net to see how a field framework can inform both our understanding of current or previous events of states of economic practice and the range of questions and analytic tools in our field of vision.[7]

One departure from the usual use of fields is our methods. Rather than use a quantitative analysis with chi-squared or log-linear methods—as Fligstein and others do to test timing and structure of organizational mergers and acquisitions or other manifestations of organizational strategies and structures—we use qualitative techniques to investigate compulsion of field relations in practice. We would not be the first to do so (cf. Stark and Bruszt 1998). Some authors use a method we could call "historical ethnography"—using actors' own experiences, practices, and categories of description to induce alignment of perceptions, interests, and responses. Another departure is combining a field framework with other theories: systems theory (Chap. 8), interpretive frameworks from area studies (Chap. 6), particular economic theory (Chap. 9), and usual political economy (Chaps. 11 and 12). Other chapters invoke a field logic to redraw the structures and boundaries of historiography rather than theory proper. And so, another departure is not to juxtapose a field framework with other frameworks—for example, fields versus markets (Fligstein 1990)—but rather to see how components of a field framework improve on existing explanations using economics and political economy. Further, we have set ourselves different goals. Some authors use

Russia to test and refine field theory, or to test and refine our narratives of Russia's economic history. Other authors use Russia's case to explore dimensions of field theory and economic theory, using Russia not to test theory, but to illustrate possible new ideas.

Further, a field framework need not only posit causal relations and connections. It can also be a framework for orienting the narrative of analysis by alerting the author to focus on particular dynamics. In particular, a field framework is a *relational* framework (Martin 2011). Not actors or institutions as rules that shape costs, but the relations between actors as actors *and* positions are what field frameworks are about. In their own ways, each chapter here is oriented to examining such relations, even in cases where the focus is on a particular individual or set of individuals. In the emergence of economics in Russia, as a field of discourse, a professional field, and a field of institutions, we see important actors navigating those relations with their asymmetric distributions of capital or power, whether knowledge or institutional authority. Yet those actors were also shaped by those relations in which they were embedded. The stories of post-socialism, for example, are not only stories of actors making use of institutions for personal gain; they were also about contradictions in webs of relations, some of which crystallized into communities of identity and meaning. In sum, we used a field framework not only to posit causation, but also as a tool to guide what we examine.

Yet while each author had autonomy in employing a field framework and choosing empirical cases for their narratives, all chapters still are not isolated from each other. In fact, common themes began to emerge as their chapters took shape. Chapters in Part I, which address intellectual history and historiography, begin to paint a common picture of how economic theory came to Russia; how that discourse shaped some Russian intellectuals; and how those actors' own intellectual contributions and practices contributed to shifts in institutions, communities of discourse, and discourses themselves. Our authors point out that Russian economics as discourse and discipline was a child not only of particular institutions (the state, universities, and after 1917 the Communist Party), but also of an "atmosphere" of relations, meanings and discourses, and events. We see throughout a sense of orientation driven not only or purely by calculations or institutional routines, but also by actors' active engagement with a complex set of rules, other actors, and categories. Actors and contexts intertwined, so that it becomes difficult to separate the two entirely. Economists in the late nineteenth century or post-war Leningrad were not atomized individuals calculating their way through academic politics and beyond (although this was not absent), nor were they reflections of contextual discourses and structures (as postmodern scholars often suggest). They were active carriers of those fields: the categories and logics and even "atmospheres" of institutions and discourses beyond the contextual domains of the latter. This is the heart of a field framework: institutions and relations matter but do not overdetermine, and fields are oriented agency within contexts of rules that constrain and enable.

The same is true in Part II, where we move from the economics discipline and its founding fields, actors, and discourses to relations between economics discourses, institutions, and policies: economics in action, rather than economics in theory. Here, too, we see chapters converging on interesting dynamics: unforeseen consequences of post-socialism that are understood only incompletely if we focus on institutions and interests and not broader fields as communities of practices and identities. One further emerging lesson of Part II is that much post-socialist confusion and contention was borne not only of competing interests and quests for wealth and power. Rather, much contention was also over resolving contradictions or uncertainty about what a *normal and legitimate* post-socialist economy should look like. That is, we cannot fully understand post-socialist Russia without taking into account that actors were not just defending or expanding interests and gain and shaping individual institutions to their benefit, but they were also trying to figure out what the entire complex of post-socialist institutions and practices should look like. Inside state, business, and other organizations, actors were implicitly taking a holistic view of the economy as they tried to reconstruct social reality.

As the organizer and editor of this volume, I leave a final word for myself. I admit to a bad habit of wanting to micromanage everything in my path, and that tendency crept in at the beginning of this project. However, that would have risked a volume reflecting my interests and interpretations, which was not our intention. The chapters are united in engaging field theory, Russia, economic phenomena, and economic theory. Beyond that, the authors approached this project with their own ideas, interests, and interpretations of the place and application and of a field framework. Each chapter begins an investigation but does not exhaust the possible analysis or narrative. The reader will find variation as well as unifying themes across chapters. Our goal here has not been to marshal data to prove a point beyond possible dissent. Rather, our aim here has been to be creative, engaging, and even provocative: probing, rather than proving. If this volume can generate spirited discussion, pro and con, among those who read it, then we have fulfilled the first stage of our task. Finally, we set out to contribute to a dialog not only between disciplines (economics and economic sociology), but also between countries. I like to think of this as an intellectual "meeting at the Elbe," an opportunity for Western (or at least English-reading) audiences to catch a glimpse of diverse and creative ideas among Russian economists. Having escaped the dogma of Marxism–Leninism, my colleagues have not entirely left Marx behind—and neither have they rushed to subordinate themselves to mainstream economics increasingly at odds with empirical reality (as the rise of behavioral economics suggests). In sum, I hope this manages to provide not only a glimpse into the richness of field theory and of Russian economic thought and history, but into the richness of Russian economics and economists today.

Note on Translation

I have translated all the chapters in this book (except my own) from Russian. The biggest challenge was maintaining a balance between accessible English and keeping each individual author's style; I did not want this book to be *my* prose because it is not my solitary creation. Errors in translation are mine alone. I have mostly followed the Library of Congress system for transliteration, with exceptions, such as leaving out representation of the Russian soft sign. I have also used some more common English spellings (Kondratieff and not Kondratiev), and in cases when the surname was foreign in origin, I used the original foreign spelling when this seemed appropriate (e.g. Hermann and not German, pronounced with a hard "g").

Notes

1. This is an important insight of Melville Dalton's (1959) classic ethnography of an industrial enterprise. Actors orient to the formal organizational chart for an initial sense of decision-making, but to get things done, they also orient to how others *really* operate—e.g. turning to an assistant manager, rather than his superior, for advice and permission to get anything done. The formal organizational chart provides the basic structure for practice, but actors construct a field of how those rules are used in business life.
2. This might sound like game theory, but field theory can subsume game theory. Both involve actors oriented to shared rules and goals, and who can act strategically. The difference is that the stereotypical (and often real) "game" of "game theory" involves actors making calculated decisions vis-à-vis common rules, goals, and each other. Fields also involve shared oriented (willfully or compelled)—except a field encompasses a wider set of decisions and interactions, as well as ritualistic action. Further, fields involve strategies regarding power and breaking the rules of the game (Martin 2011)—often missing in game theory.
3. In this sense, we understand the scope of "fields" in the same way physicists do. The scope of a gravitational field is a function of distance (decreasing inversely to distance squared) and the mass of the object at the center of the gravitational field. Stars and planets have gravitational fields—and so do people (whose gravitational fields are so small, because of small mass, as to be unmeasurable).
4. An example might be a professor embedded in an educational field (organizational), with its rules for status, promotion, sanction, and so on. She might also be embedded in a network field of blog-based activists operating outside formal organizations and who have rules about status, sanction, and so on. This might mean contradictions arise: a "professor" is presumed to act dispassionately and objectively, while the same person as "activist" must act with passion and emotional commitment.
5. This leaves aside the question of when fields are at work or are merely a potential. I will make the bold claim that whenever there is more than one actor interacting, then there already is a field of structured signals, perceptions, and practices (i.e. a field), but that field strength is variable. Discovering determinants of that variation remains the next important step. (This is similar to the claim that even I have a gravitational field,

which incidentally has increased as my lifestyle has become more sedentary. While that gravitational field is so weak that it might as well not exist, that it does exist—if current laws of physics are valid—points to important dynamics of the physical world. The same, I suggest, is true for the social world; we have much more work to do to make sense of fields.)

6. Even DiMaggio and Powell (1983) posited *correlations* between particular contexts and field effects, although they did suggest that isomorphic mechanisms (which they did not measure) were causes.
7. For this reason, we are less concerned with prediction for now. To predict requires sufficiently developed causal relations. While DiMaggio, Powell, Fligstein, and others have begun such work, this is only a first step—and usually restricted to organizational fields, which unfairly limits the scope of inquiry. Martin (2011) suggests the real power of a field framework is not necessarily formal predictions (as nice as these would be); it is whether its insights would allow us to enter a field and successfully play by its rules.

Works Cited

Bourdieu, Pierre. 1984. *Distinction*. Cambridge: Harvard University Press.

———. 1998. *The State Nobility*. Stanford: Stanford University Press.

Dalton, Melville. 1959. *Men Who Manage*. New York: John Wiley and Sons.

DiMaggio, Paul, and Walter Powell. 1983. The Iron Cage Revisited: Institutional Isomorphism and Collective Rationality in Organizational Fields. *American Sociological Review* 48: 147–160.

———. 1991. Introduction. In *The New Institutionalism in Organizational Analysis*, ed. Walter Powell and Paul J. DiMaggio, 1–38. Chicago: Chicago University Press.

Fligstein, Neil. 1990. *The Transformation of Corporate Control*. Cambridge: Harvard University Press.

———. 2001. *The Architecture of Markets*. Princeton: Princeton University Press.

Fligstein, Neil, and Iona Mara-Drita. 1996. How to Make a Market: Reflections on the Attempt to Create a Single Market in the European Union. *American Journal of Sociology* 102: 1–33.

Fligstein, Neil, and Doug McAdam. 2012. *A Theory of Fields*. New York: Oxford University Press.

Giddens, Anthony. 1984. *The Constitution of Society*. New York: Cambridge University Press.

Hass, Jeffrey K. 1999. The Great Transition: The Dynamics of Market Transitions and the Case of Russia, 1991–1995. *Theory and Society* 28: 383–424.

———. 2011a. *Power, Culture, and Economic Change in Russia 1988–2008: To the Undiscovered Country of Post-socialism*. New York and Abingdon, UK: Routledge.

———. 2011b. *Rethinking the Post-soviet Experience. Markets, Moral Economies, and Cultural Contradictions of Post-socialist Russia*. New York and Basingstoke, UK: Palgrave Macmillan.

Herrera, Yoshiko. 2005. *Imagined Economies: The Sources of Russian Regionalism*. New York: Cambridge University Press.

Martin, John Levi. 2009. *Social Structures*. Princeton: Princeton University Press.

———. 2011. *The Explanation of Social Action*. New York: Oxford University Press.

Radaev, Vadim. 2003. *Sotsiologiia rynkov: k formirovaniiu novogo napravleniia*. Moscow: Higher School of Economics.

Roy, William. 1997. *Socializing Capital*. Princeton: Princeton University Press.

Stark, David, and László Bruszt. 1998. *Postsocialiust Pathways*. New York: Cambridge University Press.

Zucker, Lynne. 1977. The Role of Institutionalization in Cultural Persistence. *American Sociological Review* 42: 726–743.

Part I

Fields of Discourses and Theory: Economics and Russia

Introduction to Part I

Part I uses insights and logics of a field framework to explore the intellectual history of Russian economics as discourse and discipline—a "community of discourse," to use Robert Wuthnow's phrase. While our authors employ fields in a subtle way, focusing on actual events, biographies, and actions (including production of knowledge), fields are certainly part of their stories. The core to these fields were institutions, personal networks (friends, colleagues), and more virtual networks of communication (journals, discussion circles of intellectual society). "Economics" as a particular discourse about the material world and the nature of wealth arose in Western Europe in the contexts of discussions and analyses of production and trade, states and policies, and statistics as a new methodology for measuring and analyzing social practices. In some countries (e.g. Great Britain), economics arose among circles of private intellectuals who were making appeals to their state and other intellectuals, but they were still writing from their own volition. In other countries (e.g. France and Germany), economics arose in the context of state institutions and emerging universities. Economics and statistical methods were to be harnessed so that states could craft policies that would serve state power (and perhaps their nations as well)—economists more as technocrats than as independent intellectuals. Overall, economics emerged initially as a community of individuals in a variety of institutional contexts pontificating on a wide range of subjects. Over the nineteenth century, "economics" as its own discourse and discipline began to emerge, albeit not without some conflict and controversy: for example, the "battle of the methods" (historical versus quantitative analyses) or the boundaries and content of the discipline (as witnessed in the separation of "economics" proper from its parent, "political economy").

One key to a field is a sense of collective orientation or "organized striving," which is compelled by different kinds of forces. Following DiMaggio and Powell, there can be the perceived need to mimic other actors and collective practices because they address a similar challenge or seem legitimate in broader contexts, the need to enact particular practices and routines as a condition of receiving some kind of resource, or common socialization, so that shared practices come more naturally as legitimate and natural than other innovations. All of these dynamics emerged even in the eighteenth century and accelerated over time, as competition in the field of emerging nation-states grew and institutional fields of intellectual activity (especially universities) expanded as well. While Russia was still a latecomer, it was not far behind in such institutional development. Russia had its own elite universities in St. Petersburg and Moscow, the Academy of Sciences, and universities or other educational institutions in other major cities. State officials were increasingly keen to strengthen Russia's capacities as a great power. And actors in these institutions quickly understood that Russia was embedded in international fields of discourse, embryonic as they were: while there were no national or international disciplinary conferences that we have today, there were still circles of networks and local institutions (especially universities) that produced knowledge. And so, in a tradition going back to Peter the Great, who visited Europe to learn what his neighbors had to offer, Russian intellectuals and officials turned to the West again. Sometimes the state brought Europeans to Russia to educate their own in nascent economic science; later, the state sent some promising young scholars abroad to further their educations. More importantly, Russian intellectuals themselves interested in various topics, including political economy and related issues, turned to Western publications and intellectuals.

This international field of institutions and other social groups—aligned in a sense of intellectual endeavor and identity vis-à-vis each other as well as their own populations—existed alongside domestic Russian social fields that shaped how economics came to be in Russia. In the eighteenth and nineteenth centuries, a civil society was developing in Russia as well: in particular, circles of educated elites and educated and talented Russians from lower classes, and a growing sense of corporate identity embedded in a shared sense of intellectual interest and mission, a "proto-discipline," as it were. In the second half of the nineteenth century, Russian universities began to establish formal positions and programs for political economy and statistics, creating the formal foundation for "economics" as a potential profession, rather than personal calling facilitated by networks of like-minded intellectuals and local societies of educated and curious elite and middle-class participants. The creation of these formal positions was the work of state and university officials and scholars themselves, some of whom took it on themselves to add ideas and energy to ground economics not only in individual talents, but also in formal occupations with status and income. Of course, not all burgeoning economists had access to such positions; some continued to rely on networks and local societies, some of which were imbibing Marxism and existing in the shadows to escape the watchful eye of suspicious state police. By the dawn of the twentieth century, "economists" as we understand them today—a technocratic practice combining institutional position (i.e. in universities), social role (i.e. legitimate producer of particular knowledge), and relation (i.e. individual networked to similar individuals sharing corporate and intellectual identity)—was taking shape. Revolutions in 1905 and 1917 would disturb the development of these fields of institutions, positions, and relations, after 1917, giving birth to a new type of modernity, that of Soviet socialism.

Authors in Part I explore various facets of this general field story. *Shirokorad* provides a general overview of the twin sources of Russian economics: the Russian state and foreign fields of actors and discourses, primarily German. Russian "economics" ended up caught in between these two forces that sometimes were complementary, and sometimes contradictory: German scholars and institutions (universities) offered the human capital the Russian state desperately needed, yet that knowledge could threaten the status quo on which the Russian state depended, from serfdom to autocracy. Put differently, Shirokorad's story is that the field of economics in Russia—again, a combination of discourse, institutions, and professional roles and identity—is a story of two other fields (European, and especially German, economics, and the Russian state) whose overlap was fraught with tension. Following this, *Ryazanov* explores the logic of economics as a particular form of research. Again, we see the tension between foreign influence—as higher status and a source of skills and knowledge for Russia—and domestic political needs. Ryazanov begins by noting that economics has been torn between claims to universality (i.e. a method for seeking Truth outside historical experiences) and the influence of local historical experiences on the questions, methods, and aims. The creation of a "national school" of economics in Russia was not unique, and it was driven by the desires of state officials and intellectuals oriented to Russia itself (e.g. Slavophiles) to create a science that could address Russia's economic issues on its own terms, and without subordinating Russian intellectual thought to foreign fields. The influence of German economics—not only as an intellectual effect in itself, but also as concrete actors bringing their own ideas to Russia—becomes clear once again. However, Russian economists could apply new knowledge to their own context and create their own national school—an economics attuned to their own institutions and political culture. Russian economics, like Russian identity, has been torn between affinity with Europe and sense of belonging to a European intellectual community, and a desire for status from uniqueness.

Dmitriev sharpens the history of economics by examining how statistics, and a quantitative methodology more generally, took root in Russian economics. K. F. Hermann played an important role, revealing how actors with particular institutional capacities, skills, and capital can influence the operating logics of fields and discourses. Both British and German approaches were available, and ultimately the German approach won out in no small part because of Hermann and his colleagues. Statistics was not the inevitable core of economics methodology: Hermann had to make a case that statistics not only provided important data for the state, as well as economists, to have a true sense of the state of Russian economy and society; it was also a true form of "science." *Raskov* turns our attention to another individual, Nikolai Sieber, who played an important role in introducing Marx to Russian economics. In telling part of Sieber's story (drawn from a broader project on Sieber's life and work), Raskov problematizes the personal biography as more than one individual's personality, knowledge, and tastes—rather, that biography is the sedimentation of experiences in different fields that the actor carriers forward. For Sieber, as for many young scholars, three fields of experience were important: the university setting as experiences of intellectual challenge, growth, and mentoring; travel abroad for purposes of study, and thus exposure to fields of intellectual discourse and practice different from those in Russia; and publication as a means for shaping discourses, raising one's status, and making a living. While Raskov guides us through part of Sieber's life, he is also guiding us through different field relations that came together through Sieber.

The contradictions of tsarist political economy and emerging modernity (influenced in part by ideas from Europe, including Marxism) came to a head in the revolution of 1905. *Markov* provides a brief excursion through the institutional turmoil of that period, and how various facets of the field of economics were touched. The 1905 revolution was a complicated affair, and not even one book can do it justice, but Markov manages to reveal the challenges economists faced. Students challenged their authority on dual grounds: their institutional authority as professors and their cultural authority as carriers of knowledge. In terms of a field framework, the revolution challenged older rules of the value and legitimacy of capital, and economists' cultural capital (particular theoretical and methodological knowledge) and social capital (institutional positions) no longer had the same value as before 1905. *Melnik* takes us to another case of institutional and field turmoil: high Stalinism after World War II. The Blockade of Leningrad had claimed more than one million victims and disrupted the work of economists, especially those at Leningrad State University. Adjusting to post-war life was its own challenge, but by 1948, the Leningrad Affair heralded a new wave of Stalinist repression aimed at Leningrad elites who led the city through the Blockade. Melnik takes us later to public "discussions" as a tool to discipline economists and professors to make sure their "science" did not challenge the authority of elite or ideology. The threat to power, it seemed, was local-level fields: a profession grounded in the search for Truth and intimately linked to Marxism–Leninism, an institution (the university), and "science" as practice and identity that was supposed to transcend social reality. High Stalinism was not only a matter of a suspicious elite bent on rooting out possible competition; it had a complex dynamic that ran not only through institutions, but also through combinations of institutions that, in this case, came together in the university.

2

Global Fields and Economic Theory: The Impact of German Scholarship on Russian Political Economy in the Eighteenth and Nineteenth Century

Leonid Shirokorad

The development of economic theory in any country is a product of the level of economic development and uniqueness of that country's economy, historical traditions of economic thought, and investment in research and education, among various factors. However, no small role is played by forces outside that country, especially economic discourses and ideas from other countries with more mature institutions—even more from those countries that seem to have similar institutions and economic challenges. It is not unreasonable to expect that older, more developed countries will exert some intellectual influence over those up-and-coming. Classical political economy in Great Britain had a powerful impact on economic thinking in continental Europe, including Germany and Russia, in the first half of the nineteenth century.

L. Shirokorad (✉)
Faculty of Economics, Department of Economic Theory,
St. Petersburg State University, St. Petersburg, Russia

J. K. Hass (ed.), *Re-Examining the History of the Russian Economy*,
https://doi.org/10.1007/978-3-319-75414-7_2

However, for particular historical reasons, Russia had relatively closer economic, political, cultural, and intellectual ties with Germany. As a result, the impact of German economic science on Russian economic thought was especially significant. This chapter provides an overview of how this influence spread in the eighteenth and nineteenth centuries. In particular, we see some mimicry resulting from a combination of network relations between key actors in both countries, as well as not only geographic but also political and institutional proximity. That is, field dynamics operated in several simultaneous ways to make Russian economic discourse an outgrowth of that in Germany, even if Russian scholars added their own innovations stemming from their own institutional contexts.

One theme of this chapter is that the rise of Russian fields of economics—as fields of discourse, professional roles and norms, and institutional positions—was a product of state policies, and private actors (especially scholars) further built upon this template. For the state, engaging foreign scholars and experts—whether by inviting them to Russia to teach or set up educational institutions, or by sending younger Russian scholars to learn in Germany from the brightest and best that German universities could offer—could improve Russia's state capacity and image abroad (an issue of legitimacy and what today we might call "soft power"), while also addressing various other problems at home. (This said, sometimes economics discourse could contribute to creating problems or drawing further attention to them, as we will see in the case of serfdom.) Russia's latecomer status vis-à-vis other European Great Powers meant a stronger role for the state in creating educational institutions (among others). This also meant that the impetus for creating new fields of economic discourse and study could also come into contradiction with the state elite's own impetus for defending power. This contradiction would rear its head repeatedly, moving the development of Russian economics between greater integration with European fields and creating their own domestic fields insulated from potentially troublesome discourses (e.g. political liberalization) from abroad.

Universities Arise: The (Russian) State and (German) Scholars

The first universities appeared in Russia in the eighteenth century, around 500 years after the birth of university education in Western Europe, and in an environment where developed primary and secondary education was essentially absent. After his direct experience of the achievements of Western European science during his travels abroad, Peter the Great created the Academy of Sciences and Academic University in 1724, primarily under the influence and with support of prominent German scholars G. W. Leibniz and H. Wolff. The first university lecturers were foreigners recruited from abroad, and there turned out to be more of them than of students recruited (often compelled) from theological academies and seminaries. Opening the Academy of Sciences and Academic University was easier than starting a modern system of education with deep and costly reforms of primary and secondary education.[1] The Academy and Academic University were created by Peter in no small part "for glory before the foreigners," in his own words.[2] As P. P. Pekarskii, a noted scholar of Russian science, remarked, "in the upper estates there was only the striving to assimilate in oneself the veneer of a European community, rather than education, which, frankly, it would be, as well as a needless weight at that position in which Russia was in the old days" (Pekarskii 1862: 33). To get an idea of how severe were the conditions and activities of the Academy of Sciences and Academic University in the eighteenth century, and how were their limited opportunities in this period, one need only read Lomonosov.

Although some of Peter's important initiatives in science and education were not sufficiently thought out, he did a great deal to acquaint Russians with Europe's best achievements, especially of German science. This was manifested primarily in the fact that, as Pekarskii noted, Russian "scholarly celebrities," who appeared on the scientific horizon at the beginning of the eighteenth century, "were formed in foreign lands," including Germany (Pekarskii 1862: 140). Suffice it to mention the educational activities of Wolff, an honorary member of the Academy of Sciences in St. Petersburg: "Lomonosov attended his lectures in Marburg, and he always remembered with awe his mentor, who spared neither time nor labor for Russian students in Marburg assigned to his supervision, and he met the expectations of the Academy of Sciences" (Pekarskii 1862: 39). It is also important that "the best of the first members of our Academy—Bernoulli, Bilfinger, Martini and some others—came to St. Petersburg by his [Wolff 's] recommendation and representation" (Pekarskii 1862: 39). In general, there were many foreign, especially German, scholars in the service of Russia during the reign of Peter. Finally, under Peter a massive number of scholarly publications, especially by Germans, were translated into Russian.

Western influence on Russian bookishness, predominantly Little Russian and Polish, manifested itself in the seventeenth century. Even then, the state exerted pressure "in favor of further rapprochement of Russia with European culture, strengthened in the era of the Reformation" (Lappo-Danilevskii 1990: 238). However, at that time this influence was exclusively religious. Pupils of existing academies and colleges "were still almost alien to the characteristic beginnings of a new education, more closely associated with the Reformation than with Catholic school wisdom. They did not yet know Descartes, Leibniz, or Wolff, nor Grotius and Pufendorf" (Lappo-Danilevskii 1990: 80). Under Peter I, this religious culture, often imbued with a scholastic spirit, gave way to "that direction that was becoming more and more intensified in Europe after the Reformation, and had at heart practical applications of results obtained by science for the benefit of man" (Pekarskii 1862: 479). As Pekarskii noted, "in Russia under Peter, European civilization not only received free access, but its establishment into Russian life became mandatory due to compulsory measures, decrees, and resolutions coming from a government that did not know the limits of its power. Peter's reform was not only in the introduction of Western education, which managed to penetrate into Russia before him, but also in taking decisive measures to spread it" (Pekarskii 1862: 479). That Peter could employ some Russians with status or connections was useful for enticing Germans to take seriously Peter's efforts. For example, one of the main leaders of Peter's reforms in science and education was Feofan Prokopovich, who, according to Pekarskii, "especially … enjoyed fame among contemporary German scientists" (Pekarskii 1862: 488).

Another feature of the new school, including at the highest level, created by Peter I, as opposed to the ancient Russian school, was that it was aimed at serving the interests of the state first and foremost (and not the church, as was the case earlier). Peter needed educated people: without them it was impossible to carry out large-scale reforms he had outlined. The liberation of Russian science from ecclesiastical captivity and putting it into service of the state in the most direct sense was an important step in this development. This was true of nascent economics as well, which was one of the new sciences that began to form under Peter I. It too had a purely practical focus; that is, it was directly aimed at solving tasks the state faced in carrying out large-scale reforms. Such was the guiding logic, for example, of I. T. Pososhkov's *Kniga o skudosti i bogatstve* (The Book of Poverty and Wealth), which was one of the most outstanding works of Russian economics in Peter's time. However, Western economic science at that time (e.g. mercantilism) was characterized by the same features. In the first quarter of the eighteenth century, works of Western scholars translated into Russian were located at the intersection of economics and other studies, primarily law, and had a pronounced practical focus (e.g. on trade law).

Full freedom for scientific creativity remained in the distance, and science was considered only a tool for solving external tasks. All this "undoubtedly contained something accidental and fragile ... and it necessarily affected adversely the further spread of knowledge in Russia, as this direction brought capricious arbitrariness, as well as superficiality, hence frivolity and a disrespectful attitude to its success, unless it was immediately understandable for many applications in practice—all this could be phenomena that directly resulted from such a view of enlightenment" (Pekarskii 1870: xxvii). Yet even in such circumstances, education made progress. As N. I. Turgenev noted,

> During the years separating the reign of Peter from the reign of Catherine II, rightly called by the famous historian Karamzin the Saturnalia of despotism, years as long as they were shameful, in those years nothing worthy of attention was done for progress, except, perhaps, the establishment of the Cadet Corps in the reign of Empress Anna ... We hasten to note, however, one more act, much more remarkable, which decorated the reign of Empress

> Elizabeth: with this sovereign in Russia, the University was opened. I am not afraid to say that never and nowhere has any institution brought so much good and benefit as Elizabeth's creation. (Turgenev 2001: 306)

Turgenev had in mind the opening of Moscow University in 1755. He noted that "the staff of professors was replenished from the very beginning by both domestic and foreign (mainly German) scholars" (Turgenev 2001: 307). Among them was I. G. Reichel, who was discharged from Leipzig in 1757.[3] He was a professor of general history, secretary of the Conference of Moscow University, and the first librarian of Moscow University. In presenting new history, Reichel always added a statistical overview of new European states. In 1764–1765, for the first time in Russia, a course with statistical content was offered, and in 1772–1773—again for the first time in Russia—a course on statistics proper was offered.[4] Reichel was an excellent lecturer. According to Prince I. M. Dolgorukii, "the gift of speech belonged to him in superiority." Reichel also exerted great influence on D. I. Fonvizin: "the professor, with his irony and his ability to grasp the unreasonable side of life … could have some influence on the moral development of our brilliant comedian" (Michatek 1913: 29–30). In general, in its first decades of its existence, Moscow University represented the best of German universities (Ikonnikov 1876a: 513).

Not only did the Russian state bring foreign scholars, especially Germans, to Russia; they sent Russia to European, especially German, scholars. Already in the first half of the eighteenth century, Russian students were sent to study at German universities, in particular Leipzig, Strasbourg, Göttingen, Heidelberg, and Halle (Vishnitser 1908: 186). Under Catherine the Great, the influence of Western ideas on Russian science and education expanded significantly. As the distinguished historian of law V. I. Sergeevich noted, "By birth and education she [Catherine] was not so much of Russia, as of Western Europe" (Sergeevich 1878: 190). Now state elites and members of educated society were interested not only in achievements of Western science and technology, from applications of which it was possible to draw direct practical benefits. The Empress was fond of French Enlightenment philosophy, and at the beginning of her reign she tried to use them to guide her state activities. She knew that in Germany, the Göttingen and Leipzig universities were penetrated with a new spirit, and so at the beginning of her reign, she sent Russian students there: in 1766–1770, 15 students were sent to the University of Leipzig, among them A. N. Radishchev. In the years 1766–1772, five Russian students were at the University of Göttingen (Ikonnikov 1876a: 500).

In the eighteenth century, German universities became important channels through which English liberalism and French enlightenment penetrated into Russia.[5] Even then they became "hotbeds of not only German, but also of world science and literature" (Kovalevskii 1915: 133). The leading place was occupied by the University of Göttingen, founded in 1734. "Unrelated to legends and medieval scholasticism, gravitating over the majority of German universities, the new university, thanks to the freedom of teaching (word and press) granted to it, in short time managed to yield rich results in the field of science. Its work became a proverb; professors' textbooks were recognized as exemplary; the influence of the theological faculty in general did not burden it" (Ikonnikov 1876a: 500–501). Russian students were already arriving to Göttingen by the middle of the 1740s. As its glory and fame increased, so did this flow: from 1780 to 1815, 81 Russian students took lectures there. Explaining the university's high rating in Russia at this time, M. Vishnitser wrote in 1908, "an extensive plan of studies, a rich selection of teaching materials, a cosmopolitan character of the university alien to any narrow local influences; the care of the University curator, Baron von Munchausen, for attracting as many listeners as possible from all countries, and especially from Russia—these are the main points that we should keep in mind when we ask why we meet so many Russian names on university lists at the end of the eighteenth and beginning of the nineteenth centuries" (Vishnitser 1908: 186–187). It should be especially emphasized that the University of Göttingen contributed in every possible way to the influx of Russian students. In particular, as M. Vishnitser noted, "the invitation from Russia to Schlözer to the history department was motivated by the desire to have more Russian listeners. Schlözer had personal connections with the academic world and with Russian society" (Vishnitser 1908: 189).[6] Johann Beckman, a professor of political economy at the University of Göttingen who once taught at the Petrovskii School in St. Petersburg, according to a Russian student, "forever kept warm feelings for Russia" and "was always glad to show his disposition towards Russian listeners" (Vishnitser 1908: 189).

Russia, Germans, and Uncertain Economics: The Political Economy of Serfdom

In the eighteenth century, the abolition of serfdom, perhaps the central and most painful social and political issue that caused sharp polemics in political and scholarly circles, was on the agenda. The belatedness and half-heartedness of reforms under Aleksandr II to no small extent predetermined the uniqueness of Russia's economic and political development in the second half of the nineteenth century and revolutionary events in the early twentieth century. Given how important this issue was, it is of interest to see what positions German economists working in Russia and Russian graduates of German universities thought about this issue. As the famous Russian historian of the nineteenth century, V. I. Semevskii, noted, "For the first time the idea of liberating the peasants in Russia was expressed at the end of the eighteenth century by Prince V. V. Golitsyn, a famous statesman who played the leading role in the reign of Princess Sophia" and who was "one of the few European-educated people of his time in Russia" (Semevskii 1888: 1).

In the eighteenth century, while Russian serfdom was expanding, some among Russia's educated elite were writing about the advisability of restricting (but not abolishing) it. These included, among others: the early Russian economist I. T. Pososhkov; Catherine the Great's Count P. I. Panin; I. P. Elagin; and Prince D. A. Golitsyn, who was educated in Germany, was married to a German woman, spent the last ten years of his life in Braunschweig (where he died), and was a leading figure in the Masonic movement. Catherine II, inspired by the ideals of an enlightened monarchy, also was inclined early in her reign to limit and mitigate serfdom.[7] Immediately after her accession to the throne, she allowed pursuit of a famous case against landowner D. Saltykova for committing atrocities against her serfs (mostly women and girls) which had not been prosecuted Elizabeth I and Peter III.[8] This scandal might have helped propel the Free Economic Society (which included Prince G. G. Orlov, who was close to Catherine) to raise the question of serfdom's fate. It was also no accident that in 1764 Catherine II permitted the publication (albeit in German) of articles by Pastor Eisen von Schwarzenberg, who arrived in Russia at the invitation of Peter III and studied the situation of the peasants in Livonia. This was the first article on the peasant question (von Schwarzenberg 1764), and its author argued that the situation of the Livonian peasantry was even more difficult than that of the peasants in the interior regions of the empire, and proposed a number of measures to alleviate it. In particular, von Schwarzenberg argued that in some countries serfdom had been abolished 100 years earlier, and that there were no lingering regrets. Apparently, this article prompted Catherine II to entrust the Lieutenant Governor General Count Iu. Iu. Brown to end landed despotism in territory under his jurisdiction.

Realizing all the complexities of the issues related to restricting serfdom, and the ambiguous response to proposals of reforms by various social strata, Catherine II decided to involve the newly formed Free Economic Society in solving this issue. This Society declared in print in 1766: "What is more useful for society, that the peasant owned land or only a movable estate, and how far should his rights extend to this or that estate?" (cited in Semevskii 1888: 48). Overall, the Free Economic Society received 162 answers to this question—120 of them were German and sent from different corners of Europe, mostly from Germany. Fifteen essays were selected for a competition on this theme, of which nine were German. The winner of the competition was a doctor of laws from Aachen, Bearde de Labelle, who proposed postponing the solution of the peasant question. He argued that "it is necessary to prepare slaves for taking liberties before they are given what kind of property they will be given" (cited in Polenov 1865: 286). In addition to his answer, it was decided to print three more works, including that of a future. All German authors, including those whose answers were not printed, opposed serfdom and recognized its inefficiency. At the same time, they were concerned that the interests of the landowners should not suffer. Therefore, they proposed that land should be transferred to peasants not as property, but only for permanent use, while at the same time increasing peasant duties.

According to V. I. Semevskii and other researchers, the best answer to the question posed by the Free Economic Society was that of A. Ia. Polenov, who for four years (1762–1766) studied law at Strasbourg and Göttingen universities. Returning to St. Petersburg in May 1767, he submitted his answer in February 1768. Polenov argued that only land ownership can interest the peasant in higher labor productivity and thus was a necessary condition for providing industry with an abundant supply of cheap raw materials, and the urban population with cheap food and consumer goods. All this would create high demand for labor that accordingly would serve "to exterminate idle life among the people" (Polenov 1865: 290). The growth of incomes would contribute to an increase in the state's incomes, and thus to the growth of its power. Maintaining an oppressed state was "not only harmful but also dangerous" for society, for it generates constant popular unrest (Polenov 1865: 291–292). In Polenov's opinion, enslaving peasants contradicted natural law, and only violence could lead people into such a deplorable situation. He emphasized that peasants provided society with its means for life and noble estates with leisure, and they defended the Fatherland when it was in danger. As possibly the most useful estate, the peasantry deserved property. In reality, the position of the peasants in Russia was truly tragic. "I do not find people who are so poor as our peasants, who having little legal protection, are subject to … resentment and undergo incessant impudence, torture, and violence" (Polenov 1865: 298).

Polenov developed something like a program for reviving the Russian peasantry. He proposed establishing primary schools in each village; setting up physicians in larger villages; and organizing police, elementary fire safety, and protection of forests, rivers and lakes in various villages. Although in principle Polenov recognized that introducing private peasant ownership of land would have beneficial effects for peasants and all of society, he also took into account that such reforms would contradict interests of the nobility. Proceeding from the need to reconcile these opposing interests, he proposed transferring land to peasants not as property, but for permanent and hereditary use. While the peasant "will regularly observe all his duties," the landowner could not take this land from him, unless the peasant neglected those duties and a court decided against that peasant. Not being a full owner of the land, the peasant would have no right to sell it or to divide between children. "Thus, the landlord will always retain his right, and the peasant will freely enjoy benefits to which he is entitled" (Polenov 1865: 307). To prevent agrarian overpopulation, local authorities would have to conduct an effective resettlement policy within the relevant region. In particular, the hope was to resettle peasant families not only with free land, but also with houses, and initially to supply them with agricultural implements, seeds, livestock, and to exempt them from duties and taxes. To encourage population growth, Polenov proposed taking into account peasants' family situations in securing "services and taxes to the sovereign and master." The peasants' rights to the products of their labor must be complete and assured (minus those parts legally set aside for the state and nobility) so that landowners could not encroach on them. To facilitate the legal status of the peasants, Polenov's proposal was sent to establish peasant courts to resolve disputes between peasants themselves, as well as between peasants and landlords. It would be dangerous, he thought, to impose these freedoms immediately, and Polenov stressed that these reforms must be preceded by a long educational work among the peasants "under the leadership of good churchmen." In addition, "to provide an example to the nobility," reforms should begin with court and state peasants (Polenov 1865: 314).

Polenov's writing, with his sharp criticism of the feudal order reigning in Russia, as well as the radicalism and thoroughness of measures he proposed to liberate Russia from this ulcer, frightened not only the Commission, which saw it as so dangerous that it did not dare publish it. The Empress herself, knowing about this work, "not only did not affect the election" of A. Ia. Polenov "to the Academy (which was no more 'free,' of course, than the Economic Society), but also failed to use his abilities to greater advantage" (Semevskii 1888: 82). Since foreign replies to the request of the Free Economic Society were unavailable not only to Russian society, but even to most deputies of the Commission that was drafting the new Code—because they were written in foreign languages—Polenov's work was the most influential of all essays, but the Economic Society feared this influence. Catherine might not have wanted to make the full negative nature of serfdom public, perhaps not only to avoid annoying or riling up serfs, but also to prevent further bad press for Russia vis-à-vis Europe, especially since Catherine had written a pamphlet refuting the less-than-flattering picture Jean-Baptiste Chappe d'Auteroche had painted about Russia (Semevskii 1888: 95–96).

Unfortunately, this work by Polenov, a graduate of German universities, was published in Russia only 100 years after its writing and the abolition of serfdom. In comments on this work, J. Borzov wrote, "It is very remarkable that the author for an entire century foresaw the necessity and, even more, the impossibility of liberating the peasants without endowing them with land; he points out with striking clarity the harmful and pernicious consequences for a mass of people who have no property; a mass that, according to the author, *knows that whatever changes occur, he has nothing to lose* … The fidelity of the author's picture of the peasants' poor state in his time and bad consequences that might occur was confirmed by Pugachev" (Borzov 1865: 316; italics in original). Borzov pointed out that Polenov's proposals made in the eighteenth century turned out so radical that they could not be realized even during the reforms of the 1860s. For example, Polenov said nothing about the peasant commune or envision the obligatory establishment of schools in villages and their state support (Borzov 1865: 316–318). In his criticism of serfdom and the position of the peasantry, Polenov was not alone. In the same year of 1767, similar ideas were expressed by deputies of Catherine's Legislative Commission, set up to discuss a new Code. Commenting on relevant statements of deputies of various classes (including some noble deputies), historian of Russian law I. I. Ditiatin wrote: "Could we have spoken more categorically, can we speak more clearly for the liberation of the peasants in the second half of the eighteenth century?" (Ditiatin 1905: 81).[9] Nevertheless, Polenov's analysis differed in depth, clarity of thought, and comprehensiveness. All this criticism at that time led to no practical results: "The activities of the Commission of 1767–68 led Catherine II to the conclusion that it was impossible to reconcile the interests of various classes, and to relax the feudal order without the risk of losing the throne … She learned firmly that the main support of her power was the nobility alone" (Kamenskii 1996: 128).

Emelian Pugachev's peasant uprising, Radishchev's *A Journey from St. Petersburg to Moscow* (which led Catherine II to describe the author as "a rebel, worse than Pugachev"), and, finally, the French Revolution frightened the ruling upper echelons of the Russian Empire and marked the beginning of a period of reaction toward the end of Catherine's reign and that continued under Paul I. In the late eighteenth and early nineteenth centuries, private printing houses were closed down, censorship was strengthened, and the importation of books from abroad was banned. By decree of April 9, 1798, Paul I forbade his subjects from visiting foreign universities, "because of the malicious rules that have arisen in them for inflaming immature minds to unbridled and depraved reasoning of those who incite" (cited in Kovalevskii 1915: 135). All this led to the curtailment of scholarly ties with Western Europe, including Germany.

Reaction and the "Foreign": Economics Insulated

The situation significantly changed after Aleksandr I became tsar. A Swiss teacher named Laharp, chosen by Catherine II as a man educated in the spirit of the Enlightenment to tutor the young Aleksandr, exerted great influence on the formation of the future tsar's personality. Having need of unlimited power and motivated only by noblest motives, Aleksandr I embarked on liberal reforms whose complexity he could not yet imagine when he took the throne at age 23. Some of these reforms affected the sphere of science and culture. In March 1801, when his father was killed, Aleksandr signed decrees abolishing Paul's prohibition on importing books from abroad and printing books and magazines in private printing houses. Censorship was substantially relaxed. Exactly one month after the death of Paul I, a decree was signed to allocate 5000 rubles annually to the Free Economic Society. An extensive program of publishing important works of great European thinkers of the eighteenth century was developed, with Adam Smith's *Wealth of Nations* being one of the first. Especially important was the cancellation of Paul's decree prohibiting study at foreign universities. "The Russian youth again went to German universities," wrote M. Kovalevskii, "among them, Göttingen again began to attract students from Russia. Professors from Göttingen were eagerly invited to the Holy See ... In response to its cosmopolitan direction, the University of Göttingen, at which in 1801 there were 456 foreigners per 701 students, tried to avoid all sorts of nationalistic biases in teaching. The connection between the Hanoverian electorate and England, in the person of the same ruler, largely explains the reason why the University of Göttingen, located at the Electorate, became the conductor of not only German, but also English science" (Kovalevskii 1915: 135).

In particular, one professor at the University of Göttingen, Georg Sartorius, expounded on the teachings of Adam Smith in his lectures. "Among the listeners at the University of Göttingen," wrote M. Kovalevskii, "we meet such subsequently-known people as A. I. Mihailovskii-Danilevskii, Nikolai Turgenev, and Kaisarov, who wrote a dissertation on the liberation of peasants in Latin and analyzed in it a note by Livonian landowner Ungern-Sternberg, in whose opinion slavery is rooted in human nature and meets principles of reason. Kaisarov publicly defended his thesis in Göttingen in 1804. In Sartorius' seminars, Danilevskii read abstracts on financial law that were imbued with the views of Adam Smith; later, when he was directing choices of reading for one participant of the Decembrist conspiracy (A. Fon-Briggen), Danilevskii had his pupil read *Wealth of Nations*" (Kovalevskii 1915: 135–137). Nikolai Ivanovich Turgenev, a future Decembrist and the author of well-known books (*Opyt o nalogakh, Rossiia i russkie*), was particularly interested in economics. While still in Moscow, Turgenev had been educated at a Moscow boarding school for the nobility opened in 1789. In 1807, he attended lectures on statistics and related subjects. Turgenev went to Göttingen with fellow students, some of whom later became professors.

According to the famous Russian historian A. A. Kornilov, "the first years of Aleksandr's reign can be recognized as the most brilliant period in the history of Russian enlightenment in the nineteenth century" (Kornilov 1993: 131). Already in January 1803, the Ministry of Education was established, which developed a broad program for creating educational institutions of various levels throughout the country, from parochial schools to universities. In 1804, Kazan and Kharkov universities were opened, and in May 1803, this Ministry transformed the Teacher's Seminary in St. Petersburg into the Teachers' Gymnasium, "in accordance with new requirements for new educational institutions and with the state of science in Europe."[10] This gymnasium required teachers with such training that no Russian teachers had. N. N. Novosiltsev, the Trustee of the St. Petersburg Academic District, was one of Aleksandr's closest friends and was part of an ad hoc secret committee formed a few months after his accession to discuss projects of social transformation, and he invited foreign scholars to study these issues. About 60 scholars from abroad were invited to Russia (Kornilov 1993: 114).

In 1804, the Teachers' Gymnasium was transformed into the Pedagogical Institute, then in 1816, into the Main Pedagogical Institute. On the basis of the latter in 1819, the Imperial St. Petersburg University was established, which immediately became the central link in the new system of educational institutions taking shape at the beginning of the nineteenth century. The initiator of the University—S. S. Uvarov, a Trustee of the St. Petersburg Academic District—was a student of German humanists. Goethe called him an international scholar (Zhukovskaia 1998: 58). He sought to attract European educated scholars to Petersburg, drew on the experience of German universities in his draft university statute (for which he was severely criticized by reactionaries such as M. L. Magnitskii and D. P. Runich), and energetically opposed recommendations by the director of St. Petersburg University, D. A. Kavelin, to base the teaching of political economy on Holy Scripture (Zhukovskaia 1998: 59, 61–62, 64).

The first rector of St. Petersburg University was a native of Hungary, M. A. Balugianskii (Baludianskii), a professor at the University of Pécs who graduated from the Košice Royal Academy (it was similar to university law faculty) and the Faculty of Law of the University of Vienna. In the 1780s, when Balugianskii studied at higher educational institutions of Austria-Hungary, an intensive process of demolishing the feudal structures in economics and politics was completed, culminating in the French Revolution. (Balugianskii graduated from the University of Vienna in 1789.) The ideological basis of this process of fundamental economic and political transformations, taking place not only in revolutionary France, but also in a more regulated fashion in Austria-Hungary, was the doctrine of natural law—which had been influenced by German scholars and professors, for example, Pufendorf, Leibniz, Wolff, and others. This approach was set against older, more medieval forms of governance (Fateev 1931: 14). In Petersburg, Balugianskii taught natural, public, and popular law to the Grand Dukes Nikolai Pavlovich (future tsar Nicholas I) and Mikhail Pavlovich (*Russkii biograficheskii slovar, T. II* 1900: 453). In St. Petersburg University, in particular, he lectured on political economy. His role in the formation of university traditions cannot be overemphasized: it was he who laid the foundations for the teaching of political economy, formulated its terminology (Fateev 1931: 33), and "held purely academic receptions of an enlightened German scholar, and sowed simplicity and comradely relations that ran counter to Magnitskii's policy" (*Entsiklopedicheskii slovar* 1890: 833).

Although Balugianskii was treated kindly by the authorities, it was difficult for him, a person brought up in European traditions, to get used to manifestations of wild ignorance and obscurantism he felt he often encountered in Russia, even in the academic environment and at the highest levels of the state.[11] In 1821, the new Trustee of the Petersburg Academic District, the arch-reactionary D. P. Runich, accused the best professors of St. Petersburg University—E. V. S. Raupakh, A. I. Galich, K. F. Hermann, and K. I. Arseniev (the last two were outstanding Russian statisticians)—of "preaching a deceptively obvious system of unbelief," including "maratism and robbespierreism" and other "isms" (*Russkii biograficheskii slovar, T. II* 1900: Pritvits-Reis: 501). Balugianskii as rector did everything to dismiss these accusations and protect his colleagues, but contrary to his will, these professors were still dismissed from the university. The Minister of Spiritual Affairs and National Education, A. I. Golitsyn, offered to send Raupakh and Hermann abroad as foreigners (they were Germans) and warn governments of the Holy Alliance about the dangerous nature of their scholarly and pedagogical activity, but the professors had influential defenders. One historian of St. Petersburg University, V. V. Grigoriev noted, "It seems to us that the unseemly behavior of some Russian professors of the university (they were 7 out of 20) who took the side of Runich and Kavelin, happened partly from their national dislike of Raupakh and Hermann as Germans" (Grigoriev 1870: 14). This vitriol made it difficult and unpleasant to work in such an environment, so in 1822 Raupakh left Russia for good, depriving the university of a major specialist in general history and literature. Raupakh enjoyed deep respect at the university, evidenced by the fact that during elections for the first rector, votes were distributed equally between Balugianskii and Raupakh. The University Council decided this issue by lots, but the Committee of Ministers recognized that solving such an important issue in this way contradicted existing rules for electing the rector, and they appointed Balugianskii to the position, at the suggestion of the Minister of Public Education.

This "University Affair" caused a sharp confrontation between ruling spheres, which contributed to its prolongation. However, when Nicholas I came to power, he ordered the dismissed professors to be declared innocent, and the case closed. The fact that he considered this scandal to be frivolous is evidenced by the fact that, in 1821, Grand Duke Nicholas Pavlovich, as the inspector general in engineering, thanked D. P. Runich "for driving out Arseniev, who could now devote his entire time to the engineering school, and asked for the expulsion from the university of a few more people like him, so that they could use them for their own benefit" (*Russkii biograficheskii slovar, T. II* 1900: 318). This moment aside, the atmosphere of obscurantism and persecution of large scientists, which reigned in the St. Petersburg University at the beginning of the 1820s, forced Balugianskii first to leave the position of rector (1821) and then as professor (1824).[12] In 1822, Balugianskii's former student from the Pedagogical Institute (where Balugianskii was a professor before the formation of the university), M. G. Plisov, was dismissed from the university; since May 1820, Plisov had taken Balugianskii's place in offering a course on political economy and speaking in defense of the abovementioned professors during the University Affair. In 1808, Plisov, as part of a group of 12 students from the St. Petersburg Pedagogical Institute, had been sent to study abroad. At the same time, "they had to spend two years in Germany (Heidelberg and Göttingen), and the third year was scheduled for trips to Paris, Rome, Vienna and other cities of France, Italy and Germany, as well as to England. The main attention was paid to political and philological sciences in Göttingen … Those sent for political economy were required to study: political history, especially new [history], the philosophical rights, natural and the people's, Roman law, statistics of European states, political economy in all systems" (Ikonnikov 1876a: 529–530). The purpose of this action was timely replacement of departments by Russian scholars. Possibly some of those who might have felt the heat of this new generation educated abroad decided to strike first—not a unique event in Russian history, alas.

The debacle at St. Petersburg University in 1821 led to a sharp drop in the level of teaching: "The consequence of the thunderstorm … was the replacement of the best professors by a generation of perfect insignificance" (Miliukov 1994: 287). N. I. Butyrskii, a specialist in philology and aesthetics but not in anything related to economics, began to teach political economy in place of the brilliant M. A Balugian and the young, highly educated, and promising Plisov. As V. I. Ikonnikov expressed it so vividly and precisely, at St. Petersburg University, "historical and political sciences before the new charter (1835) were moving in reverse" (Ikonnikov 1876b: 83). According to A. A. Kornilov, "the people's … enlightenment, which moved readily forward at the beginning of the reign [of Aleksandr I], was now suppressed, distorted and disfigured by obscurantist and reactionary measures of clerics and mystics" (Kornilov 1993: 130). According to V. V. Grigoriev, "Many professors, even the celebrities, would now be appointed teachers in decent gymnasia. For three or four people gifted and knowledgeable, there are twenty, thirty teachers who had neither knowledge nor vocation for the professorship" (Grigoriev 1856: 31). As noted by M. A. Bakunin, "[N]one of the Russian universities can still provide a classical academic education. There is only one way to achieve this: to go to Berlin" (Kovalevskii 1915: 149). The situation at Moscow University was somewhat better than in St. Petersburg. Political economy was taught according to German textbooks of Schlözer, Shtorkh, and Rau, and statistics according to the German textbook by Heim and the Russian textbook by Ziablovskii (Ikonnikov 1876b: 84).

Routinization of the Hybrid: Field Relations Insulated and Normalized, Somewhat

The period of reaction was also manifested in changes to the system of training professorial personnel. At first, it was "equally forbidden … to listen to lectures at some German universities" (Ikonnikov 1876b: 81). Soon there were almost no qualified specialists in the social sciences in universities:

> To meet the urgent need for replacing departments (from 1819–1828 there was no travel abroad), we again had to appeal to students (1828) from theological academies (six from St. Petersburg and Moscow), for their preparation at universities and sending them abroad, and to restore the Main Pedagogical Institute (mainly from pupils of theological seminaries)

> for the preparation of professors. Finally, from 1828–1838, the preparation of Russian professors was served by the University of Dorpat,[13] where up to thirty students were sent during this time. For ten years, the University of Dorpat received 22 professors. Thus, Dorpat became the hotbed of intellectual forces for Russia, and it owed its prosperity to the fact that it stood outside the catastrophes that affected Russian universities and direct relations with German universities ... And afterwards the University of Dorpat continued to serve as a school for Russian universities. (Ikonnikov 1876b: 86–87)

Measures taken to train professors, and the approval in 1835 of a new university charter that was more liberal than even German statutes at that time (Ikonnikov 1876b: 90), contributed to the improvement of the social sciences. According to T. N. Granovskii, Moscow University had excellent professors whose courses were no worse than those in the best German universities (Ikonnikov 1876b: 105–106). However, revolutionary events in Europe in the late 1840s meant that philosophy of history and political economy once again were suspect subjects in the eyes of the state (Ikonnikov 1876b: 98).

Accordingly, various restrictive measures were again introduced for universities: in March 1848 it was forbidden to send employees of the Ministry of Public Education abroad; in 1852 it was forbidden to invite foreign scholars, which led to the decline of the University of Dorpat; the rights of members of the Academy of Sciences and universities to write books or for periodicals from abroad without censorship was limited; and censorship generally became tougher (Ikonnikov 1876b: 102–103). In 1876, Professor V. I. Ikonnikov wrote of essays that could not be printed because of censors' decisions: "Reading now these works, one can only marvel at the ingenuity of events of that time and regret that so much was lost for the golden age of development of independent science. Hence the eternal, hopeless, historically-inevitable dependence of Russian thought on foreign authorities. In reports on the academic activity of universities, one cannot help noticing the significant preponderance of works on practical knowledge and loud promises of preparing the publication of major works on other sciences that have not yet appeared in the world" (Ikonnikov 1876b: 101). Even the official organ of the Ministry of Education, after the advent of another liberal era under Aleksandr II, was forced to admit the negative impact of these isolationist policies on the development of Russian science since the end of the 1840s: "Who does not know that in Russia the university reform of 1835 laid firm foundations for university education, reforms, one might say, produced under the exclusive influence of German science" (cited in Ikonnikov 1876b: 112).

Thus, the process of the formation of political economy, as of other social sciences, in Russia in the eighteenth and nineteenth centuries was under the enormous influence of a more mature German science. Strengthening ties with German universities for many decades contributed to the acceleration of the development of Russian economic science, the gradual creation of prerequisites for its transition to an independent path of development, and at the same time, for its ever-deeper integration with European economic science. This process, however, was constantly interrupted, as many economic ideas spreading in the West seemed to the powerful in Russia—often with good reason—to be politically dangerous. Outstanding German economists who worked in Russia in the first half of the nineteenth century not only taught political economy, financial law, and statistics to Russian students. They also used economic theory to study the Russian economy, making a significant contribution to this area. In 1805, Christian Schlözer in his famous work published by Moscow University, in Russian, provided many valuable recommendations on economic policy for the Russian state: the advantages of free labor rather than serf labor, the profitability of cultivating in Siberia using free labor, economic advantages of a transition from exporting raw wood to exporting processed wood products, ensuring national security by organizing production of military products within the country, overcoming dependence on imports, the economic feasibility of expanding such crops as flax and hemp, and so on. The achievements of German and European economic theory were widely used for analyzing the Russian economy and for elaborating an optimal course of economic policy by A. K. Shtorkh, E. F. Kankrin, N. Kh. Bung, and other outstanding German-Russian scholars and politicians. A similar form of hybridity was manifested in the works of nineteenth-century Russian economists who received professional training in Germany (I. Ia. Gorlov, E. R. Vreden, I. M. Kulisher, V. A. Lebedev, A. Posnikov, and others). Finally, in the process of translating European scholarship (including work of German economists) into Russian, the conceptual apparatus and terminology of political economy and other economic sciences in Russia took shape. Ultimately, the field of Russian economics—as discourse and as profession—was built on a hybrid foundation.

In conclusion, the field of Russian economics—as a field of discourse, of professional roles and norms, and of institutional relations and positions—grew out of a paradoxical relation of the Russian state to its geopolitical competitors. Initially, the state encouraged interactions with other European scholars, from sending students abroad to inviting foreigners to work as scholars, professors, and even state officials. Germans tended to predominate in this exercise in mimicry and borrowing. However, increasing such relations with the West also risked the same political tensions and even earthquakes felt abroad, leading reactionaries within the state and society to try to turn back the clock and institutionally insulate various academic fields (and others) from possible foreign influences. This proved too counterproductive, and so much of the nineteenth century witnessed the Russian state walking a fine line between engaging foreign scholars and keeping them at some distance. This hybrid model—neither mimicking perfectly, nor establishing ideal independence—did allow some development of social sciences, including economics, and also provided possibilities for home-grown innovations.

Notes

1. Criticizing educational reforms at the beginning of the nineteenth century, and noting that the bulk of funds were allocated to creating secondary schools rather than supporting existing parish schools, M. M. Speranskii told Aleksandr I that the whole system "contradicted common sense, because common sense demands beginning things from their base and working towards perfection gradually, and therefore, it would have been better to begin with folk school and end with academies" (Miliukov 1994: 284). One could have made a similar comment about Peter I—as did the ideologist of the Decembrist movement, N. I. Turgenev: "No matter how strange and sometimes absurd attempts by

Peter I to push the people towards enlightenment might have been, they could not provide a new impetus for the intellectual movement. Unfortunately, he cared more about appearances than essences, more about external splendor than content. True, he founded the Academy of Sciences and added to it something like a high school that would train teachers; but, in fact, he did little for public education; the establishment of schools at different levels of education for the masses clearly did not occupy him" (Turgenev 2001: 305–306).

2. Ultimately, these institutions played an important role in the rise of the Russian *intelligentsia*.
3. Miller himself studied at the University of Leipzig in 1724–1725. After 1725, he served the St. Petersburg Academy of Sciences. After 1754, he was the conference secretary for the St. Petersburg Academy of Sciences. He was the founder (in 1732) of the first Russian historical journal (in German), *Sammlung Russischer Geschichte*. He was also editor of *St. Petersburg Vedomosti*.
4. Professor N. Karataev of Moscow University noted in his book, written for the 200th anniversary of university, that Reichel defined statistics "as a science describing the state of various branches of the economy and culture of the state" (Karataev 1956: 12).
5. "With Radishchev's example, we were not surprised to be convinced that, despite residing at German universities, Russians returned home not only with German, but also with French and English culture, and as followers of that philosophy of Enlightenment that from England began to penetrate into France at the time of the appearance of Voltaire's *Philosophical Letters*, and came into full bloom at the time of the publication of the famous d'Alembert and Diderot's *Encyclopaedia*, Montesquieu's *Spirit of Laws*, and the first *Discourse* of Rousseau and his famous *Emile*" (Kovalevskii 1915: 132–133).
6. From 1761 to 1767, the outstanding German scholar August-Ludwig Schlözer worked in St. Petersburg. In 1769, he occupied the chair of Statistics, Politics and Political History of European States at the University of Göttingen, where he had passed his student years. He devoted himself to the study of statistics and exerted important influence on the development and acceptance of statistics as a real science. While working in St. Petersburg, Schlözer initiated the decree on the delivery of parish lists of the population using a form that he compiled, which led to statistical study of Russia's population (Maikov 1911: 340–342).

7. V. I. Sergeevich claimed the original version of the Empress' Decree to the Legislative Commission on drafting of a new Code "explicitly discussed the emancipation of the peasants" (Sergeevich 1878: 252).
8. Testimony of more than 200 witnesses supported claims that she killed at least 38 people; possibly another 26 could be attributed to her. Victims were not just killed but also tortured. "Soon ... after the death of her husband, from 1756, rumors began about her cruelties; over six years at least 21 times her peasants filed complaints about intolerable cruelties of their landowner; but every time ... cases of complaints did not make further progress and complainants themselves were extradited to the landowner" (*Russkii biograficheskii slovar* 1904; *Sabaneev-Smyslov*: 69).
9. V. I. Sergeevich noted that "among representatives of the Russian land of the eighteenth century there were people who correctly understood the needs of state of their Fatherland, who knew how to appreciate the good things in Peter's reforms and had enough courage to fight such an old ulcer as serfdom" (Sergeevich 1878: 259).
10. The Teachers' Seminary was opened in 1782 to train teachers for public schools. The formation of public schools began in 1786, when 26 major folk schools were opened in provincial cities (cf. "O nachale i postepennom vozrastanii Imperatorskogo S-Peterburgskogo universiteta" 1838: 15).
11. Illustrative in this respect is one professor of Russian history, N. Ia. Aristov: President of the Russian Academy of Sciences (in different years he also worked as a Secretary of State, member of the State Council, Minister of Education, and Chief of Spiritual Affairs of Foreign Confessions). A. S. Shishkov "assured that statistics are obliged to inform only of good deeds, and those such as murder and suicide must sink into eternal oblivion, and so one should not work on such empty things" (Aristov 1879: 83).
12. Balugianskii left his mark not only in the formation of Russia's university system, but also in the reforms of M. M. Speranskii, in the Commission for drafting laws, and in the transformation of Russian finance. As A. N. Fateev noted, "Balugianskii laid the first stone under the building of Russian financial reforms, and Speranskii's skillful and always original pen gave him a harmonious legal form applicable to Russia" (Fateev 1931: 38). Balugianskii was Speranskii's "right hand" and his most reliable support. These outstanding Russian statesmen highly valued each other. Yet, as Balugianskii's daughter, Baroness Medem, recalled,

Speranskii did not have enough knowledge of German culture or training. "Father," she wrote, "recognized Speranskii's virtues and great abilities and often said that Speranskii would be an absolutely great statesman if he knew the German element, which, unfortunately, he entirely ignored (he loved French and English, but did not have the faintest idea of German). This was an important shortcoming in his development as a statesman" (Medem 1885: 431).

13. Now the University of Tartu.

Works Cited

Aristov, Nikolai. 1879. Sostoianie obrazovaniia Rossii v tsarstvovanie Aleksandra I. In *Izvestiia Istoriko-filologicheskogo Instituta Kniaza Bezborodko v Nezhine, T. III*. Kiev: Tipografiia M. P. Fritza.

Borzov, Ia. 1865. Zametka k state A. Ia. Polenova ob unichtozhenii krepostnogo sostoianiia krestian v Rossii v 1767 g. *Russkii Arkhiv* 3: 316–317.

Ditiatin, Ivan. 1905. *Ekaterininskaia komissiia 1767 g. "O sochinenii proekta novogo Ulozheniia"*. Rostov-on-the-Don: Tipografiia "Donskaia rech" N. Paramonova.

Entsiklodopedicheskii slovar. 1890. St. Petersburg: F. A. Brokgauz, I. A. Efron.

Fateev, A. 1931. *Akademicheskaia i gosudarstvennaia deiatelnost M. A. Balugianskogo (Baludianskogo) v Rossii*. Uzhgorod: Tipografiia "Shkolnaia pomoshch.

Grigoriev, V. 1856. T. N. Granovskii do ego professorstva v Moskve. Razdel 'Smes'. *Russkaia Beseda* III: 17–46.

———. 1870. *S.-Peterburgskii universitet v techenie prebykh 50 let ego sushchestvovaniia*. St. Petersburg: Tipografiia V. Bezobrazova i KO.

Ikonnikov, Vladimir. 1876a. Russkie universitety v sviazi s khodom obshchestvennogo obrazovaniia. *Vestnik Evropy* 7 (Bk. 10): 492–542.

———. 1876b. Russkie universitety v sviazi s khodom obshchestvennogo obrazovaniia. *Vestnik Evropy* 6 (Bk. 11): 73–132.

Kamenskii, A. 1996. Ekaterina II. In *Otechestvennaia istoriia. Istoriia Rossii s drevneishikh vremen do 1917 goda. Entsiklopediia. Tom vtoroi*. Moscow: Bolshaia rossiiskaia entsiklopediia.

Karataev, N. 1956. *Ekonomicheskie nauki v Moskovskom universitete (1755–1955)*. Moscow: Moscow University.

Kornilov, A. 1993. *Kurs russkoi istorii XIX veka*. Moscow: Vysshaia shkola.

Kovalevskii, M. 1915. Borba nemetskogo vliianiia s frantsuzskim v kontse XVIII i v pervoi polovine XIX stoletiia. *Vestnik Evropy* (October): 121–163.

Lappo-Danilevskii, A. 1990. *Istoriia russkoi obshchestvennoi mysli i kultury. XVII–XVIII veka*. Moscow: Nauka.

Maikov, P. 1911. Shletser Avgust-Liudvig. In *Russkii biograficheskii slovar. Shebanov-Shiutts*. St. Petersburg: Tipografiia Glavnogo upravleniia udelov.

Medem, M. 1885. Moi vospominaniia ob ottse moem Mikhaile Angreeviche Balugianskom. *Russkii Arkhiv* 11: 415–432.

Michatek, N. 1913. Reikhel Iogann-Gotfrid. In *Russkii biograficheskii slovar. Reitern-Roltsberg*. St. Petersburg: Tipografiia Glavnogo upravleniia udelov.

Miliukov, P. 1994. *Ocherki po istorii russkoi kultury. Tom 2, chast 2*. Moscow: Progress.

O nachale i postepennom vozrastanii Imperatorskogo universiteta. Chitano v torzhestvennom sobranii universiteta v den otkrytiia ego v byvshem zdanii Dvenadtsati kollegii 25 marta 1838 g. Rektorom I. Shulginym. 1838. *Slovo i rechi, chitannye Rektorom i professorami Imperatorskogo S-Peterburgskogo universiteta v den otkrytiia ego v byvshem zdanii Dvenadtsati kollegii 25 marta 1838 g*. St. Petersburg: Tipografiia A. Pliushara.

Pekarskii, P. 1862. *Nauka i literature v Rossii pre Petre Velikom. Tom pervyi*. St. Petersburg: Obshchestvennaia polza.

———. 1870. *Istoriia Imperatorskoi Akademii nauk v Peterburge. Tom pervyi*. St. Petersburg: Izdatelstvo Otdeleniia russkogo iazyka i slovesnosti Imperatorskoi Akademii nauk.

Polenov, A. 1865. O krepostnom sostoianii krestian v Rossii. *Russkii arkhiv* 3: 287–315.

Russkii biograficheskii slovar. Pritvits-Reis. 1910. St. Petersburg: Tipografiia Imperatorskoi Akademii nauk.

Russkii biograficheskii slovar. Sabaneev-Smyslov. 1904. St. Petersburg: Tipografiia V. Demakova.

Russkii biograficheskii slovar. T. II. 1900. St. Petersburg: Tipografiia Glavnogo upravleniia udelov.

von Schwarzenberg, Eisen. 1764. Eines Lieflaendischen Patrioten. Beschreibung der Leibeigenschaft, wie solche in Liefland ueber die Bauern eingefuehlt ist. *Sammlung Russischer Geschichte*. Bd. IX. SPb. 1764. S: 491–527.

Semevskii, V. 1888. *Kresianskii vopros v Rossii v XVIII i pervoi polovine XIX veka. Tom I*. St. Petersburg: Obshchestvennaia polza.

Sergeevich, V. 1878. Otkuda neudachi Ekaterinskoi zakonodatelnoi komissii? *Vestnik Evropy* (January): 188–264.

Turgenev, N. 2001. *Rossiia i russkie*. Moscow: Obedinennoe gumanitarnoe izdatelstvo.

Vishnitser, Markus. 1908. Gettingenskie gody Nikolaia Ivanovicha Turgenieva. *Minuvshie Rody* (April): 184–218.

Zhukovskaia, T. 1998. S. S. Uvarov i vossozdanie S.-Peterburgskogo universiteta. In *Ocherki po istorii S.-Peterburgskogo universiteta, T.VII*, 56–74. St. Petersburg: St. Petersburg University.

3

Compulsion and Resistance: Origins of the Russian Research Tradition and Political Economy of the Special

Viktor Ryazanov

Economics as a Field of Science and Vocation in Russia

While the rise of a community of scholars and their discipline is influenced by discourses, institutions, and networks of actors who exchange ideas and compete over status, a discipline and profession can also be its own field of oriented, meanings, practices, and status conventions, even if these are embedded in or intersect with other institutions or institutional fields (cf. Wuthnow 1989). The peculiarities of the formation and development of economics as a discipline and social science in Russia is a story of competing sources, forces, and trajectories. It can only be understood by considering the interaction of these dynamics: a global discourse, with its own competing schools and converging theoretical logics; Russia's status as both a Great Power at the same table as Western counties *and* as a latecomer with lesser status vis-à-vis Western intellectual discourses; and

V. Ryazanov (✉)
Faculty of Economics, Department of Economic Theory,
St. Petersburg State University, St. Petersburg, Russia

J. K. Hass (ed.), *Re-Examining the History of the Russian Economy*,
https://doi.org/10.1007/978-3-319-75414-7_3

a sense at home of uniqueness borne of this status competition *and* of different economic circumstances. All three of these dynamics had their own "domains" in particular fields of institutions and networks, and the unique nature of Russia's research tradition in economics was inextricably linked to and interacted with all three. Global economics discourses paid great attention to identifying general principles and characteristics, although there were different camps located in different countries. In Russia, the state and educated non-economists focused on the needs and status of the country, first and foremost; "economics" was a tool for national improvement. Russian scholars working at universities and embracing "political economy" and then "economics" were in between both of these fields of actors and meanings, navigating both the universal and the specific. It was from the intersection of these three fields and their actors that Russian economics created a set of innovative tools and concepts that gave Russian economics its own national "flavor."

In the logic of field theory as a framework for making sense of social organization, the representatives of academic disciplines and schools "try to create a 'local' and stable world in which dominant actors generate meanings that enable them to reproduce their advantages. These actors create status hierarchies that determine the positions of dominant players and pretenders to domination" (Fligstein 2013: 62–63). The formation of the Russian economics research tradition reflects these dynamics. This history of Russian economics, from its beginnings to the present, is characterized by competition between different schools for status positions, including setting the main issues and parameters of "economics," and this took on an acute form during critical periods in Russian history. Exploring peculiarities of the formation and development of Russian economics has potential value-added insights relevant to issues of historical originality in the formation of scientific fields not only in Russia, but beyond discourses and into economic practices implemented and economic policies pursued.

Let us single out one remarkable detail. Economics as one scientific field "is a place (i.e. a gaming space) of competitive struggle, the specific rate in which is the monopoly on scientific authority, defined as technical ability and—simultaneously—as social power, or, if desired, a monopoly on scientific competence" (Bourdieu 2005: 474). Such was part of our story. The very birth of academic economics in Russia took place under the direct influence of the English school of classical political economy, and in subsequent periods, Russian economists bequeathed not a small number of important innovations by mastering and rethinking achievements of global economic thought in its various trends and schools. In the development of Russian economics, it is important to note two such features. First, the originality of research in the history of Russian economics was born in the process of mastering the methodology of political economy, for a time, the leading school in global economics; but at the same time, Russian economists were stimulated by the growing desire to make their own discoveries and set out new areas of research. Further, the formation and development of Russian economic thought could not but be affected by competition from different schools as they emerged in global economic thought.

Competition between different schools in Russia reflected a common property associated with the formation of a leading scientific school—orthodoxy—and alternative trends competing with it in all their diversity—heterodoxy. This is how a "discussion field" arises in the sphere of economic knowledge, in which orthodoxy and heterodoxy collide, and scientific authority, a special type of scientific capital, is established. At the same time, the scientific field, which characterizes objective relations between existing scientific positions, always acts as a place for unequal struggle between spokesmen (agents) unequally endowed with scientific capital (Bourdieu 2005: 489). The history of economics in Russia is an example of such a clash of scientific ideas, but one where domination was usually determined by borrowing leading ideas from European (Western) economics as orthodox economic knowledge; competing scientific views, the heterodoxy, reflected national originality.

The Universal and the National: "Schools" of Economic Thought

Let us analyze what Russian scientific traditions represented for economic knowledge and how they might have been linked to the formation of a "national school." I begin with a brief discussion about whether it is justifiable to single out national theoretical and economic schools and how this could help us better understand Russian economics.[1] These issues were the subject of a number of Russian conferences that led to the publication of monographs and many articles (cf. Abalkin 2003; Iakovets 2003; Iakutin 2010). There has also been no small interest in the history of Russian economic thought among foreign scholars, especially Germans. In this connection, it is enough to single out the work of economists from England (Barnett 2005) and Germany (Zweynert 2008). Italian scholar G. Carpi (2016) has also addressed the history of Marxism in Russia, and in so doing, also raised questions about the history of Russian economic thought.

The first question to be addressed is the extent to which it is valid to classify schools of economics as a "national field," versus being driven more by a global discourse and professional field, with its own schools of thought and positions. Among economists, there is one opinion that rejects the idea of a national economic school: economics, like any other science (physics, chemistry, mathematics, etc.), has no national borders, and because its subject and laws are universal, economics has to be a truly international creative activity. However, this argument oversimplifies the realities of scientific discourses. Variation in discourses persists across nations even for the natural sciences and mathematics, which have the strongest claims to universalism. As Spengler noted in 1927 (Spengler 1993: 59), "There are several number-worlds as there are several Cultures. We find an Indian, an Arabian, a Classical, a Western type of mathematical thought and, corresponding with each, a type of number—each type fundamentally peculiar and unique, an expression of a specific world-feeling, a symbol having a specific validity which is even capable of scientific definition, a principle of ordering the Become which reflects the central essence of one and only one soul, viz., the soul of that particular Culture." The analysis of Archimedes differs from the analysis of Gauss not only in form, but also in content; in Lobachevskii's geometry, parallel lines still intersect, even if in infinity.

Another argument for the universality of economic theory is the fact that the orientation to the search for truth in scientific research should remove any national coloring from the field of the economics. This attitude is significant if we consider economic knowledge to reflect universal economic processes in which an "economic person" participates with a standard set of properties and interests. In this case, economics acquires the property of universal scientific knowledge, simultaneously focused on the study of universally valid forms of organizing economic life and ensuring economic growth, while becoming a universal science of economic processes and development. This interpretation of the place and role of economics is characteristic of neoclassical orthodoxy, claiming a dominant position in the system of modern economic knowledge. It does not reject the existence of other scientific schools, but these are understood as existing in the same global field. They historically precede "mainstream" economics, or are a supplement for possible mismatches for some parameters, or act as peripheral theories that contradict the main field that shapes modern economic thought and defines professional competence. Any "national schools" are rejected as a classification for schools of economic thought, as this provides little for developing an analytical apparatus for economic research and identifying progress. To the extent the "national" matters, it is as particular variables that influence particular national traditions or trajectories in the formation and development of economics in individual countries. In other words, the national section of economic research is limited to sociocultural specifics in the manifestation of the universal for a global mainstream field.

On the other hand, there are serious arguments for singling out a national component in classifying schools of economic thought. As I have elaborated on this topic elsewhere (Ryazanov 2010), I limit myself to a few core theses. First, various currents in economic thought arise and develop with reference to a specific economic field (e.g. by country or region). This means that the initial formation of certain schools is not random or an outgrowth of some universal form. Moreover, their distribution in global economic space often occurs not through direct transfer and mimicry, but through enrichment from interaction with national differences and traditions. Thus, the simplest form of "national" in economics is the presence (or absence) of a national school in a particular country. In this case, it is either built into the general economic field or exists as an alternative process. In some cases, accumulated experience in global economic thought suffices to fulfill economics of all its basic functions in a given country, and corresponding theory is perceived as universal, explanatory, practical, and able to satisfy basic needs of that society. In other specific country-specific cases, the general and universal, with their accumulated economic knowledge, may not be sufficient to understand economic reality and to determine effective ways to develop that national economy. It is important to understand the reasons for the limited application of general and universal economic knowledge, for this is what gives rise to a strong impetus to develop a national component of economic theory as either alternative or supplement to existing universal theory.

In this context, it is worth mentioning examples of designating economic schools with a national label. The "English" and "Austrian" schools of political economy, the "German historical school," and "American institutionalism" reveal how core economic frameworks grew out of historical experiences and institutional contexts. Other similar labels refer to institutions (especially universities) in which particular individuals and networks supported competing economic frameworks, for example, the Chicago school or Cambridge school (or, in contemporary American macroeconomics, "freshwater" and "saltwater" schools). Such classifications not only indicate countries of origin, but also specific experiences, institutions, and discourses of those contexts that influence the emergence of certain traditions. National schools of economics stand out for their closer attention to the particular that is built into the general economic field.

The fact that heterogeneous links interact and interweave in economic fields, revealing the presence of the universal, the particular, and the individual, cannot but affect the formation of a complex structure of global economic thought. This means that not only the universal, but also the particular in economic discourses can induce the development of special economic theories, the subjects of which are national economic systems. This circumstance is perhaps the most important methodological and practical basis for taking national schools as valid: when peculiarities are significant and require the creation of their own economic model given these economic specifics. Perhaps this is not surprising: as institutional approaches across many disciplines have noted, diversity in economic practices across national models reflects various historical ways of organizing capitalist economies. If this is the case, then there might be universal variables that are specific to space and time and cannot be fully integrated into a general, universal model. Further, the more diverse and broader the national specifics, the higher the need for a special or contextual component to general economic theory. The latter can take two forms: an adaptation of general forms of governance and policy in a specific country, taking into account that country's specifics; or as a conceptual justification for creating a special economic system whose parameters are provided by forces of the corresponding school of economic thought. These more contextual models can draw on insights of other schools, including those that claim to unveil universal laws, but their main purpose is to understand and implement nationally special economic policies that correspond to concrete needs in those places and times.

Another reason for the emergence of national economic schools is that economic knowledge from each country cannot but reflect not only national experiences, but also national identities and dispositions. While mainstream theory uses generic *homo economicus* for reasons of convenience, an economic entity in a real economy cannot be recognized as an abstract "economic person," for in each case she or he acts as a concrete person, and actors embedded in national institutions and discourses most likely carry some facet of those phenomena in their dispositions and practices. In modernity, nations play significant roles in situating individuals in social time and political space, including in national and global economic fields (Greenfeld 1992). Common languages, institutions, and traditions that make up nations inculcate particular identities in individuals (whether or not they want to embrace them) that can be as strong as social class or similar forms of community. Further, organization of the economy, and actors' experiences in that context, also contribute to dispositions and worldviews. It is not accidental that basic concepts of the economy and its role in social life, as well as views on economic categories and methods of governance, differ not only across countries but also within them, for example, along ethnic bases. Emerging national economic schools that have the potential affect the formation of these economic national self-awareness and worldview—just as experiences in developing economic contexts shape those emerging economic worldviews and schools of thought.

Finally, we must consider that behind the real economy, there are not only general principles of how economic systems operate, but also national economic interests, including competition for profitable positions in the international division of labor. Therefore, economic theory can fulfill not only a narrowly understood ideological function associated with social and class interests, but also an even more significant role—to reflect and protect national economic interests in the competitive struggle in the global economic field. The national character of schools of economic thought is precisely connected with their importance for ascertaining and justifying a country's competitive advantages in the international division of labor. Since these can arise from unique economic structures, the task of any national school of economic thought is not simply to make sense of those elements of that economic structure, but also to look for ways to transform them into geopolitical advantages.

Thus, there are good reasons to take seriously the idea of whether economics as a science can have a specific national dimension in addition to the universal. Perhaps this should not be surprising, given that one of the main objects of economic analysis, even in conditions of modern globalization, remains the national economy. This in turn creates the need for economic theory that takes into account national specifics as an integral component.[2] If nations with their own institutions and cultural traditions have important explanatory force for economic processes, theory should take these into account, rather than leaving them as random variation exogenous to the model. If economics as a science plays an important role in justifying the transformation of what is nationally unique into a competitive advantage and thus protecting national economic interests, then national variation matters as something in and of itself, something possibly *sui generis*.

The Russian School of Economic Thought: Three Beginnings, Three Sources

Let us now turn to Russian economics against the background of the development of the global economic discourse. This is a broad topic, and so I will focus on one facet: the claim in Russian economics to have the status of a "national school," grounded in its own economic and discursive fields. This requires exploring the degree of independence of processes of formation and development of economics in Russia, and the extent of Western influence.

Academic economics emerged in Russia under the influence of this emerging young science in Europe. In particular, political economy provided a formal reason for thinking that Russian economic thought could not develop independently of its context. Russian scholars studying the economy mimicked economic and political forms and knowledge elsewhere. This orientation to European economic thought was largely explained by the desire to rely on the European experience of designing an industrial market economy. Russian scholars viewed this as an opportunity to accelerate their own development and to overcome economic backwardness.[3] Those who claim that Russian economics was an offshoot of foreign ideas concluded that its very appearance occurred under the direct impact of emerging economic schools in Europe, originally the English classical school, and then others (French and German).[4] This assumption about the secondary nature of Russian economic views was also argued by making reference to the synchronization of changes in these views with changes in European intellectual fashions. V. V. Sviatlovskii, who devoted part of his research to studying "foreign influences" on Russian economists and politicians, did not ignore "synchronization" (Pokidchenko and Kalmyzhkova 2003: 66–88). Up to the present, a considerable number of Russian authors (and foreign scholars as well) link the development of Russian economics with shifts that occurred and are taking place in the global field of economic thought.

That "foreign influences" were significant historically in Russia is not in doubt. Starting with Peter the Great, the overall development strategy was aimed at massive Westernization of the country, and borrowing Western economic ideas played a key role. This practically predetermined the secondary status of Russian versions in discursive and professional fields. At the same time, the growing fashion for political economic knowledge by analogy with European countries spread to educated Russian society. The upbringing and education of members of the Imperial household and most of the nobility, which formed the basis of the ruling elite, were under the dense and long-lasting impact of Western European culture, including lifestyles and languages, that made up the appropriate lifestyle and nature of thinking. This could not help but facilitate the concrete specifics of the formation of the Russian nobility, almost half of which, by some estimates, was of non-Russian, European origin (Alekseeva 2007: 73).

Because of the priorities of existing international economic, political, and cultural ties of Russia, immigrants from Germany played a special role in the history of Russian science, including economics. It is well known that the first Russian academician in political economy, Andrei Karlovich Shtorkh, was German by birth, and the first professor of political economy in Russia was Christian von Schlözer, who headed the department of "diplomacy and political economy," which was created in 1804 at Moscow University, until 1825; he also published the first textbook in Russia on political economy, a two-volume work in 1805 and 1806 titled *The Initial Foundations of the State Economy, or the Science of the National Economy* (*Nachalnye osnovaniia gosudarstvennogo khoziaistva ili Nauka o narodnom khoziaistve*), which was also translated into German and French. German economists also actively participated in Russian state administration: for example, for 69 of the 115 years of its existence (1802–1917), the Ministry of Finance of the Russian Empire was headed by an ethnic German (7 of 18 ministers).[5]

The increased and sustained interest in European science and education was manifested in different ways and forms. One such form was inviting foreign scientists and economists to work in Russia,[6] while also sending representatives of educated Russian society to study in Europe. According to rough estimates, during the eighteenth and nineteenth centuries, more than 20,000 people from Russia were trained in European universities; for comparison, according to available data, from 1859 to 1900 Russian universities trained only 58,000 specialists. Russian specialists who received their education and professional skills in Europe created channels to disseminating ideas in Russia's economic field; however, having landed in an excellent economic environment, they were forced to focus on the need to adapt this knowledge to the Russian context, which provided an impetus for the emergence of independent scientific research. It is also important to note that in many cases, Russia's higher authorities played an active role repeating and disseminating European economic ideas and European economic and administrative experiences. The authorities saw in Russians mastering theoretical knowledge not only the prospect for familiarizing the Russian *intelligentsia* with Western scientific traditions with corresponding knowledge and skills, but also an important social and educational mechanism that could enable a possible reformation. Therefore, it is not surprising that the state actively popularized new economic knowledge in official publications and organized translations of foreign works on the economy. In many cases, such transfers were carried out at the direct request of the state with financial support. This is how the first Russian edition of Adam Smith's *Wealth of Nations* appeared in 1802–1806, with other classics soon to follow.

No less significant for the spread of European economic knowledge was the creation of departments of political economy[7] and the inclusion of political economy and statistics into a number of compulsory disciplines taught in universities. Just as in Europe, Russian universities became the most important institutions for the development of scientific knowledge and the formation of professional skills. At the same time, competition between scientific directions developed in leading Russian universities created its own field of discourse, reflecting the struggle for status positions and authority in economics. In this case, it was university professors that were most attracted to the European choice. We can assume that their ideological and theoretical searches, being under the strong influence of volatile European scientific fashions, took place within the framework of a general global field of economic discourse. Therefore, it is natural that in the pre-Revolutionary period, professors actively mastered both liberal and Marxist ideas.[8]

All this could not but create the impression of an almost exclusive role of European economic thought in the formation of Russian economics. As V. M. Shtein wrote, led to the fact that "the pages of our economists' works were filled with arguments about phenomena that played almost no role in the Russian economy … For them, the main subject is the capitalist enterprise and the relationship between entrepreneurs and workers, developed according to the book, and not the lifeblood" (Shtein 1948: 86). Without denying the importance of foreign influences on the formation of the content of Russian economics, we should still pay attention to other roots. Relevant views evolved not only in universities and academic environments, but also in the Free Economic Society (established in 1765 "to promote farming and saving" and to study and generalize foreign business experience), which regularly organized contests around pressing economic issues. No less significant were numerous social and political circles and journals in which key issues of Russia's development were debated, including the abolition of serfdom and the fate of capitalism in Russia. Discussions of the Russian economy were extensive. For example, in many of his novels, Fyodor Dostoevskii discussed economic and financial problems, especially money and the strengthening of finance in Russian society, or problems of ownership and new ways of management after the 1861 land reform and liberation of the peasants. It is no coincidence that Dostoevskii's economic views, as well as those of other Russian writers (e.g. Nikolai Gogol and M. E. Saltykov-Shchedrin) were often the subjects of special studies (e.g. Carpi 2012). In such a multifaceted environment, a broad range of views on economic issues existed, and original ideas and practical recommendations not dependent on foreign authorities were advanced and honed. As V. M. Shtein noted, "Therefore, historians of political economy for a long time did not notice in Russia independent thought that was embodied not in thick volumes written according to the requirements of the Western European methodological canons, but in journal articles and notes that smack of the 'illegal,' or even in critical reviews about literary works" (Shtein 1948: 86).

That non-academic branch of economic discourse developing in parallel with academic economics in the nineteenth century had its own foundations in earlier ideas of its predecessors. These, in turn, reflected the peculiarity of the economic structure and way of life as they historically evolved in Russia. It is not by chance that in the first article devoted to the history of economics in Russia was by a Moscow University professor, V. N. Leshkov, called "Medieval Russian Science of National Wealth and Welfare." Published in 1855, this work was an analysis and comparison of three of the earliest and most notable economic works of Muscovite Rus and the Russian Empire: *Domostroi* (the mid-sixteenth century), *Books on Poverty and Wealth*, by I. T. Pososhkov (1724), and A. Volynskii's *Instructions* (1724). The professor, highlighting the peculiarities of these authors' economic views, drew attention to a significant difference between Western and Russian thinking about wealth as a key economic concept in emerging economic conditions. "the Western science of wealth," Leshkov emphasized, "cannot make a poor man rich; the Russian does not undertake this task at all; but he does more, learning and being content with what is, rooted in the belief that wealth is not a goal, but a means" (Leshkov 1855: 14). Considering economic science as a science of national wealth, Leshkov reasonably attributed its origin in Russia to the sixteenth through the eighteenth century as an original scholarly project.

Such a starting point for the emergence of economics in Russia, with some reservations, corresponded to a general pattern borne of the need to study a new and complex economic reality, and the need to develop practical recommendations. In this connection, it is worth mentioning the work of I. T. Poshoshkov. It is no accident that M. P. Pogodin, who published this work in 1842, assessed it as a political economy of "common sense" that was "a complete treatise on the state of Russia." In a biographical sketch devoted to this Russian thinker in 1898, A. N. Miklashevskii described Pososhkov as "the first Russian economist," whose work "is not of a strictly economic nature," representing "an original combination of mercantile ideas with canonical Western ideas—a combination all the more curious in that it was created beyond all literary Western European influences."[9] In 1899, M. Filippov, while accepting the existence of parallels in the development of political economy in Russia and the West, argued that these parallels "did not exclude, of course, the individual traits of Russian science in communication with the peculiarities of our economic life and the special conditions for the development of social thought in our country" (*Istoriia russkoi politicheskoi mysli* 1955: 26).

In the early works of Russian thinkers, the possibility of the emergence and consolidation of one's own original approach to the analysis of economic life, fundamentally different from the Western view of the economy, was laid down. It was characterized by special attention to the moral component of economic activity, an emphasis on the problem of "correct distribution" in the interests of multiplying social wealth and rejecting "unjust wealth," and the priority of labor on the basis of Orthodox Christian values. Equally significant was the interpretation of the place of economic activity in the life of society and its individual members. One of the prominent representatives of Slavophilism, I. V. Kireevskii, around the middle of the nineteenth century compared political economy as a science in the West with the historical situation in Russia:

> The Western man was looking for the development of extra means to alleviate the gravity of domestic insufficiencies. The Russian person strove with an inner elevation over external needs to avoid the gravity of external needs. If the science of political economy existed then, without any doubt it would not be understandable to the Russian. He could not agree with the integrity of his view of life—a special science of wealth. He could not understand how one can vex the sensitivity of people to external needs with the intention to multiply their efforts to material production. He knew that the development of wealth is one of the secondary conditions of social life and must therefore not only be in close connection with higher conditions, but also in the subordination that has been accomplished by him. (Kireevskii 2007: 216)

This emerging alternative conception of economies and economics—the idea of the integrity of the human world as key—led to an expansionary version of the original economic concept of wealth. It was not only translated from a goal into a means of economic activity, but also had its own essential difference in content and composition. Not reducible to the material (monetary) component, wealth included a "non-material" component. For I. T. Pososhkov, for example, this was defined as the desire for "true truth." And the idea of integrity of the world as a unity of economic, social, and moral spaces was further developed, becoming a life-giving source not only for domestic economic thought, but also for the unique formation of the Russian school of philosophy with its special attention to the human problem, and its cognitive capabilities. This was embodied in the original development of philosophical anthropology—the science of studying the place of man in the world and society. Naturally, the economic ideas that arise in each country draw inspiration from prevailing philosophical concepts that lay the methodological framework in cognitive practice and shape research agendas for all branches of social science. It was natural that views on Russia's special historical path were originally born and developed in sharp discussions within the framework of general historiosophical and philosophical ideas, and then relayed to other fields of scientific knowledge, including economics. This is how the domestic theme was consistently developed, which found its complete embodiment in the "Russian idea" and in posing the question of the original trajectory for Russian development. Its full-scale development required both creative mastering of foreign theoretical knowledge about society and its laws of development, and direct study of Russia's own experience, with close attention to the Orthodox spiritual practice of the Russian people, reflected in its national culture and mentality. At the same time, the very promotion of the "Russian idea" bred a powerful creative impetus for research. This is exactly what Russian philosopher I. A. Ilin meant by regarding it as a "creative idea" expressing "the Russian historical *originality* and at the same time the Russian historical *calling*" (Ilin 1992: 436, italics in original).

It is important to emphasize that the gradual shift of Russian economists' attention, from abstract theoretical topics to concrete problems facing the Russian economy, led to a turning point in academic and university economics. It was increasingly reoriented from purely educational activities, based mainly on foreign sources, to the study of Russia's real economy as a kind of economic field—not merely a set of universal and abstract laws that happened to be on Russian soil, but tendencies and dynamics that were embedded in Russian experiences and institutions. This created currents that overcame the sense of apprenticeship and imitation of fashionable economic theories from the West. There are good reasons to believe this turn took place in the middle of the nineteenth century, when the country entered the era of fateful reforms in the economy and society (Ryazanov 1998). In this new period in the development of academic economics, there was a disengagement of two lines of thought. Conditionally, we can suggest one was oriented to studying the general and universal in the economy, while the other focused on the national and special. Moreover, the development of this second line of thought could not but interact with other researchers who were not professional economists, but among whom there were outstanding scholars, publicists, politicians, and writers. It was understandable that at some points their ideas intersected, while in other ways they differed significantly. One could speak of continuing research within existing schools in the global economy and focusing on purely scientific interests, while achieving innovative results in a number of areas (e.g.in the theory of cycles and crises or in agrarian topics). In the other case, the formation of a national (Russian) school of economic thought in a narrow sense with its special attention to the economic problems of Russia stimulated independence and a creative approach in research.

Thus, the emergence of an economic science in Russia was not solely related to Western status and influence. Although at the initial stages, it dominated the academic and university contexts, the adoption and development of its national version took place in other spheres of society, not directly related to pure science and education. Therefore, the history of economic thought in Russia is not limited only to academic spheres. In this regard, there were least three sources that explain diverse reasons for the emergence of economics as an independent branch of knowledge. The first was the need to understand the surrounding economic world. The second was the need to formulate the economic policy for the country, to develop practical recommendations for organizing economic activities. The third was the need to study and use scientific ideas and economic experience of other countries. Accordingly, sources for the formation and development of economics in Russia were: Western economic schools, which played a leading role in the emergence of university and academic science; early economic views of Russian thinkers, which contained embryonic ideas about their own model of governance and reflected features of economic mentality and practice, which they developed in writings of the national school; and Russian philosophy, which formed its methodological and general theoretical platform in the study of Russian civilization, including its economic sphere and mental practice.

Political Economy of the Special: Russian Economics

The unequal sources of the origin and development of Russian economics served as the basis for the gradual formation of two dissimilar poles, competing for influence and status in this field of discourse. Although their common object was the Russian economy, the existing antagonism between them broke the single space of economics into two independent and detached parts. One was guided by the study of the general and universal in the economy, and therefore largely was immersed in the global context of economics, reproducing new trends in its development and making its own contributions. The second part arose under the influence of the need to comprehend Russia's own economic peculiarities, which found their expression in the search and justification for a Russian path of economic development. This bred a certain stiffness in the confrontation between theoretical positions and ideas about the future development of Russia's economy and society, which was most clearly manifested in the severity of discussions between Slavophiles with Westerners (1840–1850), and between Narodniks and Social Democrats and liberals (1880–1900). To a greater or lesser degree, such opposition was reproduced in subsequent periods, and its echoes reverberate even today.

Over time, such a sharp disengagement in Russian economics gradually became outdated, and most likely this trend will continue in the future. But nevertheless, these two poles survived until today. On the one hand, their presence reflects the discursive nature of scholarly activity itself and the usual competition of different schools for status. Therefore, the diversity of economic approaches is a standard situation. On the other hand, constant difficulties and disruptions from reconstructing the economy using outside models and claims gave rise to doubts about the validity of this approach, and this encouraged efforts to put forward original economic ideas and to develop alternative projects for Russia's economic development, all oriented to the specificity of its economic institutions and experiences. This laid the foundation for the development of a national economic school, we can define it as a *Russian school of national economy* or a *political economy of the special*. The development of one's variant of political economy became the main direction for development of a national scientific tradition. This school included different trends, such as Slavophilism, Narodism, the Orthodox-autocratic branch, the philosophy of economy, Eurasianism, the economic theory of socialism, and many specific names of economists not formally included (e.g. D. I. Mendeleev, S. Iu. Witte).

The Russian school of national economy (or political economy of the special) can and should be interpreted by its position in the field of global economics, placing it in that part that previously was represented by historical schools of the nineteenth century and in recent decades by institutionalism. Such proximity to economic schools of other countries does not invalidate the design of the national school. Once again, it should be emphasized that its status does not imply that something exists entirely outside the global field, which is the totality of schools and currents competing for scientific leadership and status positions. In addition, shared economic challenges can spur parallel ideas—just as the internal logic of the development of economics can lead to the emergence of the same theoretical concept, not excluding common theoretical problems economists of different countries confront. This said, let us highlight a few key ideas illustrating the novelty of theoretical postulates advanced by Russian economists in the heyday of this school, the last third of the nineteenth century and the beginning of the twentieth century.

First, one of the main achievements of Russian economic thought is the idea of diversity in economic structures and development. This presents one of the most important aspects associated with the possibility of a mixed, multi-sectoral economy as sustainable organization of social production. This was born under the influence of accumulated experience of reforming Russia and was a response to numerous failures that arose with liberal market reforms. At present, the interpretation of most modern economic systems as mixed seems to be elementary and is presented in many textbooks of theoretical economics. Corresponding models of the economy with a heterogeneous structure were worked out. But at this earlier time, when the concept of a multi-sectoral economy was emerging, opposing views dominated global economic thought, especially the possibility of "pure capitalism" or "pure socialism," with appropriate generalizations. Moreover, even today neoclassical analyses almost entirely exclude non-market modes that have not yet been overcome in developed market economies, limiting them to the interaction of state and market, to which the entire content of the mixed economy is reduced. Second, the Russian school of national economy developed a concept of economic reconstruction and development on its own, which made it possible to explore a "Russian way." This concept made it possible to avoid depending on imitating the West in theory and practice, to wager on creative independence, and to lay foundations for national self-awareness and worldviews. This concept rejected the interpretation that non-Western institutions, peculiarities of culture and mentality as economic anomalies, created obstacles to successful social and economic development that must be replaced by Western institutions as universal constructs. Note that objective reasons for choosing an independent economic path seemed to be supported by historically significant achievements unique in world history. For 400 years, Russia expanded its territory by 40 times and increased its population by more than 60 times. All this required colossal effort and the ability to overcome this exorbitant institutional burden. For 537 years—from the Battle of Kulikovo (1380) until 1917—Russia spent 334 years engaged in war, that is, almost two-thirds of this period. Equally important, even in such geopolitical conditions, economic results were impressive. In the 1880–1970s, the GDP per capita in comparable prices increased almost 7.5 times.

Third, the Russian economic school proposed an alternative option for implementing programs for developing Russia: using a non-liberal market and nationally oriented reforms. The concept of the Russian way of reforming and improving the economy was logically completed with a reform program. This rejected the option of denying changes in the economic and political structure of Russia, and at the same time proposed a course of reforms based not on achieving abstract goals, but on real opportunities and prioritizing national interests.

The emerging Russian research tradition in economics had its own peculiar features. These include: criticism of "pure economism" and promoting the idea of the integrity of the economy with all spheres of society; positive relations of economic activity and religious ethics to create a Christian Orthodox economy; reliance on an interdisciplinary approach to economic phenomena; orientation to the real economy, with special attention to the human factor and the role of statehood as the embodiment of a community of interests for achieving creative goals in economic development; and special attention to social consequences in governance and to meeting social priorities. Another important circumstance was that, unlike in Western political economy, which expressed the interests of the bourgeoisie or working class, the Russian school of national economy and, above all, its populist cousin, acted as spokesmen for the interests of the peasantry. Although on the periphery of European economic thought one can also find a similar modes of thought (e.g. Sismondi's work), its role, and the nature of ideas advanced were fundamentally different. It defended interests of the peasantry as a class of petty bourgeoisie, suffering from dictates of big capital. In turn, the populist school in Russia, approaching the peasantry as a laboring class, ultimately went on to develop its own distinctive model of "peasant socialism."

In Place of a Conclusion: Lessons and Significance of the Russian Research Tradition

The domestic research tradition in the field of economic knowledge formed with reference to the Russian economy of the nineteenth and early twentieth centuries. Specific theoretical assessments and practical recommendations belong to history. But is there nothing to revive from "old" traditions, and should we wager on diverse scientific developments in global economic science? Does this mean the design of political economy is special and its methodological grounds are outdated and in applicable to Russian economic realities of the coming twenty-first century?

We are doubtful. Russia's modern economy has its own field of economics that differs in its originality, no matter how it is compared to standards and not leveled to eliminate "dissimilarity." Hence, there is still an impulse to preserve special theories in the emergence of development strategies and transformative activities. In fact, we are talking about the need to turn economic theory to problems of the real Russian economy. This is primarily an appeal to study humanity and systems of economic relations in which they are embedded. The subject of the study should not be limited only to functioning markets and the movement of goods. After all, "markets are structured by multiple social relations within societies and between societies" (Fligstein 2013: 121–122), and some of these realities emerge as theories that reflect those varying social conditions. Therefore, a productive ontology must be fully supplemented by an anthropological ontology. Such a reorientation is a direct inheritance of the Russian political economy tradition, which had its own competitive advantage in the field of the global economic discourse.

A new, perspective view of the economy is increasingly based on the concept of human integrity and the inseparability of all its qualities, and this characterized the key idea of the Russian research tradition. The school of behavioral economics that has gained status recently provides an example of meaningful analysis based not solely on axiomatic premises and mathematical (quantitative) correlations, but on a qualitative analysis of economic processes as human interaction. This approach assumes that economists must master and use tools of anthropological and social–psychological analyses, understand mechanisms of evolutionary biology, and be able to extract that of value that was present in past economic experiences. Important lessons of this research tradition for modern economics are associated with its departure from the absolutization of economism and technocracy, with emphasis instead on interdisciplinarity and synergy, with the acquisition of this trait of "philosophical knowledge." This change also shows the need to change from a natural–mathematical orientation of economics to a social and humanitarian emphasis, which is significant not only in theory, but also in business practices. This cannot but change the key criteria of economic scholarship, which can no longer be only mathematical rigor of calculations and modeling, the schematics and theorems of knowledge, the discovery of the supposedly miraculous that is capable of explaining all those formulas. The facts are that these science-based standards only give rise to excessive ambitions and temptations, which are not confirmed in practice by analytical and prognostic achievements.

The story here seems to foreshadow similar trends later. The development of political economy in the USSR, not so directly connected with global economics, can nevertheless be viewed as reproducing established traditions. The dominance of political economy in that period reflected the leading role of the classical school in its Marxist version, which in its early historical time acted as the main trend in economics. In this regard, the rapid break with the tradition of political economy in post-Soviet Russia was not accidental. The mass transition by economists to neoclassical positions formally returned Russian economics to a common global discourse in which neoclassical orthodoxy predominated—and again, as "pupils" mastering new true teaching and acquiring professional competencies. Russian economics began as an apprentice to older and established Western schools and discourses, and how long this "apprenticeship" would last was not obvious—and much has been the same since 1991. Such a protracted period of development of neoclassical orthodoxy can turn the peripheral state of modern Russian economics into a stagnant state. To a greater extent, this is facilitated by the assertion of the neoclassical monopoly that replaced Marxist orthodoxy. Dictatorship in science, like monopoly in economic life, inevitably leads to stagnation, depriving it of its ability for self-development and for responding to changing economic relations and to new challenges. The long-standing statement by one of the ideologists of Russian Slavophilism, A. S. Khomiakov, remains instructive: "Science needs not only freedom of opinion, but also freedom to doubt." This means that the search for a new theoretical paradigm in economic science should certainly stimulate access to new areas of research, but this also does not exclude the importance of restoring lost scholarly traditions and the opportunity to return to the analysis of contemporary problems of Russia's economic development. As Pierre Bourdieu rightly noted, "real economic theory can be built only through a break with antigenetic prejudice and the assertion of oneself as a historical science" (Bourdieu 2005: 131).

The Russian research tradition, in a sense, created its own fields in Russia before the revolutions of 1917, and the sense of uniqueness persisted even when Marxism was imported and enforced in Soviet-era economics departments. While Western ideas did flourish because of the status of Western ideas and actors—a status only enhanced by the position of the Great Powers geopolitically—Russian economics had its own "home base," with a state confronting sometimes unique economic contexts and challenges, and a competing class of "theorists" outside the university system that could propose alternative ways of thinking about what "economics" and "economies" are and should be. The rule of "pure economism," which was the object of close critical attention in the Russian school of national economy, today has acquired a hegemonic status, and such "global economism" has had arguably devastating consequences demonstrated by the last world crisis. More importantly, the very idea of economics as a science, with methods and tasks developed in the past, can remain relevant for revealing the nature of Russia's current economic system and determining its prospects for domestic economics—a topic of clear desire and interest.

Notes

1. In this chapter, I equate the ethnic label "Russian" (*russkii*) and the legal label "Russian" (*rossiiskii*) when I refer to schools of economic thought. Strictly speaking, there are differences between the two, given the complex ethnic composition of the country's population and possible discrepancies between different cultures. While this point warrants its own discussion, for sake of space, I elide it here. Note that I also take for granted the idea of a "nation" and "national," when nations are political constructions. A national school of economic thought presumes a "nation," and nation-building is obviously related to economy-building and theory-building. But this is a complex theme for another work.
2. Again, it might be possible to incorporate contextual specifics (of social space and time) into a general theory, as a set of variables. However, if context matters in this way, then those particular variables and causal relations are no longer invariant. Prediction would only work if those contextual specifics themselves are fairly constant. So, while some issues might be addressed invariant "laws" (e.g. regarding inflation), others might not (e.g. the effect of state activity in the economy).
3. Thus, the process of creating a Russian school of economic thought paralleled the process of designing industrial policies: "latecomers" could copy earlier modernizers (Gerschenkron 1962).
4. Zweynert (2008) chose 1805 as the starting point of his analysis, noting that in this year the first textbook on political economy was prepared by Christian von Schlözer—dismissing earlier originality.
5. Ministers of Finance for the Russian Empire who were German by birth included E. F. Kankrin, P. F. Brokh, M. Kh. Reitern, S. A. Greig, N. Kh. Bunge, S. Iu. Witte, and P. L. Bark.
6. From its creation in 1725 up to the end of the eighteenth century, the Academy of Sciences was made up of more foreigners than Russians. In 1799, of 111 members of the Academy, 76 were foreigners, of whom 68 were Germans. Only 26 were Russian. Russians began to dominate the Academy only in the second half of the nineteenth century.
7. The first departments of political economy were created at Moscow University and Kazan University (1804), and then at St. Petersburg University (1819).
8. As an example, Marx's *Das Kapital* enjoyed great success in Russia, which Marx himself noted in a letter to Sorge in November 1880. In Russia he

was "more read and appreciated than anywhere else…" (Marx and Engels 1964: 380). At the same time, during Marx's life, *Kapital* was not translated into English, even though he wrote it while living in Great Britain and using Britain as his empirical case. The first volume was translated into Russian in 1872, and it was his first translation into a foreign language. The English edition appeared only in 1886 after his death (1883).

9. Cf. *Bolshaia biographicheskaia entsiklopediia* (Moscow, 2009), at http://dic.academic.ru/dic.nsf/enc_biography/.

Works Cited

Abalkin, L.I. 2003. *Ocherki istorii rossiiskoi ekonomicheskoi mysli*. Moscow: Nauka.

Alekseeva, E.V. 2007. *Diffuziia evropeiskikh innovatsii v Rossii (XVIII–nachalo XX v.)*. Moscow: Rosspen.

Barnett, V. 2005. *A History of Economic Russian Thought*. New York: Routledge.

Bourdieu, P. 2005. *Sotsialnoe prostranstsvo: polia i praktika*. Moscow: Institut eksperimentalnoi psikhologii.

Carpi, G. 2012. *Dostoevskii—ekonomist. Ocherki po sotsiologii literatury*. Moscow: Falanster.

———. 2016. *Istoriia russkogo marksizma*. Moscow: Common Place.

Fligstein, N. 2013. *Arkhitektura rynkov: ekonomicheskaia sotsiologiia kapitalisticheskikh obshchestv v XXI veke*. Moscow: Higher School of Economics.

Gerschenkron, A. 1962. *Economic Backwardness in Historical Perspective*. Cambridge: Harvard University Press.

Greenfeld, Liah. 1992. *Nationalism: Five Roads to Modernity*. Cambridge: Harvard University Press.

Iakovets, Iu.V. 2003. *Rossiiskie ekonomicheskie shkoly*. Moscow: MFK.

Iakutin, Iu.V. 2010. *Rossiiskaia shkola sotsialno-ekonomicheskoi mysli: istoki, printsipy, perspektivy*. Moscow: Ekonomicheskaia gazeta.

Ilin, I.A. 1992. O russkoi idee. In *Russkaia ideia*, ed. M.A. Maslin, 436–443. Moscow: Respublika.

Istoriia russkoi politicheskoi mysli, vol. 1. 1955. Mosacow: Politizdat.

Kireevskii, I.V. 2007. *Dukhovnye osnovy russkoi zhizni*. Moscow: Institut russkoi tsivilizatsii.

Leshkov, V.N. 1855. *Iz vospominanii 12-ro ianvaria 1855 goda: Uchebno-literaturnye stati professorov i prepodavatelei Imperatorskogo Moskovskogo Universiteta, izdannye po sluchaiu ego stoletnego iubileia*. Moscow: Universitetskaia tipografiia.

Marx, K., and F. Engels. 1964. *Sochineniia*. Vol. 34. Moscow: Politizdat.

Pokidchenko, M.G., and E.N. Kalmyzhkova. 2003. *Istoriki ekonomicheskoi mysli v Rossii: V. V. Sviatlovskii, M. I. Tugan-Baranovskii, V. Ia. Zheleznov*. Moscow: Nauka.

Ryazanov, V. 1998. *Ekonomicheskoe razvitie Rossii. Reformy i rossiiskoe khoziaistvo v XIX–XX vv*. St. Petersburg: Nauka.

———. 2010. Russkaia shkola ekonomicheskoi mysli: universalno-vseobshchee i natsionalno-osobennoe. *Vestnik SPbGU*. Seriia Ekonomika 3: 66–84.

Shtein, V.M. 1948. *Ocherki razvitiia russkoi obshchestvenno-ekonomicheskoi mysli XIX–XX vekov*. Leningrad: Leningradskii gosudarstvennye universitet.

Spengler, O. 1993. *The Decline of the West. Form and Actuality*. New York: Knopf.

Wuthnow, R. 1989. *Communities of Discourse*. Cambridge: Harvard University Press.

Zweynert, Joachin. 2008. *Eine Geschichte des ökonomischen Denkens in Rußland 1805–1905*. Marburg: Metropolis Verbag.

4

Statistics Comes to Russia: Science, Quantitative Analysis, and Shifts in Economic Thinking

Anton Leonidovich Dmitriev

Statistics as a scientific method and discipline appeared in Russia at the end of the eighteenth century—but it did not arrive on the force of its own insights into Truth. The history of the origin of statistics as a key branch of knowledge and its acceptance in the West, first as a descriptive tool for state power and policy, has been the subject of several studies (Ptukha 1955, 1959; Kaufman 1922; Gozulov 1972; Ploshko 1955; Ploshko and Eliseeva 1990). This is in addition to an extensive literature on individual personalities and trends in Russia. The expansion of statistics in Russian economic thought involved a combination of individuals interested in and working on this subject, institutions shaping and implementing decisions that brought statistics into Russia in a particular way, and a loose field of intellectual discourse and status located outside Russia

A. L. Dmitriev (✉)
Faculty of Economics and Finance, Department of Economic Theory and History of Economic Thought, St. Petersburg State University of Economics, St. Petersburg, Russia

Faculty of Economics, Department of Economic Cybernetics, St. Petersburg State University, St. Petersburg, Russia

J. K. Hass (ed.), *Re-Examining the History of the Russian Economy*,
https://doi.org/10.1007/978-3-319-75414-7_4

proper but enveloping scholars and institutions of Russia and other European countries. It might seem that the spread of statistics as a science in its own right, and as a component of objective scientific analysis, follows a natural evolutionary logic of discovery and enlightenment. Yet, however much the validity of science and scholarship is grounded in reality (e.g. disconfirmation by empirical facts), validity and legitimacy also depend on institutions that can impose at least consideration, if not adoption, of particular categories and forms of knowledge; on fields of actors situated in positions of status and power; and on networks and actors' strategies, especially to use positions and relations to advance particular claims of knowledge and validity.

Even since the "battle of the methods," quantitative methods and statistics have enjoyed a central place in judging "valid" economic knowledge. Russia was no exception to this trend and raises interesting questions about fields, institutions, and knowledge that I cannot definitively answer in one chapter but for which I will provide a narrative laying out a wider discussion. Namely, these issues are: how statistics penetrated Russia, what ideas had the most significance and why, and what institutional configurations and transformations shaped these processes. At the outset, let me emphasize that for several decades there was no clear distinction between statistics, political economy, and geography in Russia. A clear division began only in the 1830s, and was finally realized when quantitative methods began to penetrate statistical methodology.

First Steps

As the historian of political economy and statistics V. V. Sviatlovskii (1906: 163) noted, the first information about teaching statistics in Russia dates back to the 1760s. In June 1762, August Ludwig von Schlözer taught a basic course in statistics at a private boarding school set up for children of Hetman K. G. Razumovskii. By Schlözer's own admission, "everything was borrowed from foreigners," and Russian materials used were of poor quality and obsolete (von Schlözer 1875: 119). Von Schlözer intended to create a system of Russian statistics in the spirit of "descriptive government."[1] This reveals an important early impulse (not only in Russia) for adoption and expansion of statistics: part of the emerging state-led project of rationalizing, studying, and shaping "society." Perhaps unsurprisingly, this descriptive school of statistics was called "state studies" (*gosudarstvovedenie*), implying the state's "gaze" of those under its power. Its representatives believed statistics was a science tasked with a systematic description of entities of interest to the state (Ploshko and Eliseeva 1990: 21). Such entities included anything that could affect the welfare of the state, and statistics was considered to be a science dedicated to social, legal, and public facets of state activity. The most famous intellectuals arguing for the importance of statistics in this endeavor were the Germans Hermann Conring, Gottfried Achenwall,[2] and Anton Friedrich Büsching—whose ideas would have a significant impact on statistics in Russia. While German intellectuals developed this state-centered idea of statistics, a competing approach took root, one from seventeenth-century England: "political arithmetic." At its origins stood William Petty, Gregory King, and Charles Davenant. Unlike state studies, political arithmetic decentered the state and focused instead on a broader plurality of social fields with an affinity for "nature"—that is, political arithmetic was designed to reveal laws of social life, and methods for measurement were considered an indispensable condition for studying mass data. In fact, the term "political arithmetic" testified to the application of mathematics to politics, that is, measuring the field of socioeconomic life (Ploshko and Eliseeva 1990: 14),[3] only not as a hierarchical structure revolving around a sovereign, but rather around a community of the realm.

In his main work on statistics, *Theorie der Statistik* (1804), Schlözer claimed that quantitative characteristics were the preferable form of data, and he even added a liberal logic to his project, calling statistics a "barometer of civil freedom" (Schlözer 1804: 52) that was incompatible with despotism. According to Schlözer, statistics revealed how order and system are mandatory, implying regular social fields whose rules and rhythms were beyond elite control. He claimed that history is statistics in motion, and a statistical account is a frozen story (Schlözer 1804: 86). His followers considered this to be the true definition of statistics and, in spite of his further ideas, refused to study causes and consequences (Sheinin 2014: 137). Later, a representative of the descriptive trend in statistics, A. G.

Obodovskii, modified Schlözer's formula: history is to statistics as poetry is to painting (Obodovskii 1839: 48). Statistical data, according to Schlözer's logic, should be collected by officials, which scholars would then interpret. For convenience, they offered the idea of *veresunitaeagunt* (territory or population; connection or device; to control; cf. Sviatlovskii 1906: 169–170). On some issues, Schlözer sided with the state, but he did not reject quantitative data, and instead gave them significant meaning.

As Iu. Ianson noted in the 1880s, "… next to repetition of the basic principles of the Achenwell school in a new way, we find in Schlözer's statistics, as in everything he did, many thoughts far ahead of the scientific level of contemporaries. Comparing these thoughts with the routine of government, we are convinced that even Schlözer's talent could not breathe a living spirit into the old statistical school" (Ianson 1887: 12). In the beginning of the twentieth century, V. V. Sviatlovskii stressed that "in the Achenwall-Schlözer school, there are still few elements of true scientific spirit," and it is "only improved and somewhat ennobled and profound, yet still state-oriented." As a result, Russian statistical science, blindly following the footsteps of the German school in the third quarter of the eighteenth century, established this trend under the name of "descriptive statistics" (Sviatlovskii 1906: 170). Sviatlovskii also concluded categorically that "the introduction of a descriptive school was a great misfortune for a young Russian statistical science that needed fresh water" (Sviatlovskii 1906: 170).

The first manual on statistics, as a doctrine on "entities of state interest," was published by Johann-Gottfried Reichel, professor at Moscow University, who came to Russia in 1757. He initially taught German language and literature, and then began to teach universal history, to which he added statistics of modern states. Reichel gave lectures on state studies in Latin, and one student, A. D. Baibakov, transcribed and published his lectures in Russian under the unwieldy title *A Short Guide to the Knowledge of the Natural, Ecclesiastical, Political, Economic and Educational Status of Some of the Most Eminent European States, Collected from the Public Teaching of History in 1773 at the Imperial University of Moscow and its Continuation to Modern Times*. He wrote at the beginning of the work, "*Statistics*, as generally understood, is the knowledge of the condition of states and republics. In particular, this means an *accurate and well-known explanation for the current state of Europe's most eminent possessions.* This includes in itself all the centuries and the whole world of empires, states, and republics; and this reasoning is about the special possessions of our times" (Reikhel 1775: 316; emphasis in original).[4] Reichel's work had a geographical basis, not uncorrelated with empires: his book included chapters on Spain, Portugal, France, Great Britain, the Netherlands, Denmark, Sweden, and Poland. He stressed that geography, supplied with statistical information, had great significance: prosperous European countries had accumulated data on wealth (or its lack), population, manufacturing, trade, income, military capacity, and the like of the homelands and possessions (empires), which would provide the foundation for statistical science (Reikhel 1775: 7). His work, according to Ptukha (1955: 221), was notable for employing a concept of "statistics" broader than that of Schlözer, Conring, or Achenwall. Under the influence of Anton Friedrich Büsching, Reichel conducted a comparative analysis, although he paid little attention to material forces.

Statistical works of the first academician-economist Andrei Karlovich Shtorkh deserve close attention, as they have been neglected, even in Russia, and only the demographer and historian of statistics M. V. Ptukha paid much attention to him. In the first volume of his study of statistical scholarship, Ptukha gave Shtorkh prominent billing, noting that his first statistical work was aimed at a mass audience. This work was in the vein of "entities of state interest"—that is, centered on a paternalistic state guarding its possessions—and it came to Russia from Prussia and was associated with Achenwall and Büsching, among others. However, Ptukha noted that political arithmetic slipped into Shtorkh's work: Shtorkh provided much statistical information on the population of the St. Petersburg, placing emphasis on the demographics of the development of the city. Shtorkh also drew on a Prussian form of political arithmetic, as well as on other statisticians. In the nine-volume tome *Historisch-Statistisches Gemälde des Russischen Reichs am Ende des achtzehnten Jahrhunderts* (1797–1803), Shtorkh produced a number of "calculations typical for political arithmetic," that is, marriages, births, deaths, the period of doubling of the population of Russia, and so on (Ptukha 1955: 368). For source material he used data on the natural movement of the population for 1793 for nine dioceses and the province of Riga. Describing this work, Ptukha noted that Shtorkh created the most complete, albeit unfinished, historical, and statistical description of Russia at the end of the eighteenth century. Shtorkh also published a collection of articles on history and statistics at the end of the eighteenth century.

Unfortunately, Shtorkh's work, devoted to a statistical description of Russia, was interrupted in the beginning of the nineteenth century. As Ptukha (1955: 368) noted, a detailed historical review took a very long time, but Shtorkh ended up with other important tasks. The decisive role in changing the position of statistics was played by Shtorkh's election to the Academy of Sciences (where he would serve as vice president) and his duties associated with teaching political economy. Ultimately, it was less Shtorkh's publications, as the preparation he did, that would be important (Ptukha 1955: 125). In the process of such data collection, Shtorkh came to the conclusion that state officials would have an easier time making sense of entities of interest if data were easier to read, especially in tabular form. In the end, Russian and foreign state officials looked to Shtorkh as a scholar whose works provided the most complete and reliable statistical description of Russia in the late eighteenth and early nineteenth centuries. His synthesis of state-centered statistics and more pluralist political arithmetic, while still embryonic, was novel and important for Russia.

K. F. Hermann and Statistical Science

The beginning of the nineteenth century, with tsar Aleksandr I on the throne, the development of the Russian state continued apace, and statistics played a role in this process—in particular, with the labor of K. F. Hermann and K. I. Arseniev. Perhaps the key role belongs to the former, Karl Theodor Hermann,[5] who moved to St. Petersburg in 1795 after graduating from the University of Göttingen. He began his career teaching history, geography, and statistics at the Naval and First Cadet Corps. In 1798 he was appointed rector of the Academic Gymnasium, in 1806 as professor of statistics at the Pedagogical Institute (after 1816, the Main Pedagogical Institute), and after 1819 an ordinary professor in the department of statistics. In 1811, Hermann was appointed head of the statistical department of the Ministry of Police, Russia's first central state statistical agency. After the merger of the Ministry of Police and the Ministry of Internal Affairs (1819), he headed that statistical department until 1835. In 1816–1817, Hermann gave private lectures to future Decembrists on political economy and statistics (Syroechkovskii 1954: 162). In 1819, Hermann gave a special course of lectures on these subjects open to the public. In 1821 D. P. Runich initiated a criminal investigation against Hermann and three other professors at St. Petersburg University (K. I. Arseniev, A. N. Galich, and E. Raupakh) for crimes against the state and religion. Hermann was suspended from teaching, and his textbooks on statistics were banned; a court was held at a conference of professors, at which he behaved in a dignified manner. In 1824, he was fired and prohibited from teaching in the Ministry of Education. His case was closed in 1827 and, by order of Nicholas I in 1828, the case was expunged from his record.

An important feature of Hermann's theoretical work that came to occupy a prominent place in Russian statistical science of the early nineteenth century was the connection between methodology and theory of statistics as a special scientific discipline. Hermann considered challenges of methodology to be part of constructing a theory that "would contain rules and samples which, when processing each particular statistical subject, could be made useful in state reasoning about the economy and politics" (Hermann 1809: 26). This approach contributed to understanding and resolving various issues, while also significantly expanding statistical theory, contributing to its transformation from a scholastic "clarification" and descriptive endeavor to genuine science in its modern sense. The main work containing a complete and systematic presentation of the theory of statistics was *A General Theory of Statistics for Students of this Science*. How contemporaries evaluated this work can be judged by the response of K. A. Arseniev: "this small work in all fairness can be considered a classical creation in this genius" (Arseniev 1818: 11). Almost 100 years later, A. A. Kaufman claimed Hermann's work contained "a number of extremely valuable ideas of a methodological and even theoretical nature" (Kaufman 1922: 8).

The initial part of Hermann's work was devoted to the history of statistics. He argued that "tables about the state of the army and finances were the first beginnings of Statistics. Later, information about state structure and administration was added. And so the power and strength of government were the first subjects of statistical research" (Hermann 1809: 1). Hermann believed progress in statistics was related to demands of society, as revealed in economic and political development in England and France. In England, "growing national wealth and demands of merchants ... produced new knowledge, under the name of the State Economy, and separated from the political sciences, Stuart, Arthur Jung, Adam Smith, by means of which they acquired immortality. But the science of State economy is the Philosophy of Statistics" (Hermann 1809: 6). In France, "the bad state of public finances did the same thing as a *coup* and people's wealth in England. They began to talk about many important subjects of the state economy" (Hermann 1809: 7). The result of economic, political, and cultural changes was a political arithmetic with rudiments of statistical science. "But before Statistics, as a special science, became known under a special name," he wrote, "this discovery was preceded by another, in the reasoning of its pre-preparatory science, that is, Political arithmetic... Economists attached Political arithmetic to agriculture, and at the base of all their studies laid down their agricultural calculations (*calcusagricolas*); the English, between whom Arthur Jung, Preuss Priestley, and Thomas Wick distinguished themselves, applied this to all areas of national industry" (Hermann 1809: 9–10).

The emergence of the statistical science Hermann associated with Achenwall's work, but in his teaching, he embedded statistics in historical contexts and the completion of a long period of historical development:

> And so, when Theologians in Germany lost their excessive influence, when Law Natural and Civil and Political Economy and Arithmetic became special sciences; much statistical material was accumulated, and their importance in the political sciences was recognized so universally that one German scientist, Achenwall in Göttingen (the place through which political enlightenment from England moved to Germany), then known entities of state interest, under the name of Statistics, gathered in One whole, in the year 1749. He issued the first educational book sowing new sciences, and introduced it to schools and universities. (Hermann 1809: 11)

Hermann agreed with critical remarks of British statistician Sinclair about Achenwall's approach, in connection with the prevalence of law in that discourse, explaining this as due to German specifics: "it was very natural that German statistics should have had the character of State Law, since shortly before this, science dominated in universities, and was well processed. But the science of political economy, which deals especially with objects, to the state of the people concerned, was only then discovered and brought to perfection in England and France, and in Germany, on the contrary, little was known" (Hermann 1809: 12–13).

Hermann explained attempts to deny the independent significance of statistics as a science, as in Büching's works (where he linked statistics with political geography), as due to the predominance of state law and a general state-centered logic. As a result, he paid much attention to revealing its connection with other "political" sciences, first of all with political economy. Hermann stressed the need to develop a special statistical theory designed to develop rules that would make statistics useful for study of economics and politics. Its foundations, he claimed, were political sciences, which needed to be brought together to find first principles. While this progress resembled the philosophy of Descartes and Kant, a theoretical foundation was necessary to make statistics the foundation of all political sciences (Hermann 1809: 25–26). Hermann's work correctly outlined how statistics would develop. In particular, Hermann discovered the main forces that determined the development of statistics: he linked its emergence and development with practical needs of the state and with the state of socioeconomic life. When he identified the importance of England's social and economic development, demonstrating successes achieved in economic science and the impact of the latter on statistics, he showed the stimulating effect of the maturation of the pre-revolutionary crisis in France on statistics. This was valuable not only for explaining the history of statistics, but also for a deeper understanding of the very nature of statistics. He showed a direct link between political arithmetic and its influence on the formation of statistics, as well as the fact that political economy directly could directly affect the latter.

In his section on theory, Hermann cited general principles of cognition, grounding his argument in the philosophy of Immanuel Kant. Hermann's use of Kant's contrast between the sensual and reason was in the claim that "the necessity of theory is based on properties of the human mind, seeking to everywhere find unity" (Hermann 1809: 28–29). For the mass of statistical works at the time, representatives of the descriptive school and political arithmetic, there was no proper theory of statistics that could uncover and scientifically comprehend processes of statistical cognition. The pure empiricism of the descriptive school, which reduced statistics to description and derived causal explanations beyond its limits, excluded the possibility of a satisfactory statement of the question of statistical theory. There was no "theory" in the German descriptive school. The influence of Kant's philosophy on Hermann's development of statistics had progressive significance. In addition, Hermann linked creation of a theory of statistics with requirements of practice, and this ensured the reality of his theoretical constructs. Hermann also recognized the leading role of social sciences, in particular political economy, in relation to the statistics (Hermann 1809: 25), and "the science of the State economy is the philosophy of statistics" (Hermann 1809: 6). These two circumstances created a solid basis for Hermann's theoretical constructions, which raised possibilities for developing a theoretical foundation for statistics.

Hermann proposed a succinct, formal definition of the theory of statistics: "The theory of statistics is the science of the initial concepts serving the knowledge of its materiality, and the main grounds necessary for the possible perfect presentation of its parts" (Hermann 1809: 32). This allowed Hermann to proceed to the development of his theory, a first in the history of Russian statistics. Hermann noted a set of issues making up this theory: "The general theory of statistics is a science: (1) the concept of Statistics, (2) its limits, (3) its parts and types, (4) its sources, and (5) its usefulness" (Hermann 1809: 33). This promised to cover a broad swath of conceptual space, even if much was still in the bud. Importantly, Hermann noted that the usual belief was that statistics was counting objects—but this was incorrect. Rather, numbers had contextual meaning, and "so that each [statistical] table can be intelligible, we also need a historical-political interpretation" (Hermann 1809: 16–17).

Let us also note one more important circumstance connected with Hermann's activity. In 1806–1808, he published in the *Statistical Journal* (of which there were only four issues), the purpose of which was processing and publishing information from reports by the Ministry of Internal Affairs, and to disseminate statistical data for various reading circles. Although only 175 copies of the journal were printed, and there were only 110 subscribers (50 of whom were the Main Administration of Schools), the journal played a positive and progressive role (Sviatlovskii 1906: 175). In this journal, Hermann wrote a remarkable article (Hermann 1806) distinguishing "statistics in the spatial sense" as "a thorough knowledge of the condition of the state at some known time," and statistics "in the closest sense" as "a thorough knowledge of all that has an obvious influence on the welfare of the state at some known time" (Hermann 1806: 16). "The philosophy of statistics," he wrote, is "rules of state economy and politics that require statistical information, and statistics should be used by all sciences to respond to requests made by them that they must 'statistically process'" (Hermann 1806: 9). The largest part of *Statistical Journal* was various materials for domestic and foreign consumption, understood in the spirit of the descriptive school. The journal printed articles on political economy, and the first issue featured a piece by future Rector of St. Petersburg University, M. A. Balugianskii, entitled "National wealth. The image of various economic systems." Thus, Hermann's journal was the conduit for advanced European ideas to enter into Russian discourse, such that A. A. Kaufman called this issue of this journal Hermann's "greatest merit" for Russian statistics (Kaufman 1922: 6).

After Hermann

In 1797, Evdokim Filippovich Ziablovskii was appointed professor of history and geography, and later of statistics, at St. Petersburg Teachers' Gymnasium, which was transformed in 1803 into the Pedagogical Institute. In 1805, he was awarded the title of extraordinary professor, and in 1809, of ordinary professor of geography. After the removal of Hermann from St. Petersburg University, Ziablovskii took over the department of statistics. From 1821 to 1825, Ziablovskii was rector of the university. Much of his scientific and educational work was devoted to geography, although he did devote some work to statistical descriptions of Russia and Europe. In the first of his works on the subject,

Ziablovskii, expounding on Russia's own vision or perception, adhered to the scheme of German scientist I. G. Meisel, and in terms of theory, he stood behind Achenwall, even giving him the title "father of statistics": "Achenwall's concept of statistics, in my opinion, has an advantage over others; statistics is thorough knowledge of the real vision of any state. It is much clearer, more definite, and more fruitful; it is confirmed directly on that foundation, and concludes that perfection, according to which useless, unimportant, and pointless circumstances are separated easily and conveniently; only a sensible choice of materials is required" (Ziablovskii 1808: 9). Making statistics a science required establishing two preliminary concepts: states and vision. Under state vision, Ziablovskii understood what other state scholars understood: they were based on the people's well-being.

> Under *general* statistics, of course, is what all the known lands of our globe describe. *Private*, or *special*, statistics is that which speaks of the most remarkable possessions on the globe by position and various other subjects. The latter statistics are again divided into *general* and *particular*. The first contains the most remarkable news about the entire state of any people, which is the statistics of the Russian Mr. Shtorkh, or the statistics of Shvartnerov about the kingdom of Hungary, and the second includes some part of the state, which is statistics as used in France, [a state] divided into departments. (Ziablovskii 1808: 26–27; emphasis in original)

Thus, Ziablovskii was an antagonist to Hermann and Arseniev; Sviatlovskii characterized him as "a typical figure of a bureaucrat-professor of the pre-reform time" (Sviatlovskii 1906: 181).[6]

Konstantin Ivanovich Arseniev was a disciple of Hermann. In 1817, he was approved as an adjunct professor at the Pedagogical Institute for Geography and Statistics. In 1818–1819, two parts of his work on statistics and the Russian State were published (Arseniev 1818, 1819). In this work, Arseniev distinguished the theory of statistics, or statistics as science, and paid much attention to history and theory of the state. By 1821 he acknowledged the views of progressive elements in the state, Hermann and Schlözer. Arseniev's career corresponded with a decline in the prestige of statistics: recall that in 1821, Hermann and Arseniev were brought to justice for "insulting" religion and the state, and their work was withdrawn from circulation. This led to a period of degradation in Russian statistical scholarship, best seen in the number of statistical publications. In 1821 Ivan Andreevich Geim, who studied at Helmsted and Göttingen Universities and taught for 20 years at Moscow University, completed the last important work on statistics from this earlier dynamic phase. Again, the role of the state was crucial: "Statistics is a thorough knowledge of the state's real vision… Statistics describe the state in its present form and condition, i.e. first the *state* in its own meaning, then the people or the society united in it" (Geim 1821: 1–2; emphasis in original). However, after this no real work on statistics by Geim appeared for ten years, except for small brochures or secondary school textbooks.[7]

However, an official in the Debt Commission, T. F. Stepanov, who defended his dissertation in 1824 at Kharkov University on political economy and statistics, published an article on statistics (Stepanov 1831). The starting point of his theoretical research was rejecting a descriptive understanding of science in favor of an attempt to discover objective laws. Political economy and politics give a general definition of laws governing public life, while statistics would aid with their concrete realization. Statistics would also serve as a guide for politics and political economy—a positive view of science and society already taking root in France. And that next period in the development of statistics in Russia began in the second half of the 1830s and was associated with the work of Quetelet. In 1838 and 1839, two works by V. S. Poroshin and A. G. Obodovskii appeared. Obodovskii borrowed his theoretical approach from V. Butte and Hermann, noting that in the nineteenth century, "Achenwall-Schlözer's idea was revealed even more in splendor, and statistics again took the place of honor in the line of political science, to which was placed the immortal Schlözer. Everyone finally became convinced that policy measures can never be true unless they are based on statistical data" (Obodovskii 1839: 2). After analyzing the various definitions of statistics, Obodovskii used Butte's definition: "Statistics is a systematic representation of data from which it is thoroughly understood to what extent the state has reached its goal at any particular moment" (Obodovskii 1839: 25–26).

Poroshin, who taught political economy and statistics at St. Petersburg University, took a different approach. Well versed in philosophy, political economy, and statistics, Poroshin sharply criticized descriptive trend in statistics in his dissertation. His provisions boiled down to the following. Science and theory are inseparable; as a chronicle is not history, so description is not statistics; statistics are comparatively pragmatic; among moral sciences, there is no place for descriptive sciences (Sviatlovskii 1906: 188). From Poroshin's point of view, "the subject of statistics is a person; it is a moral science; a new reason why it cannot be a description" (Poroshin 1838: 38). The task of science is to describe species and to put each thing in its place. Similar phenomena are also encountered in the "moral world." Statistics, in his opinion, resembles experimental sciences about nature. "Both sciences, statistics and political economy, deal with what remains outside the solemn procession of history, in the everyday and universal circle of being and action. And here nothing prevents us from considering what is given by experience as a condition for knowledge (*Währnehmung*) of truth of the unconditional (*Wahrheit*)" (Poroshin 1838: 46–47). Poroshin also raised logical constructions of Say: the connections between statistics and political economy suggested that political economy is the basis of statistics, which itself is not a science. According to Poroshin, Say destroyed a ghost, because if statistics does not express general truths, it cannot be a science. These considerations were derived from the fact that statistics notes phenomena as they occur and as a result of the fact that truths obtained via political economy in one way are verified by another (Ptukha 1959: 403). Poroshin ended his work stating that his views were shared by other European scholars, and that the desire for developing science in this intended direction was "not an individual dream," but a vision shared by Joy, Quetelet, and Ritter (Poroshin 1838: 54). Russian statistical science saw a clear departure from the descriptive trend that was finally rejected by the middle of the nineteenth century.

Conclusion

Statistical science, arriving to Russia from the West, was greatly influenced by policies of the German state—statistics as new lenses for the state's gaze and authority—further disseminated among scholars: this was the policy of giving the state a special social role. Germany already had status in Russia, going back to when Peter invited foreigners, especially

Germans, to come organize academic organizations. Despite shortcomings in the methodology of state studies, mostly the low desire to use mass quantitative data, qualitative indicators began to accumulate and contributed to the development of a comparative method of analysis. Main attention was focused on material and economic factors of the life of the state. Ideas of political arithmetic penetrated discourse later but were fraught with difficulties. Yet the state could bite those serving its interests: the "University Affair" at St. Petersburg University and the dismissal of Hermann and Arseniev slowed the development of statistics. In the end, both for good and ill, state institutions played a significant role in the spread of ideas, even those that, like economics, are supposed to owe any dissemination to their proximity to Truth.

Notes

1. O.B. Sheinin (2014: 136) believes that the history of the state has not been sufficiently studied.
2. Achenwall came up with the word "statistics," drawn from Italian *statista*—a state official.
3. Only in J. F. Melon's translated *The Political Experience of Commerce*, in the section "On Political Arithmetic," is it stated that "everything is subject to calculation: it extends to matter, only to the moral science of belonging" (Melon 1768: 225). However, the translator William Petty called him a "one-sided nationalist" and his work "written on false grounds."
4. Note "possessions" and empire—the importance of domains of natural control.
5. In Russian, "German" (but not pronounced as the nationality).
6. An analysis of Ziablovskyii's views is contained in the detailed work of L.S. Berg (1946), which includes numerous opinions about him by his contemporaries.
7. At the end of the 1820s, the only published articles about the subject of statistics and its tasks as a science were by I. Maevskii (1828), I.A. Shchedrinskii (1828) and K. F. Hermann (1829).

Works Cited

Arseniev, K. 1818. *Nachertanie statistiki Rossiiskogo gosudarstva. Ch. 1. O sostoianii naroda*. St. Petersburg: Tipografiia Imperatorskogo Vospitatelnogo Doma.

———. 1819. *Nachertanie statistiki Rossiiskogo gosudarstva. Ch. 2. O sostoianii pravitelstva*. St. Petersburg: Tipografiia Imperatorskogo Vospitatelnogo Doma.

Geim, I.A. 1821. *Opyt nachertaniia statistiki glavneishikh gosudarstv*. Moscow: Universitetskaia Tipografiia.

Gozulov, A.I. 1972. *Ocherki istorii otechestvennoi statistiki*. Moscow: Statistika.

Hermann [German], K. 1806. Teoriia statistiki. *Statiticheskii zhurnal, T. 1, Ch. 1*: 1–28.

———. 1809. *Vseobshchaia teoriia statistiki. Dlia obuchaiushchikh sei nauki*. St. Petersburg: Imperatorskaia akademiia nauk.

Hermann [German], K. F. 1829. Vzgliad na nyneshnee sostoianie Statistiki v prosveshchennykh gosudarstvakh Evropy. *Zhurnal Ministerstva vnutrennikh del* 1 (95–106): 373–384.

Ianson, Iu.E. 1887. *Teoriia statistiki*. 2nd ed. St. Petersburg: Tipografiia Shredera.

Kaufman, A.A. 1922. *Statisticheskaia nauka v Rossii. Teoriia i metodologiia. 1806–1917*. Moscow: Tsentralnoe statisticheskoe upravlenie.

Maevskii, I. 1828. Vzgliad na statistiku kak na politicheskuiu nauku. *Vestnik Evropy* #10 (June): 273–285.

Melon, Zh.F. 1768. *Politicheskii opyt o kommertsii*. St. Petersburg: Imperatorskaia Akademiia nauk.

Obodovskii, A.G. 1839. *Teoriia statistiki nastoiashchem sostoianii, s prisovokupleniem kratkoi istorii statistiki*. St. Petersburg: Tipografiia K. Vingebera.

Ploshko, B.G. 1955. Russkaia sotsialno-ekonomicheskaia statistika nachala XIX veka (1800–1820 gg.). *Nauchnye zapiski Leningradskogo finansovo-ekonomicheskogo instituta* 9: 55–93.

Ploshko, B.G., and I.I. Eliseeva. 1990. *Istoriia statistiki*. Moscow: Finansy i statistika.

Poroshin, V. 1838. *Kriticheskie issledovaniia ob osnovaniiakh statistiki*. St. Petersburg: Tipografiia Akademii nauk.

Ptukha, M.V. 1955. *Ocherki po istoriia statistiki v SSSR, T. 1*. Moscow: Izdatelstvo AN SSSR.

———. 1959. *Ocherki po istoriia statistiki v SSSR, T. 2*. Moscow: Izdatelstvo AN SSSR.

Reikhel, I.G. 1775. *Kratkoe rukovodstvo k posnaniiu naturalnogo, tserkovnogo, politicheskogo, ekonomicheskogo i uchebnogo sostoianiia nekotorykh znatneishikh evropeiskikh gosudarstv, sobrannoe iz publichnogo v 1773 godu obucheniia istorii v imperatorskom Moskovskom universitete i do samykh noveishikh vremen prodolzhennoe*. Moscow: Pechatano pri Imperatorskom Moskovskom universitete.

Shchedrinskii, I.A. 1828. Vstunitelnoe chtenie o sushchnosti, predelakh i polze statistiki. *Vestnik Evropy* #16 (August): 271–310.

Sheinin, O.B. 2014. K istorii gosudarstvovedeniia. *Finansy i biznes* 1: 136–156.

Sviatlovskii, V.V. 1906. *K istorii politicheskoi ekonomii i statistiki v Rossii*. St. Petersburg: Nachalo.

Stepanov, T.O. 1831. O statistike voobshche. *Literaturnaia gazeta* 3 (20): 159–164.

Syroechkovskii, B.E. 1954. Pestel i K. F. German. (K voprosu o rannikh politicheskikh vozzreniiakh Pestelia). *Uchenye zapiski Moskovskogo universiteta* 167 (Kafedry istorii SSSR): 151–178.

von Schlözer, A.L. 1804. *Theorie der Statistik: nebst Ideen über das Studium der Politik überhaupt*. Göttingen: Vandenhoeck & Ruprecht.

———. 1875. Obshchestvennaia i chastnaia zhizn Avgusta Liudviga Shletser, im samim opisannaia. Prebyvanie v Rossii ot 1761 po 1765 g. Izvestiia o togdashnei literature. In *Sbornik Otdeleniia russkogo iazyka i slovesnosti Iperatorskoi Akademii nauk*. St. Petersburg: Tipografiia Imperatorskoi Akademii nauk.

Ziablovskii, E.F. 1808. *Statisticheskoe opisanie Rossiiskoi imperii v nyneshnem ee sostoianii, s predvaritelnymi poniatiiami o statistike i o Evrope voobshche v statisticheskom vide. Ch 1*. St. Petersburg: Tipografiia Imperatorskogo teatra.

5

Networks, Fields, and Political Economy in Fin-De-Siècle Russia: The Life and Work of Nikolai Sieber

Danila Raskov

A scholar's intellectual biography is incomplete without its historical, social, and political contexts, since these contexts give meaning to the scholar's work and contributions. That scholar's biography *is* that context, just as the unfolding of biographies constitutes contexts. Political economy, like any scholarly endeavor, refers to a space of collective practices and depends on a combination of consensus, rules of the game, and institutions that reproduce practices through competition, power, and those very games. The field concept proposed by Pierre Bourdieu (1976) and subsequently developed by Neil Fligstein and Doug McAdam (2012) offers an adequate approach to the social phenomenon of an individual intellectual biography. On the one hand, the biography is an intersection of various field experiences crystallized in that individual's *habitus*. On the other hand, a scholarly or scientific field involves a symbolic space in

The work on the chapter was supported by a grant from the Russian Foundation for Basic Research, RFFI, #16-23-41003. I am grateful to François Allisson, Federico d'Onofrio, and Leonid Dmitrievich Schirokorad for valuable discussions and comments.

D. Raskov (✉)
Faculty of Economics, Department of Economic Theory,
St. Petersburg State University, St. Petersburg, Russia

J. K. Hass (ed.), *Re-Examining the History of the Russian Economy*,
https://doi.org/10.1007/978-3-319-75414-7_5

which there is competition for authority and the capacity to act legitimately on behalf of that science. As Bourdieu writes, "The 'pure' universe of the 'cleanest' science is the same social field as any other, with its inherent correlation of forces and monopolies, struggles and strategies, interests and profits, but in this field all these invariants are clothed in specific form" (Bourdieu 1976: 89). The intertwining of various fields, which enable us to speak about science, gain credibility, recognition, and symbolic capital, which are the main entities that interest us in discussing an intellectual biography. A certain aspect of this problem is the evolution of views on a specific branch of knowledge—in this case, political economy in the Russian Empire—and the Russian émigré academic environment that formed in the 1870s and 1880s in Switzerland and Germany, countries that were Russia's most significant scientific partners at that time.[1]

Nikolai Ivanovich Sieber (1844–1888) is known as the translator of David Ricardo, a Russian Marxist, and one of the last representatives of the classical school of political economy. The main treatise in its most complete form that represent his contribution to political economy is *David Ricardo and Karl Marx in Their Socio-Economic Studies* (Sieber 1885), compiled from his dissertation on Ricardo's theory (Sieber 1871), and later journal articles on Marxist political economy. Sieber's life and works at various times attracted the attention of various scholars (Kleinbort 1923; Tkachenko 1928; Naumov 1930; Rezul 1931; Reuel 1956; Raskov 2016; Guelfat 1970; Scazzieri 1987; Kapeller 1989; White 2009, 2011; Smith 2001; Allisson 2015). While various explorations of his life provide a plethora of materials that bring his life to light, there remains too little on the social relations that made up his life and thought. This chapter aims to rethink his intellectual biography from the interaction of different fields, with their rules, hierarchies, and types of reproduction of collective practice.

Sieber's biography is replete with mysteries and questions. Why did the young, talented professor leave Kiev University in 1875, having only just begun to teach there? Was this out of solidarity with M.P. Dragomanov, who was fired? Was it because of family circumstances, as Sieber's wife Nadezhda Olimpovna Shumova, later a well-known chemist, decided to obtain her education in Switzerland? Did Sieber identify with the radical movement of Russian Marxists—and if not, then why did he communicate with prominent figures of underground revolutionary organizations and was afraid to return to Russia? What was his relationship and attitude to Ukrainian nationalism or Ukrainian movements? How did he support himself while living in Switzerland? While we could try to answer these questions with reference to Sieber the individual, this would miss what shaped his interests and identities, and what triggered these and other interesting events in his life. That is, biography reveals relations as well as individual dispositions, as the two are never far apart. As such, an intellectual biography forces us to turn our efforts to understanding contexts in which the formation and activity of Nikolai, Mikola, and Niclaus (Kapeller 1989), as he was called in different contexts, took place.

The task for this chapter is thus restricted to examining the fields and institutions through which Sieber's life passed and, consequently, endowed his life with contacts and networks: authorities and teachers, associates and friends, publishers and editors, and so on. I will set out to establish and define contexts, fields, and connections that will make Sieber's intellectual biography become clearer. We will concentrate primarily on three venues where Sieber made most of his new contacts, bearing in mind that he was not very fond of publicity and left little behind about his private life: these three institutional and field contexts are university work and life, travels abroad into a community of intellectuals and émigrés, and his work in the field of academic and economics publishing.

Sieber's intellectual development began at St. Vladimir Imperial University of Kiev, with which he was associated from 1863 until 1875 (with a minor break). During this period he was an undergraduate, a graduate student, assistant professor, and finally professor. He defended his dissertation, went on a lengthy trip to Europe, participated in the publication of the newspaper *Kievskii telegraf*, and became a member of the Southwestern division of the Russian Geographical Society, where he worked on the census. It is possible that at this time he may have been a member of the Kiev Hromada—in any case, he had close friends there and Sieber was an active participant in joint undertakings with them.

The second context with its fields that had a significant influence on Sieber was his academic travels abroad for the purpose of becoming acquainted with the most recent achievements in various European universities. For many talented graduates such trips played an important role in forming their academic interests and developing contacts for future careers. Indeed, in this early period, Sieber's travels enabled him to meet people who would become lifelong friends and acquaintances (Sergey Podolinskii, Aleksandr Chuprov, Ivan Yanzhul, and Valerian Smirnov). The trip not only introduced him to life at these universities, but also placed him in a position to become involved in underground intellectual circles and to get to know the Russian emigrant community (e.g. at the Russian Library in Zurich), and even to meet his future wife.

The third facet of Sieber's story is the field of publishing, including participation in editorial work at newspapers and magazines, and regular preparation of articles. Sieber was the author of books and articles, and he did not stop writing and publishing in Russian, even though he was in Switzerland. The work of publishing houses and magazines had its own specifics. A stable circle of authors formed around each magazine. Publication included certain contractual relations that presupposed mutual obligations and often an honorarium. Such activity included illegal and semi-legal publications (*Vpered!* and *Volnoe slovo*), which had their own conspiracies and risks. Participation in publishing and the appearance of articles in public spaces inevitably led to purely scholarly articles in politics and the struggle of various groups for recognition and influence. Sieber did not refuse to participate in more radical movements—often his publications in particular outlets raised his authority in the field of political economy, and regardless of Sieber's personal views, he worked for his editors and their ideas and movements that stood behind them.

Sieber and the University

A long period of Sieber's life and creativity was associated with the University of Kiev—a period that led him to the heights of recognition in the field of political economy. Sieber entered the St. Vladimir Imperial University of Kiev as a student in 1863, and in December 1875 he wrote a petition requesting dismissal. For more than 12 years, he was affiliated with the university, with an eight-month break for his work as a world mediator in the Volyn province. In the dry records of the university, it appears that from November 1, 1868 to November 1, 1871, Sieber was a "professorial scholarship student" in the Department of Political Economy and Statistics, under the leadership of Grigorii Matveevich Tsekhanovetskii (1833–1898). Tsekhanovetskii introduced Sieber to political economy and the theory of value. At the end of 1871, Sieber defended his master's thesis on Ricardo's theory of value and was given the right to travel abroad for two years to continue his academic studies. In September 1873, he was registered as an assistant professor in the department of political economy and statistics, even though he returned from those travels only in January 1874.[2]

In this period when Sieber's scholarly interests took shape, his position was worked out, and reliable networks formed, Kiev University was, in students' recollections, relatively liberal and open. The university charter of 1863 breathed the spirit of the liberation reforms of Aleksandr II. The posts of the rector, dean, professor, and associate positions were elected rather than appointed. Lecture attendance was voluntary, and for a fee, one could attend lectures and other talks. The University was divided into four faculties: historical-philological, physical-mathematical, legal, and medical. Economics was taught in the Faculty of Law. Among 13 professors, three positions were related to economics: professor of police law (security studies and welfare studies), professor of financial law, and finally professor of political economy and statistics. Sieber learned police law from Nikolai Khristoforovich Bunge (1823–1895), the future rector of the university and finance minister. Bunge and Tsekhanovetskii were Sieber's two main teachers in political economy. Tsekhanovetskii, according to students, willingly "offered his services for managing students in choosing authors of works in political economy, for a more complete study of this science ... mostly classical economists were indicated: A. Smith, Mill, Malthus, Ricardo ..." (Chevazhevskii 1994: 194). Bunge lectured in an impromptu manner, "smoothly, vividly, with an unchanged smile on his face" (Chevazhevskii 1994: 209). He helped Sieber obtain a scholarship to prepare his dissertation.

Students' recollections of the university provide impressions of the atmosphere there. From 1870 to 1875, V. Chevazhevskii studied in the Faculty of Law (Chevazhevskii 1994: 193–215). He recalled a free, creative atmosphere, where students had much personal freedom, external control and supervision were not felt at all, and students communicated in circles distinguished by self-organization. Students created collections for mutual aid and from which one could obtain a loan; students had a self-governing library and a canteen with an elected committee. The most important event at the canteen was theoretical tournaments and incessant debates. Students with radical views formed more closed societies; they did not like casual guests, and they were singled out for their activity, high conceit, and readiness to train workers.

One of Sieber's teachers, law professor Aleksandr Vasilievich Romanovich-Slavatinskii (1832–1910)—whom students called "Colonel" behind his back, left his memories of Sieber as a student: "His was a nervous nature, deeply impressionable, with which his balanced, rather positive mind could not always cope. He was somehow quivering and feverish about phenomena of life and questions of science, which he considered a royal goddess at whose feet he was ready to prostrate himself with fasting and prayer. A thoroughbred idealist, he was not capable of any deal and compromise; he was disgusted and intolerant of everything that did not fit his purely puritanical demands. He resembled a plant signaling 'do not touch,' which was compressed and closed from every touch" (Romanovich-Slavatinskii 1903: 36–37). He attentively attended lectures, sitting in the first desk, and then checked lithographs of the lectures with the teacher. Such reminiscences by his former professor, who was keenly interested in life of the university, had already formed many years after Sieber's death, after walks with Sieber's sister in Yalta, where the seriously ill Sieber spent his last years.

Sieber quickly began to publish work and earned scholarly authority. In 1869 he published his work *Consumer Societies* (Sieber 1869), in which he was less a theorist than practitioner of political economy, as he was less interested in theoretical propositions that consumer societies could win trade mark-ups and have retail goods at wholesale prices, but rather the experience of organizing such societies in Russia and abroad. What put Sieber among the leaders in political economy was the publication of his dissertation (Sieber 1871). The university gave Sieber the opportunity to become acquainted with such teachers as Tsekhanovetskii and Bunge to join the Kiev school in economics. At the same time, even a cursory acquaintance with attempts to characterize the Kiev school of political economy shows how Sieber's position differed fundamentally from those in his context. Investigating the legacy of A. Bilimovich, A. Sušjan singled out the following characteristics of this school: (1) more liberally oriented (a doctrine of harmony), (2) opposed to socialist ideas, and (3) strongly influenced by the German historical school (Sušjan 2010). Sieber immediately became a follower of the classical school in political economy, primarily of Ricardo, and he skillfully incorporated the teachings of Karl Marx. Sieber criticized the subjectivist school of utility, and so was far from the extremes of liberalism. Regarding how radical Sieber was his socialist views, D. B. Naumov in his book about Sieber claimed Sieber was a Marxist who denied practical revolutionary activity, which he called "castrated Marxism" (Naumov 1930). Sieber's teacher N.H. Bunge was an admirer and follower of police law and the historical school. Judging by reports from his travels abroad, Sieber was more sympathetic to the German historical school than to free trade, but he did not develop his work in this direction, and he preferred to deal with more speculative problems of classical political economy.

Kiev University formed Sieber as a scholar. Even after he immigrated to Switzerland in 1875, in many documents in Russia—among publishers and various state departments, for example—he remained a former professor of Kiev University. At the university, Sieber began to speak on behalf of science and quickly gained prestige both due to his teachers, and to his strict approach and clear study of the theory of value and capital. His authority was strengthened by Marx's favorable review of his work, and he quickly moved up the ranks of the university hierarchy, eventually gaining a professorship. Within the framework of the scholarly field, the university occupied a significant place, and this in turn meant inclusion into relationship with the authorities, involvement in intellectual circles and self-governing student movements, and in the very atmosphere of freedom-loving students. This period of Sieber's association with the university was also the period of the formation of networks that Sieber would carry throughout his entire life. To a certain extent, two other fields touched university life: his travels abroad and his experiences in the world of publishing.

Study and Travel Abroad

Students began to be sent abroad to study under the reign of Peter I. This method of education preceded even schools and universities (Vladimirskii-Budanov 1874: 254–256). Once universities were founded, the best graduates were given the opportunity to attend lectures in Europe and from there to prepare their dissertations. In the 1860s and 1870s, traveling abroad was the primary means to train future professors (Masoero 1995) and accompanied the completion of the formation of scholarly interests and implied intensive contacts and meetings with the Western world and its culture, intellectual and in the broadest sense, and building relations with the Russian world abroad. In the 1870s, A. I. Chuprov, I. I. Ianzhul, A. S. Postnikov, and later V. A. Goltsev, N. A. Kablukov, I. I. Isaev, and N. A. Karyshev joined Sieber in such an intellectual journey.

Sieber was sent abroad to continue his academic work abroad in 1872–1873.[3] Professor Bunge's instructions for Sieber's trip show its expansive scope geographically and thematically: "To study political economy and statistics, Mr. Sieber should acquaint himself with how these subjects are taught in Germany, particularly in Leipzig and Berlin. Then Mr. Sieber must learn how bureaus of statistics are set up and operated, particularly in Berlin (under Engel), Vienna, Brussels, Paris, and if possible, also in London, Copenhagen, and other cities where statistical data are collected and processed under the supervision of well-known statisticians. Independent of these activities, Mr. Sieber should, to the extent time allows, study the economic institutions in Western Europe and the United States, if circumstances and financial means allow him to make such a trip. The subjects of these activities are not only the forms of industry in general, but also those establishments whose goal it is to increase the level of public welfare."[4]

Sieber's European journey covered a vast expanse, as demonstrated by letters he sent to the Ministry of National Education requesting funds: Heidelberg, Zurich, Leipzig, Florence, and Paris. According to a published report, most of Sieber's time and attention was spent in Switzerland and Germany. The first part of his trip was spent in Heidelberg and Zurich (Sieber 1873), where he critically appraised university libraries and lectures he attended. In the growing confrontation between the Manchester school of free trade and the German historical school, Sieber was more sympathetic to the latter, which still did not prevent him from disagreeing with their methodology. One gets the impression that he supported the ideas of Brentano, Rocher, and Knies, that economic life is not limited to personal interests. Thus, the study of associations, communities, and other collective forms of economic and social organization acquires special importance. In this case, Sieber preferred the general and homogeneous over the local and particular, so as to separate the essential from the nonessential. Sieber devoted much time to visiting factories and studying organizations for raising public welfare than he did to attending lectures. He was particularly captivated by consumer societies and associations of craftsmen and small businessmen. He personally observed public discussions over revising the federal constitution in the Swiss canton of Glarus and studied elements of community organizations in Switzerland.

Sieber also began to make what would become long-lasting relationships with other scholars: Aleksandr Ivanovich Chuprov (1842–1908), Sergei Andreevich Podolinskii (1850–1891), Ivan Ivanovich Ianzhul (1846–1914), and Maksim Maksimovich Kovalevskii (1851–1916). Sieber was also attracted to activity connected with underground organizations, struggles, and the production of periodicals; his contacts here included Valerian Nikolaevich Smirnov (1849–1900), Rosaliya Khristoforovna Idelson (1848–1915), Peter Lavrovich Lavrov (1823–1901), and Nikolai Grigorievich Kulyabko-Koretskii (1846–1931).

In Leipzig, he and Chuprov attended lectures by Roscher. This was the beginning of a long friendship, and throughout their lives they invariably supported one another. For the rest of his life, Sieber would remain in contact with Moscow economist Aleksandr Chuprov. Their relationship was warm and friendly, and Sieber gave Chuprov the right to conduct affairs with Kozma Soldatenkov regarding publication of his *Essays on Primitive Economic Culture* (1883). Their correspondence continued even after he was forced to relocate to his sister's residence in Yalta in 1884, despite his severe illness. After Sieber's death, his sister Sofia consulted with Chuprov, as the person he was closest to, regarding Sieber's publications, maintaining his legacy, and publishing his books and articles.

Chuprov wrote a short but tender and heartfelt obituary: "The life of this untiring overachiever was not in vain. Those who recall his deep and analytical mind, the childlike innocence of his heart, his ideal kindness, and irreproachable honesty not only of his deeds, but also his most treasured ideas, cannot help but add even more of his personal qualities to his sympathetic character" (Chuprov 2009: 325). It was through Chuprov that Sieber met and grew close to Ivan Ianzhul, who wrote in his memoirs that he had met and "became very close to" Sieber in Leipzig, where he was attending courses by Roscher and Knapp (Ianzhul 1910: 83).[5] Ianzhul ended up writing an article on Sieber for *Brockhaus and Efron Encyclopedic Dictionary*. Sieber also became acquainted with historian and sociologist Maksim Maksimovich Kovalevskii of Kharkov University, who was taking a longer journey at the time. Their interrelations were manifested in Sieber's review of Kovalevskii's book in 1879 and in their collaboration for the journal *Critical Review* Kovalevskii published.

During this time, Sieber encountered the Russian émigré community in Zurich and became an active participant in their various activities. Podolinskii dined with him often[6] and unsuccessfully tried to recruit him to write for the journal *Vpered*, which was published by Valerian Smirnov. In his memoirs, Nikolai Kulyabko-Koretskii, an ordinary participant in the revolutionary movement, provided some information about this relationship (Kulyabko-Koretskii 1931). He related information about Pyotr Lavrov, leader of an entire section of the movement, and described activities occurring in the Russian émigré community in Zurich in 1872 and 1873, as well as his work at the publisher that printed *Vpered*.

A central institution that brought together the Russian émigré community and also brought them into contact with other European intellectuals was the Russian Library, located on the wide Platte Street. As Kulyabko-Koretskii noted in his memoirs, one could recognize Russian students on the street "by their careless dress, loud voices, lively and frequent gesticulations, the long hair of most of the men, and in contrast, the short hair of many of the young women" (Kulyabko-Koretskii 1931: 12). The space recalled the freedom-loving student circles in Kiev: "'Reading' in the library without being used to the circumstances was difficult, since from morning to evening there was a constant flow of people and loud conversations and heated theoretical debates about all kinds of social and philosophical matters, all in a thick cloud of tobacco smoke." Therefore, this was no unimportant site: in fact, Sieber took the side of Pyotr Lavrov and Sergei Podolinskii[7] in an incident regarding the Russian Library in Zurich.[8] The library was founded by an associate of Mikhail Bakunin, Mikhail Sazhin, who went by the pseudonym Arman Ross. Many people confirmed that the library was rich in contemporary books, including tomes on political economy and sociology. According to Sazhin, it was as if regular tea had turned into a library oriented to the ideals of "Michel" Bakunin's anarchism (Sazhin 1932). Membership was for an initial fee of 10–15 francs and then 3 francs per month thereafter, and this source of income provided the library with a degree of independence (Sazhin 1925: 41–42). The library's secretary was Valerian Smirnov, and the librarian was his wife Rozaliya Idelson—"a young, beautiful, and very elegant brunette of small stature, extremely friendly and happy to help the library's readers" (Kulyabko-Koretskii 1931: 15).

Sieber's archive did not survive, and so it is necessary to infer his personal relations from archives of those with whom he exchanged letters. Three letters in the GARF archive in Moscow from Sieber to Idelson (as librarian of the Russian Library) are mostly devoted to books Sieber requested.[9] On November 22, 1872 he reported from Leipzig on the dispatch of Stein and threatened to expel Marx. In a letter from March 3, 1873, sent from Munich, he asked for more books: "I come to you with the humble request to send me the following three books from the Zurich library (in the case, of course, if the latter at the end of the seizure case … was again placed at the disposal of those persons whose legal property it constitutes) my dissertation, the posthumous works of Herzen (1st edition), and the January edition of *Otechestvenye zapiski*. If the January edition is in your hands, then will you be so kind as to make for me an excerpt from Mikhailovskii's 'Literary and Journal Notes,' and place in it *everything* that is said about my dissertation. The books will be returned very soon, and along with them I will send you a monthly fee for the right to use them (5 francs, it seems?)."[10] On March 31, 1873, Sieber wrote from Munich instructing Chuprov to send books and money, and stating that he was moving to Florence. This correspondence makes clear how important the exchange of books was and the role played by the Russian Library of Zurich in 1872–1873 for Sieber and an entire intellectual community that was developing important strands of economics and socialist thought (cf. Bankowski-Züllig 1991). The internal conflict between Lavrov's allies and the Bakuninists coincided with the Russian state's prohibition on students staying in Zurich in 1873, after which many moved to Paris, Bern, and Geneva. Following this, the Russian colony in Zurich lost its importance.

Publications and Fields of Political Discourses

Like discussion circles, publications were one of the most important components of intellectual, social, and political life. In the eighteenth century they were already starting to play an important role in education and enlightenment, and the printed word created expanded communities of discourses and identities (Wuthnow 1989). In creating media for disseminating intellectual, political, or other ideas, actors create discourses with particular core categories, arguments, and identities that serve to orient further discourses. As actors orient to these discourses and related identities, they interact more with others sharing that communication, creating fields. Those fields of discourse, in turn, continue to shape the identities of actors who enter those fields and participate in discussions. An actor's intellectual dispositions draw her into particular fields, through which she hones and develops her own interests and arguments and contributes to that field. Sieber's intellectual life, like that of many of his contemporaries, was closely tied up in publication activities—from his own creations, to facilitating the publication of other people's works, to helping guide broader policies for which authors and subjects to champion (or, conversely, to ignore). The growth in production of written materials and of publics that consumed them contributed to the rise of various intellectual endeavors that crystallized into academic disciplines and political movements, for example, and Sieber was no exception to this phenomenon. Further, by the nineteenth century, it became possible to make a living through publishing, whether as an author or as an editor (or both). Some markets could also create competition over popular authors, enhancing their status and networks as well as income, allowing them to shape those fields even more.

Sieber's publication record was at its most active between 1870 and 1886. His articles were written in Russian and most were published inside Russia. He was particularly active in his Swiss period (1875–1884), when he lived with his wife in Berne, when he published from 4 to 12 articles each year, for an average about 8 articles of different lengths (cf. Rezul 1931). He contributed actively to the journals *Universitetskie izvestiia* (University News), *Znanie* (Knowledge), *Slovo* (Word), *Otechestvennye zapiski* (Fatherland Notes), *Vestnik Evropy* (European Herald), *Russkaia mysl* (Russian Idea), *Iuridicheskii vestnik* (Juridical Gazette), *Kriticheskoe obozrenie* (Critical Review), and *Volnoe slovo* (Free Word). Sieber's first articles were published in *Universitetskie izvestiia*, the monthly journal of the University of Kiev, where in 1871, he published his master's thesis Ricardo's theory of value and capital, which was followed by a report on his stay abroad and a sketch about Ricardo (1873). From 1874 to 1877, Sieber published a series of articles on Marx's economic theory of Marx "cleansed of its metaphysical subtleties" in the Petersburg journal *Znanie*, and in the political and economic section of that journal his articles on urban economy and reviews of various books also appeared. After *Znanie* was closed by the censors, Sieber continued to publish a series of articles on Marxist theory in the journal *Slovo*. In 1878, Sieber published five articles on Marx dedicated to the theory of public cooperation, large-scale industry, machine theory, the theory of capital accumulation, and population theory. (Subsequently, Sieber added most of this material to his book in 1885.) Along with multiple reviews in 1879, his article "Chicherin *contra* Marx (Criticism of Criticism)" came to light and he published a series of articles on the relationship between social economy and law.

In 1877, Sieber made his debut into the St. Petersburg literary journal *Otechestvennye zapiski* with a review in defense of Marx, "Several remarks on G. Iu. Zhukovskii's article 'Karl Marx and his book on capital.'" The journal, encyclopedic in scope magazine and published at that time by A. A. Kraievskii and M.E. Saltykov-Shchedrin was increasing in popularity and in 1881 had already 10,000 subscribers (Dementiev et al. 1959: 284). Under the heading "Economic sketches," Sieber's articles were published on a variety of topics: the peasant commune, agriculture in England, Holland, and Belgium, the status of working classes, successes of a large-scale farming, reports by Swiss factory inspectors, and the housing issue. In the 1880s, he began to cooperate with *Vestnik Evropy*, the St. Petersburg journal on historical and political sciences published by M. M. Stasiulevich. After 1882, Sieber placed a number of his articles in the Moscow-based journal *Russkaia mysl*: for example, on laws of the global economy, the distribution of landed property in Germany, and Henry George's book *Progress and Poverty*. In 1880 Sieber's review of M. Kovalevskii's book on communal land ownership appeared in *Iuridicheskii vestnik*, a journal with a liberal slant—Kovalevskii was a member of the journal's editorial board—and in 1881 they published Sieber's article on the work of Karl Rodbertus-Jagetzow. Sieber not only published in *Iuridicheskii vestnik*; he also became a member of its permanent staff (Dementiev et al. 1959: 305). In the pages of the journals he analyzed Charles Letorneau's book on sociology and ethnology, explored the agrarian question in Ireland, compared farming and law, and discussed Albert Post's view on primitive law. (A. I. Chuprov, I.I. Ianzhul, and N.A. Kablukov also published work in this journal.) Sieber's last article, on the housing issue in Berlin, Paris, and London, was published in *Iuridicheskii vestnik*.

Sieber's work was not confined to these various journals; he also wrote for and worked with newspaper publishers, a history that was sometimes exciting and dramatic. During a trip abroad in Zurich, Sieber was involved in a dispute over different approaches to societal development. On one side, to which Sieber leaned, was Lavrov, who advocated a more moderate movement and who did not deny the principle of statehood as such. On the other side, associates of the anarchist Mikhail Bakunin held more radical views. Not unimportantly, Sieber's friend S.A. Podolinskii actively helped Lavrov organizationally and financially. The organizers of a new newspaper literally had to hunt down potential authors to write for them. In a letter dated January 2, 1873, Podolinskii reported to Lavrov from Leipzig, who decided not to promise print anything, but rather to report that the publisher was waiting for work without any promises—although he also complained that he had to listen to long and boring conversations about the program to organize the newspaper *Vpered!* (Lavrov 1974). Kulyabko-Koretskii wrote of failures to persuade Sieber: "Unfortunately, I was not able to convince Sieber to cooperate in with the journal *Vpered!*—he was dissuaded by the lack of time and the abundance of scholarship that he had already began and that demanded a conclusion" (Kulyabko-Koretskii 1931: 187).

This was only one part of Sieber's experiences with newspapers. In 1875, he began to cooperate with the newspaper *Kievskii telegraf*, a partnership that lasted until June 18, 1876, when the newspaper was closed by order of the government after the issuance of the Ems ukaz (Dementiev et al. 1959: 381). At the end of 1874 the newspaper reported that from January 1, *Kievskii telegraf* "will be completely transformed, both in terms of content and appearance. Having organized a significant number of permanent employees, we hope to meet the long-recognized need for the existence of a private literary and political newspaper in a city like Kiev that would fully reflect current life and would treat it with freedom to critique. With constant participation of K. I. Kibalchich in the editorial office, we were promised cooperation of the following: Prof. V. B. Antonovich, Prof. I. G. Borshchev, Prof. F. K. Volkov, Prof. V. G. Demchenko, Prof. M. P. Dragomanov, N. I. Zhitetskii, Prof. N. I. Sieber, A.V. Klassovskii, Prof. I.V. Luchitskii, I. P. Novitskii, Dr. K.R. Ovsianyi, Prof. Romanovich-Slavatinskii, A. D. Ushinskii, E. E. Tsvetkovskii, P. P. Chubinskii, and others" (*Kievskii telegraf*, December 18, 1874). In the year and a half that this newspaper existed, Sieber worked with the same group of Kiev intellectuals with whom he collaborated in the Geographical Society and other enterprises supported by the old Kiev Hromada.

If earlier the newspaper printed feuilletons, general information, domestic and foreign news, and reprints from the central press, after 1875 the new publishers stated that the newspaper would be free, critical, and political. Neighborly relations in one editorial board and Sieber's friendship with M. P. Dragomanov, V. B. Antonovich, and S. A. Podolinskii spoke for themselves. D. B. Naumov and M. Grushevskii stressed that Sieber at least did not act as a Ukrainian and did not write in Ukrainian, but still stood "nearer in spirit" to the old Kiev Hromada: "Sieber took an active part in *Kievskii telegraf*, around which members of the 'Hromada,' and became a member of the South-Western Division of the Russian Geographical Society, which included prominent figures of the Ukrainophile movement. He took part in organizing, conducting, and processing statistical work undertaken by the Hromada. Sieber also participated in work, as a lecturer in political economy, for illegal women's courses organized by activists of the 'Hromada.' He took part in the organization and work of the Kiev consumer society, organized by the 'Hromadans'" (Naumov 1930: 17–18).

At the same time, in a programmatic statement from December 1874 on issuing a new type of publication, the Ukrainian question was bypassed, and the words "Ukrainian" or "Little Russian" were never mentioned. A high intellectual level was accentuated, and the newspaper was understood to be the voice of the South Russian *intelligentsia*, which, without prejudice, party spirit, and any exclusivity, but only proceeding from the study of the laws of life, would direct "social phenomena in accordance with conclusions of science and the common good" (*Kievskii telegraf*, December 18, 1874). One could say that this was precisely the credo of 30-year-old professor Sieber, a creed he followed hereafter, trying only to study social phenomena scientifically and to direct these toward the common good.

Despite the policy of not accepting anonymous submissions, permanent columns often went without a signature. There are reasons to believe that Sieber was one of the newspaper's permanent authors: "On the dwellings of the poor population of Kiev" (1875, No. 8, 38), "Apartment tax in Kiev" (1875, No. 119, 122, 1876, No. 14), "Our consumer societies" (1876, No. 16), "Six-Year Results of City Budgets" (1876, No. 23), "Our Industry and Speculation" (1876, No. 29, 32), and many others. Beginning with issue #60 for May 21, Sieber published a positive review of A.I. Chuprov's master's thesis, "The railroad economy, its economic characteristics, and its relationship with the interests of the country," and in #34 for 1876, he was defending Chuprov's journalistic style and conclusions in his response to a critical note in *Vestnik Evropy* by N. G. Bunge "Monopoly of the railway kingdom." The newspaper also created its own space for assessing current events, with special emphasis given to Little Russian cases. In this sense, it raised Kiev from a simply mimicking St. Petersburg news, to an independent center capable of independently assessing events and even of taking its own initiative (e.g. in matters of the census). It is no coincidence that the newspaper had the word "political" in its subheading, as the editors wanted it to play a political role primarily through the growth of its readers' intellectual demands. Sieber's participation in this newspaper shows his active civic position and his desire to serve political purposes.

The newspaper *Volnoe slovo* was published weekly in Geneva from August 1881 to May 1883, and its editors declared that it was the organ of the constitutional-liberal "Zemskii soiuz," and in reality A. P. Malshinskii, on instructions from the third branch, founded it with the aim of managing sentiments abroad. M.P. Dragomanov, an old friend of Sieber's, and Pavel Borisovich Akselrod, who graduated from Kiev University and went out to the people to engage in revolutionary activity, took up the task of publishing the newspaper. Despite his usual caution, Sieber became an active member in this overseas publication and submitting a number of articles and reviews, often under the name "PS." In issue # 41 for 1882, a review of *The Fate of Capitalism in Russia* was published (Dubianskii 2016), and the newspaper also published his reviews on books by G. D. Shcherbachev, in "The Present Position of Russia," and G. George's *Progress and Poverty*. In one letter to Akselrod, dated August 5, 1881, Sieber elaborated on his disagreement with George's basic theses: "The author proves with his new theory only that once America is open, it cannot be reopened, even in America itself."[11] In a rare occurrence, Sieber responded to an offer to publish his work in a Swiss publication. Perhaps he was impressed by the relatively moderate views there that it turned out reflected the preferences of those were funding the publication. It also turned out that the always cautious Sieber was participating in a fake magazine.[12]

During this Swiss period, Sieber also actively published works from his own pocket, and this investment contributed to his main source of earnings. We do not know much about his relations with publishers, but data from rare archival discoveries indicate that these relations involved constant tensions and misunderstandings. In a letter P. B. Akselrod in December 1881, Sieber commented about *Essays on Primitive Economic Culture*, eventually published by Kozma Soldatenkov in 1883: "The printing of my work is slowing down for an unknown reason. Everything went well, I soon found a publisher, but for two months now there has been no rumor, no spirit, no answer to my inquiries. In a few days, however, I hope to resolve this purely Russian misunderstanding."[13] By proxy, A. I. Chuprov conducted Sieber's affairs for publishing the book. From the agreement with the publishing house, we learn that *Essays* was printed with a circulation of 1200 copies, with remuneration of 75 rubles per printed sheet.[14] In total, Sieber received 1350 rubles: 300 rubles on April 6, 1882; 200 rubles on October 21; 250 rubles on November 16; and 600 rubles on March 25, 1883. There is some reason to believe that the minimum honorarium from a journal could start at 75 rubles and rise to as much as 200–250 rubles. Sieber, it seems, was beginning at the lower, starting rate.[15] His working relations with his publishers were not straightforward, either. While editors tried to involve authors in the entire process of publication, authors were also dependent on editors, especially after they had granted consent to publish. Sieber could not refrain from complaining about how helpless he felt when interacting with editors, and about how much their ideas about moral concepts differed: "It is generally difficult to conduct business with our printing houses, journals, etc., which deny the author all human rights. For example, my Ricardo was in print for *three years* … and now I learn from strangers that it has been out of print for three months. After that, I write to Antonovich, asking that he send me a few copies."[16]

In sum, publishing as a subfield was developing in the context of an emerging intellectual and professional market, and in which editors and authors had common *and* differing interests in producing discourses for that public. Core actors and statuses, boundaries, identities, and rules in this emerging, embryonic field—if there was a single—were still rough or in flux. The core structures were personal relations between intellectual authors, whose personal relations allowed them to act on each other's behalf, to hone their arguments in discussions and debates, and to provide a growing critical mass of intellectual elites who might command the attention (and thus book and journal sales) of broader publics, in particular other intellectuals and budding economics (in Russia and abroad) and educated society (again, in Russia and abroad). Another set of actors were the publishers themselves, who did not enjoy market dominance that could allow them to dictate tastes or take risks with potentially innovative (or off-the-wall) author. In the background were other no less influential fields and institutions: in particular, market economies that demanded profit for survival, thus influencing what tastes and discourses might emerge (i.e. those with enough demand to provide sufficient income to publishers); and states exercising censorship to differing degrees and contributing to themes, identities, and content of their different intellectual discourses. Transform Russia's *imperial* state into something federal (closer to the American model), for example, and notions of ethnic identity might have been less pressing (hypothetically), leaving more room for discussion of capitalism, analytically and critically.

Sieber contributed to this history in his own small way. The discourse here offered reflected the independent, critical view of a scholar well versed in modern economic theory, particularly in classical political economy and Marxism. This view appealed to current issues by explaining and interpreting them from the standpoint of advanced approaches at that time. In this sense, many of Sieber's publications added an enlightened, polemic, and critical function to the usual academic style. This was in accordance with ideals of the cosmopolitan *intelligentsia* of that time. At the same time, reality politicized Sieber's activity, and he consciously or not, he became involved both in the so-called Ukrainian question, and in underground and revolutionary activity.

Tensions of Economic and Political Fields, or, Was Sieber a Ukrainofile Socialist?

While other intellectuals followed traditions of statistics and an *analytic* strain of political economy, introducing these into Russia via writings and institutional work (e.g. setting up departments and courses in universities), a parallel *critical* tradition was emerging as well, as the conclusion to the preceding section suggests. The earlier tradition of political economy did not separate politics (institutionalized power) from economic activity, and Marx and those who followed in his stead left the tools to expand inquiry into power with a critical edge. At the same time, the "battle of the methods" and the rise of a more positivist, quantitative, and (arguably) constrained vision of "economics" created the impulse for some intellectuals to carve out institutional and discursive space to create a field of "economics" as an independent discipline, with its own domain and rules of status and legitimate inquiry. In this tradition we see the unification of statistics and economics, for example, and a focus on pragmatic suggestions for policies of economic development. At this time, however, a host of factors were driving debates and mobilization in a political direction. One was the rise of a working class, with its organizational champions (nascent trade unions and political parties or movements) and intellectual champions (especially European socialists). Another was the consolidation and expansion, physically and institutionally, of nation-states that were developing not only structural tools of control, but also political identities for control and mobilization of their populations. For all the claims and assumptions by the growing economics profession, that the "economy" was its own autonomous sphere with its own laws and thus its need for a separate profession to study it, economies are expressions of power as well (Mitchell 2002). The critical tradition in political economy thus had not only seeds from the past; it had soil in which to grow, especially in the eastern half of Europe.

The Russian state contributed to this critical discourse with its own repressive work, much of which until 1880 was carried out by the Third Department.[17] Institutionally suspicious—that was its charge, after all—Third Department cadres tried to watch carefully those who might be involved in various attempts to destabilize the power of the Russian state. They were at work not only at home, but also abroad, where various Russian and other intellectuals from the Empire were studying, or living in exile, or traveling. Sieber was not outside their field of vision. Paradoxically, their characterization of Sieber as a leader among socialist Ukrainophiles in Kiev was based on denunciations of mediocre (or worse) quality—yet this characterization also had a basis in reality. Sieber clearly held sympathies for those unwilling to consign Ukrainian identity and culture into imperial Russia's dustbin of history. He stood by those who studied Ukrainian language, history, and everyday life, and he seemed to provide some support to figures who were much more dedicated to some form of political struggle, even revolution. As the field of economic discourse could intersect with political fields and forces, we need to take a closer look at how politics and economics intersected in these particular networks and institutions.

A brief reference to Sieber in a report compiled by the Third Department noted that "Sieber fled abroad with Professor Dragomanov. Sieber headed an organization of Ukrainian socialists in Kiev. At meetings at Fedor Volkov's[18] home from 1874 to 1876, Sieber gave a course on socialist political economy according to Karl Marx, whose work Sieber translated [from German] into Russian... Sieber, who lives abroad and who maintains a permanent relationship with Russia through their relatives living in Kiev, is among prominent figures of the social-revolutionary party."[19] Each individual detail of this reference is either partially true or erroneous; his relatives lived in Yalta, he was cautious vis-à-vis the social revolutionary party, he was academically minded, he was careful by his nature, he translated Ricardo rather than Marx, he did not take part in fictitious meetings, his travel abroad was legal and formally approved (and so by no means did he "flee"), he wrote nothing that could be construed as sympathetic to the Ukrainian cause, he did not write in Ukrainian, he only communicated with or was personally (not ideologically) close to those who devoted energy to the Ukrainian question (e.g. Dragomanov and Podolinskii), and his struggle with Lavrov and Smirnov was ideological.

In commentary about the denunciations, published in the magazine *Byloe* in 1907, A. A. Rusov and F. K. Volkov—participants in these events and well-known in the Kiev Hromada—provided valuable remarks for understanding not only of those who reported and wrote these statements, but also what actually happened and how it was perceived by actors themselves.[20] Those who made these reports to the Third Department, Velednitskii and Bogoslavskii, were dreamers and chatterboxes who knew little. Further, the Kiev Hromada was a circle of people from the university and scholarly environment, who were involved in constitutional and educational activities; inclusion in this community could be blocked by a single vote of members, suggesting strict boundaries. "In the scholarly sphere, except for the work of the pedagogical-professorial and teaching staff, this circle left its traces in the form of a mass of scholarly work in history, economics, ethnography, etc., in the formation of the Southern Department of the Imperial Russian Geographical Society, the organization of the Third Archaeological Congress, the Kiev one-day census, the Proceedings of the Ethnographic Expedition of Chubinskii, the foundation of the Nestor Chronicle Society, 'Kiev Antiquity,' and so on, in the publication of a number of studies and scholarly collections in the genre of monumental Ukrainian historical songs, edited by Antonovich and Dragomanov" (Rusov and Volkov 1907: 153–154). The publication of the *Kievskii telegraf* constituted their work in the sociopolitical sphere. This comment leaves no doubt that Volkov and Rusov were confident that Sieber was part of the Kiev Hromada. At the same time, this gives us a sense of terminological nuances. To participate in the Kiev Hromada did not mean to one was a Ukrainian philosopher. This term was used more by the authorities, such as confessions enticed by the Third Department on confessions, or in the Ems ukaz of 1876 that marginalized the Ukrainian language and culture. It is no coincidence that the authors of the Third Department commentary wrote, "The frequent mention of this or that person by the 'leaders of Ukrainophiles' among people who lived in the 70s and 80s and who were members of the former Kiev Hromada can only bring a smile" (Rusov and Volkov 1907: 159).

It is also important to take into account that the real interaction between St. Petersburg and Kiev was more difficult than the struggle between the imperial and the national. As Hillis correctly demonstrated, in the context of the Polish uprising in 1863, interest in Ukrainian history as a counterweight to the Polish gentry was supported in St. Petersburg, as exemplified by the activities of Antonovich's journal *Osnova* (published in St. Petersburg), support of Ukrainian Sunday schools, the translation of the Manifesto on the emancipation of peasants into Ukrainian, and the establishment of the Geographical Society in Kiev (Hillis 2012).

It is clear that Sieber's life did not become calmer after this denunciation. In 1882 he confessed to Ovsyaniko-Kulikovskii in Paris: "Although I am a Swiss citizen, he said, that will not prevent the Russian government from putting me in jail, and maybe even hanging me ... just for his inclination, type of thought, friendship with Dragomanov, acquaintance with emigrants."[21]

A review of Sieber's publications shows that this form of participation in academic life played a leading role from the very beginning. Already in the short Kiev period, after returning from his foreign mission in 1874–1875, Sieber was also involved in a wider field—not only academic, but also popularizing and sociopolitical. The newspaper *Kievskii telegraf* was the voice of the Kiev *intelligentsia*, the source of evaluation and unbiased news and reviews of "Little Russian" life. Participation in this newspaper alongside leading representatives of the Kiev Hromada was a sign of Sieber's desire and readiness to participate in political life. The closure of the newspaper by the government only further strengthened this resolve.

In the Swiss period, Sieber continued to publish actively in Russian, mostly in influential publications out of St. Petersburg. He popularized and defended Marx, wrote on topical economic issues, responded to books and articles with reviews, and generally supported the reputation of a neutral and unbiased adherent to the scientific approach. Often the publishers themselves were trying to attract an authoritative author about economic theory. Yet the publishing world differs from the academic world: market principles play a large role that often leads to a misunderstanding between author and publisher. The review of Sieber's contacts related to publications showed that he was actively involved in this work. In this space, often in addition to the will of the author, there was a clash of the logics of academic and political fields.

Conclusion

This snippet of Sieber's biography and micro-history of various forces and relations raises a logical question: What is hidden behind these details regarding of the fate of the scholar—as individual, and as role and position in broader institutions and fields—in the Russian Empire of the 1870s–1880s, as civil society was starting to take shape? Sketching the details of the academic and political fields of the work and the life of the scholar allows us to glimpse what exactly it meant at that time to be an economist in this intellectual circle. To use a popular Russian toy that tourists seem to love, the scholarly field as like a large *matryoshka*[22] that can be decomposed into smaller subfields: university, academic travel, and publications. What did it mean to enter the university and become a student of an Imperial University at this time? It meant an intense student life, broken down into more circles and corporate entities with varying degrees of closeness and dispositions. To advance in the university, to participate in academic life, the "student-professor" relationship was quite important. In Sieber's case, we see personal support from Bunge and Tsekhanovetskii. The dissertation was symbolic capital with significant weight. Academic travel abroad was not simply a whim of fancy; it was literally an institution already incorporated into academic life and supported formally by ministries of state. Being chosen to undertake such a journey already meant likely inclusion into the company of the professoriate. This entire period in a scholar's life contained the important contexts and practices that shaped a scholarly *habitus*: entering a significant university, relations with professors, imbibing texts, cooperating with journals and newspapers in which one honed arguments and learned the art of shaping discourse, exchanging letters with colleagues, participating in various circles and discussions, writing books and other output, familiarizing oneself with practical aspects of economic and political life, and forming personal contacts that combine a high level of intellectual communication and sympathy as a foundation for relations that follow one through life. Authority and symbolic capital were just as crucial in the academic field as in others, and it contributed to a hierarchy of status: in particular, publications (then as now) were core forms of symbolic capital. A typical example is *Otechestvennye zapiski*, which had 10,000 subscribers during Sieber's collaboration. The editors literally hunted for talented authors with academic authority, and they often received refusals if the journal's program did not coincide with their positions. In the case of illegal publications, there was real political risk. At the same time, most journals paid royalties and a certain rhythm in the provision of texts. Thus, through journals it was possible to find a way to gain symbolic capital and a regular income.

From his correspondence it is clear that Sieber did not have a position in Switzerland and did not remain a loner. Often he refused to unnecessary contacts, using some illness as an excuse, but he had stable connections, some of which lasted to the end of his life. The question of Sieber's relations to Ukrainophiles and socialists was an issue of the intersection of academic and political fields. As a member of the old Kiev Hromada, he actively participated in a variety of intellectual and public projects. Educational and social activities intertwined with interests of Little Russia as a concrete site. Although Sieber did not write in Ukrainian and did not leave written statements about this topic, just as he did not consider it right for him to participate in the revolutionary struggle and somehow push history, Sieber was neither a Ukrainian philosopher, nor a revolutionary—subjectively. Objectively, by virtue of the logic of these spaces and his positions in them, he was. Belief and position might not have corresponded perfectly, but fields and actors do not map perfectly, either. Politically, belief and position were one and the same to the Third Department, as they often are for those actors whose charge is domination or revolution.

Sieber was born and died in the Crimea. A truncated marble column was erected as a monument at his Yalta grave. This image is consistent not only with how disease broke his life, but also with his intellectual biography, in which the scholarly was truncated by the political.

Notes

1. This is similar to the consistent cross-national variation in economic theory (Fourcade 2009): that there are national discursive fields and broader international fields that influence one another, but never entirely, such that distinct national traditions can persist even as they contribute to each other's fields.
2. GAK (State Archive of Kiev), f. 16, op. 314, d. 163 ("On the release of Nikolai Ziber from service as university instructor"); f. 16, op. 465, d. 4759 ("Personnel file, N. I. Ziber"). Sieber apparently had no indication of affiliation with the university in his external passport (GAK, f. 16, op. 34, d. 163, l. 20).
3. RGIA (Russian State Historical Archive), f. 733, op. 141, d. 158 (case on sending Sieber abroad for academic purposes, 1871–1873).
4. RGIA, f. 733, op. 141, d. 158, l. 22.
5. Ianzhul also related a meeting with the Shumova sisters in Heidelberg, "one of whom later married my friend Sieber" (Ianzhul 1910: 89).
6. Kulyabko-Koretskii (1931: 59) related a story about Sieber's trip with Podolinskii to Mühlhausen in German Alsace, where they shared their home with workers from the Dolphus textile factory. Sieber and Podolinskii agreed that in the end it put the workers in an even more dependent position.
7. Podlinskii was the son of a landlord and graduated from the faculty of Mathematics and Physics at Kiev University. In 1871 he went abroad to support the publication of *Vpered.*

8. This conflict was between followers of Mikhail Bakunin and their more radical anarchist agenda, with followers of Pyotr Lavrov, with a moderate program for education and change. Sieber was on Lavrov's side. For more details, see Bankowski-Züllig (1991).
9. GARF (State Archive of the Russian Federation), f. 1737, op. 1, d. 80 (three letters of N. I. Sieber to R. Kh. Idelson, November 22, 1872; March 3, 1873; March 13, 1873).
10. GARF, f. 1737, op. 1, d. 80, l. 2.
11. International Institute of Social History. Akselrod. 45a. Ziber. Letter from N. I. Ziber to P. B. Akselrod, August 5, 1881 (https://search.socialhistory.org).
12. Sieber still did not completely distance himself from the underground struggle, from more moderate socialists who raise the working class question and criticized imperial power. While Sieber often refused, his cooperation with *Volnoe slovo* reveals that this was not a fundamental refusal, but more often was connected with his inclination to caution, too many other obligations, and poor health. In the 1880s Sieber again saw the need for more active participation in this struggle.
13. International Institute of Social History. Akselrod. 45a. Ziber. Letter from N. I. Ziber to P. B. Akselrod, August 5, 1881 (https://search.socialhistory.org).
14. TsGA (Central State Archive), Moscow, f. 2244 (A. I. Chuprov), op. 1, d. 3070. A printed sheet—even today the measure of size of a publication (rather than pages or word count)—might be up to 40,000 characters, depending on the presses. So this was far less than 75 rubles per *book* page.
15. As editor of *Otechestvennye zapiski*, Saltykov-Shchedrin tried to pay new authors 200 rubles per printed sheet (Krivenko 1891).
16. TsGA, Moscow, f. 2244, op. 1, d. 1551 (letter from N. I. Ziber to A. I. Chuprov, January 24, 1883). Emphasis in the original.
17. Until 1880, the Third Department was an arm of the Russian state charged with rooting out political challenges to the crown. In 1880 the Third Department was abolished in the course of reforms restructuring the security apparatus. Its functions were transferred to the Ministry of Internal Affairs.
18. Fedor Kondratievich Volkov (1847–1918) was an ethnographer and archeologist. In 1879 he was forced to leave Russia and until 1905 lived and published in France.

19. GARF, f. 109, op. 230, d. 1741, l. 1–3. Similar concise phrases about Sieber as a "prominent figure in the social revolutionary party" are repeated in the state police report (GARF, f. 102 d-3, op. 1881, d. 753). The case was dismissed after Major Deil arrested A. M. Smirnov.
20. "Svod ukazanii, dannykh nekotorymi iz arestovannykh po delam o gosudarstvennym prestupleniiam (mai 1880)," *Byloe*, #6 (1907): 118–152.
21. Russian National Library (St. Petersburg), Manuscript Division, f. 1000, op. II, d. 996, l. 15.
22. This is the Russian figure of dolls nested one inside the other, from large to small.

Works Cited

Allisson, François. 2015. *Value and Prices in Russian Economic Thought: A Journey Inside the Russian Synthesis, 1890–1920*. London: Routledge.

Bankowski-Züllig, Monika. 1991. Die erste 'Russische Bibliothek in Zürich' (1870–1873). In *"Der Parnass liegt nicht in den Schweizer Alpen." Aspekte der Zurcher Universitatsgeschichte*, ed. Verena Stadler-Lahart, 229–259. Zürich: Rohr.

Bourdieu, Pierre. 1976. Le champ scientifique. *Actes de la recherche en sciences socials* 2 (2–3): 88–104.

Chevazhevskii, V. 1994. Iz proshlogo Kievskogo universiteta i studencheskoi zhizni (1870–1875). In *Z imenem Svyatoho Volodymyra: Kyivskyj universytet u dokumentah, materialah ta spohadah suchasnykiv: u 2 kn*, ed. V. Korotkyj and V. Ulianovskyj, 193–215. Kiev: Zapovit Kyiv.

Chuprov, A.I. 2009. Nikolai Ivanovich Sieber. In *Rossiia vchera i zavtra. Stati. Rechi. Vospominaniia*, 323–325. Moscow: Russkii Mir.

Dementiev, A.G., A.V. Zaitsev, and M.S. Cherepakhov. 1959. *Russkaia periodicheskaia pechat (1702–1894). Spravochnik*. Moscow: Gospolitizdat.

Dubianskii, A.N. 2016. Ziber i Vorontsov o kapitalisticheskom puti razvitiia Rossii. *Terra economicus* 14 (4): 107–118.

Fligstein, Neil, and Doug McAdam. 2012. *A Theory of Fields*. New York: Oxford University Press.

Fourcade, Marion. 2009. *Economists and Societies. Discipline and Profession in the United States, Britain, and France, 1890s to 1990s*. Princeton: Princeton University Press.

Guelfat, Isaac. 1970. Aux sources de la marxologie authentique: N. I. Sieber. *L'homme et la société* 17 (1): 141–148.
Hillis, Faith. 2012. Ukrainophile Activism and Imperial Governance in Russia's Southerwestern Borderlands. *Kritika* 13: 301–326.
Ianzhul, I.I. 1910. *Vospominaniia I. I. Ianzhula o perezhitom i vidennom v 1864–1909 gg*. Vol. 1. St. Petersburg: Tipografiia Stoikovoi.
Kapeller, Andreas. 1989. 'Н.И. Зибер – Микола Зібер—Niclaus Sieber.' Ein Schweizer als Wegbereiter des Marximus in Russland. In *Slavitische und slavenkundliche Beiträge für Peter Brang zum 65 Geburtstag*, ed. Carsten Goehrke, Robin Kemball, and Daniel Weiss, 671–684. Peter Lang: Bern.
Kleinbort L.M. 1923. *Nikolai Ivanovich Sieber*. Petrograd: Kolos.
Krivenko, S.N. 1891. *M. E. Saltykov-Shchedrin. Ego zhizn i literaturnaia deiatelnost*. St. Petersburg: Obshchestvennaia polza.
Kulyabko-Koretskii, N.G. 1931. *Iz davnykh let. Vospominaniia labrista*. Moscow: Vsesoiuznoe obshchestvo politkatorzhan i ssylnoposelentsev.
Lavrov, P.L. 1974. *Gody emigratsii. Arkhivnye materialy. Ot Vpered k gruppe starykh narodovoltsev*. Boston: Riedel.
Masoero, Alberto. 1995. Paysages sociaux et itinéraires de formation. Les voyages d'études des économists russes à l'étranger. Annees 1860 et 1870. *Cahiers du Monde russe. Cultures économiques et politiques économiques dans l'Empire tsariste et en URSS, 1861–1950* 36 (1–2): 7–35.
Mitchell, Timothy. 2002. *Rule of Experts: Egypt, Techno-Politics, Modernity*. Berkeley: University of California Press.
Naumov, D. 1930. *Nikolai Ivanovich Ziber. Iz istorii marksistkoi ekonomicheskoi mysli na Ukraine i v Rossii*. Kharkov: Proletarii.
Raskov, D.E. 2016. Sieber kak kabinetnyi economist-antropolog. *Terra economicus* (4): 92–106.
Reuel, A. 1956. N. I. Ziber kak nauchnyi i obshchestvennyi deiatel v Rossii 70-kh i nachalo 80-kh godov XIX veka. In *Russkaia ekonomicheskaia mysl 60–70-kh godov XIX veka i marksizm*, 287–341. Moscow: Gospolitizdat.
Rezul, Ia.G. 1931. N. I. Ziber (bibliografiia). *Kartoga i ssylka* (7): 132–174.
Ricardo D. 1873. *Sochineniia*. Kiev: Universistetskia tipografiia.
Romanovich-Slavatinskii, A.V. 1903. *Golos starogo professor. Po povodu universitetskih voprosov*. Vyp. 2. Kiev: Universitetskaia tipografiia.
Rusov, A.A., and F.K. Volkov. 1907. Kommentarii к svodu ukazanii, dannykh nekotorymi iz arestovannykh po delam o gosudarstvennyh prestupleniiah (mai 1880). 1 (6): 153–162.
Sazhin, M. P. (Arman Ross). 1925. *Vospominaniia*. Moscow: Vsesoiuznoe obshchestvo politicheskikh katorzhan i ssylno-poselentsev.

———. 1932. Russkie v Tsiurikhe (1870–1873 gg). *Katorga i ssylka. Istoriko-revoliutsionnyi vestnik* 10: 25–78.

Scazzieri, Roberto. 1987. Sieber on Ricardo. *Contributions to Political Economy* 6: 25–44.

Sieber, N.I. 1869. *Potrebitelnye obshchestva*. Kiev: Tipografiia Fedorova.

———. 1871. *Teoriia tsennosti i kapitala Rikardo s nekotorymi iz pozdneishikh dopolnenii i raziasnenii*. Kiev: Universitetskaia tipografiia.

———. 1873. Otchiot magistra politicheskoi ekonomii Nikolaia Siebera o prebyvanii s ianvaria po oktiabr' 1872 goda. *Universitetskie izvestiia*, August: 1–55.

———. 1883. *Ocherki pervobytnoi ekonomicheskoi kultury*. Moscow: K. T. Soldatenkov.

———. 1885. *David Rikardo i Karl Marks v ikh obshchestvenno-ekonomicheskikh issledovaniiakh*. St. Petersburg: M.M. Stasiulevicha.

Smith, David Norman. 2001. The Spectral Reality of Value: Sieber, Marx, and Commodity Fetishism. *Research in Political Economy* 19: 47–66.

Sušjan, Andrej. 2010. Historicism and Neoclassicism in the Kiev School of Economics: The Case of Aleksandr Bilimovich. *Journal of the History of Economic Thought* 32 (2): 199–219.

Tkachenko, M.M. 1928. Ziber u Kyivi. In *Iuvilejny sbirnik*, ed. M.S. Grushevskii, 349–358. Kiev: Ukrainian Academy of Sciences.

Vladimirskii-Budanov, M. 1874. *Gosudarstvo i narodnoe obrazovanie v Rossii XVIII-go veka. Chast 1*. Yaroslavl: Tipografiia G. V. Falk.

White, James D. 2009. Nikolai Sieber: The First Russian Marxist. *Revolutionary Russia* 22 (1): 1–20.

———. 2011. Nikolai Sieber: An Introduction to a Political Economist Approved by Marx. *Research in the History of Economic Thought and Methodology* 27: 151–154.

Wuthnow, Robert. 1989. *Communities of Discourse*. Cambridge: Harvard University Press.

6

Fields of Discourse Perturbed: The Revolution of 1905 and Economic Teaching and Thinking at St. Petersburg University

Maxim Markov

The most important aspect of the development of any science is its formation and development as a professional academic discipline. In the world of economic science, the formation of an autonomous academic field occurred in the late nineteenth to the early twentieth centuries. As Marion Fourcade (2009: 2) notes about the history of the economics profession: "In this process of 'academicization' or 'disciplinarization,' economics migrated from salons and learned societies to universities and other higher education establishments." Russia began to go through this process as well, albeit with a lag in comparison with countries of Western Europe and then a tragic interruption after the Revolution of 1917, after which the academy fell under the control of the Communist Party and Soviet state, and the only acceptable form of economic science became orthodox Marxism-Leninism. We cannot judge how Russian economics would have developed without Bolshevism in power and what role scholars at

M. Markov (✉)
Faculty of Economics, Department of Economic Theory,
St. Petersburg State University, St. Petersburg, Russia

J. K. Hass (ed.), *Re-Examining the History of the Russian Economy*,
https://doi.org/10.1007/978-3-319-75414-7_6

St. Petersburg University would have played. However, Mark Blaug, composing his list "great economists before Keynes," recognized the unrealized potential of St. Petersburg scholars when he included two of its graduates—V. I. Bortkiewicz (1868–1931) and N. D. Kondratieff (1892–1938)—on that list (cf. Blaug 1986).

Joachim Zweinert, the author of the most serious contemporary work on the history of economic thought in Russia, ends his study with the revolutionary year 1905, "since the first Russian revolution summed up the nineteenth century in Russia" (Zweinert 2008: 25). In this chapter, the 1905 Revolution is seen as the first act and dress rehearsal for the events of 1917 and is taken as a starting point for this analysis of the formation of the academic field of political economy at St. Petersburg University. Key roles were forces of external influences—political, administrative, social—as well as the ability of scholars to resist these influences and transform them. Thus, we will investigate field autonomy, using as a main indicator "the strength of its refraction and ability to transform" (Bourdieu 2001: 52).

Organizing Research and Teaching: University Autonomy and Academic Freedom

St. Petersburg University was founded by Peter the Great in 1724. In 1819, the university was reformed and its history continued uninterrupted, along with teaching and research in political economy. The development of Russia's university system ("Imperial" universities) in the nineteenth century was directed by the state, and general university statutes set structures of management and internal organization. In total, four such statutes were adopted, in 1804, 1835, 1863, and 1884. When introducing changes in university statutes, the state treated issues of academic freedom and university autonomy differently, as it set itself a dual task: the practical and political training of qualified professionals to suppress the revolutionary movement, "so that people's education conformed to our order of things and was not alien to the European spirit," erasing "the confrontation of so-called European education with our needs"

(Uvarov 1864: 4, 106).[1] In the second half of the nineteenth century, professors' struggles for university "autonomy" became an important facet of the Russian liberation movement, while, "Russian professors, to judge by many of their writings, tended to idealize the degree of academic freedom in other countries and the amount of respect with which other societies viewed academics" (Kassow 1989: 394). The general university statute of 1884 operated for the last period of Imperial Russia, with some changes introduced by the "Provisional Rules for the Management of Institutions of Higher Educational," issued in 1905.[2] Leaving little room for self-organization, the Charter regulated a wide range of issues of university life: the number and composition of departments in each faculty, procedures for paying teachers, tuition fees, and ways to ensure student discipline.

According to the Charter, the university was managed by the Rector, with participation of the University Council and Administration, assemblies and deans of faculties, and the inspector of students and his assistants. The Rector was selected by the Minister of Education from ordinary university professors and appointed by order of the Emperor for four years. For research and teaching, the main decision-making body was the Council, but such issues as selection of new professors were to be approved by the Ministry of Education. Generally, each university consisted of faculties headed by a dean and a faculty assembly. Standard faculties included historical-philological, physical-mathematical, legal, and medical. St. Petersburg University had no medical faculty, but there was a faculty of Oriental languages. The dean of the faculty was elected by a trustee of the educational district from among faculty and was confirmed in office for four years by the Minister of Education. Faculty meetings included all professors and made decisions about internal affairs, which were then approved by higher authorities.

From a formal perspective, universities had an extremely low degree of administrative autonomy. However, as E. A. Rostovtsev (2009a: 142–143) noted, "*in fact,* professorial councils (including the Council of the capital's university) were of decisive importance for the whole course of the internal life of universities... Any interference by the Ministry (despite its formal legal validity) in what professors considered to be their inseparable competence, met with the most stubborn resistance." The situation was somewhat different for academic freedom, especially teaching. Consider the teaching of economics from the organizational side: this topic was studied in the Faculty of Law, which included departments[3] of Roman law, civil law and litigation, commercial law and litigation, criminal law and procedure, history of Russian law, state law, international law, police law, financial law, and so on. Of course, economists were grouped mainly in the department of Political Economy and Statistics. At the same time, associate professors (*dotsenty*) in economics could teach courses in other departments, for example, financial law.

Independence in teaching came from long service by those approved to the position of professor. A professor who served 25 years as a teacher was awarded the title of Honored Professor and after 30 years of service earned the category of "untenured professor," who remained a member of the faculty and received a pension equal to a full salary. (By the time of full service, such professors also reached the highest level in the civil service.) Ordinary professors received good support for that time: an annual salary of 2400, plus 300 rubles for dining in university canteens and 300 rubles for housing, bringing total remuneration to 3000 rubles. This said, salaries had been set as far back as 1863 and had not changed for decades, while the cost of living had increased significantly. Criticizing this state of affairs and comparing professors' salaries with those of higher officials and railway chiefs, Professor P. I. Georgievskii asked a rhetorical question: "Is the activity of all categories of persons listed above more important for the state and society than the work of scholars (except for development of science), who prepare all those persons?" (Georgievskii 2016: 154). In addition to salary, professors received honoraria for lectures. One innovation in the Charter of 1884 was introducing a fee system in Russian universities and replacing in-house positions and junior teachers (associate professors) with *privatdozent*, instructors who were not professors and did not have salaried positions.[4] According to the Charter, students were also charged for lectures and had to pay fees to teachers as well as to the university: one ruble per week per half-year. The Charter provided that, "If the same subject is taught by several teachers, then the student may attend lectures and take part in practical exercises of whichever designated teacher he wishes."[5]

This provision of the Charter was intended to encourage competition across compulsory courses taught by teachers not on staff, especially those with *privatdozent* status. To enter the university as a *privatdozent*, one had to apply to the relevant Faculty and, having obtained consent of full faculty, could start teaching the desired course. Competition in teaching university courses could increase the degree of autonomy and independence from administrative and political pressure, allowing a variety of views to be expressed. However, there was a risk that these lecturers would try to attract more students to their courses by fawning over them and even indulging their political views. Further, such competition contributed to conflicts between lecturers. For example, the fee for teaching a course on political economy could be fairly impressive. This course included four hours of lecture per week for half a year and was compulsory in the Faculty of Law, but students of other faculties or those who simply wanted to audit the course could attend. Thus, more than 500 students might take a single course, and a rate of one ruble per week hour for the half-year provided more than 2000 rubles from that one course for that instructor. This was not always a fair system: a *privatdozent* who provided optional special courses or worked in smaller departments received less income. The University Council could provide additional remuneration to such instructors, but "as a rule, the average size of such remuneration was small and amounted to 200–300 rubles for one annual hour (600–1200 rubles per year)" (Rostovtsev and Barinov 2012: 44).

P. I. Georgievskii sharply criticized the fee system, asserting that "the charter of 1884 is imbued from a beginning with distrust towards professors and a naive (yes, forgive this word) confidence in the *privatdozent*" (Georgievskii 2016: 133). Such remarks are not surprising when we discover that Georgievskii was a professor of political economy and had to compete with other notable economists, such as A. Isaev, V. V. Sviatlovskii, V. G. Iarotskii, and others. This situation escalated in 1906, when M. I. Tugan-Baranovskii, one of the most brilliant Russian scholars of the time, became a *privatdozent* and quickly became popular among students. As Pitirim Sorokin recalled,

> If the *privatdozent* was an outstanding scholar and popular lecturer, he often had more students enrolling in his course and, accordingly, a greater income than a less famous but full (ordinary) professor. It was exactly the

> same with *privatdozent* M. Tugan-Baranovskii and Professor Georgievskii at St. Petersburg University. They both read parallel courses on political economy, but the number of students enrolled in Tugan-Baranovskii's course was many times greater than for Georgievskii. Their incomes also differed accordingly.[6] (Sorokin 1991: 69)

However, Sorokin's description of Russian academic life at the beginning of the century is fraught with inaccuracies; being a student, he could not judge his teachers' financial affairs. Tugan-Baranovskii's fee could have been higher than Georgievskii's, but the latter also received a salary. As well, the requirement that the faculty consent to a course being offered limited possibilities for competitive situations—although there were such instances in the Faculty of Law, especially for courses in political economy, as such consent could be given for various reasons. However, St. Petersburg University was the exception rather than the rule. Georgievskii noted, "It seems that the *privatdozent* did not offer compulsory courses in any Russian universities, except at St. Petersburg, and in the latter, not in any faculty except for the Faculty of Law, until recent years" (Georgievskii 2016: 131–132).

As a rule, obtaining a university position required an academic degree: a master's degree for the position of *privatdozent* and doctorate for a professor. In practice, this requirement was not always observed: in the Faculty of Law from 1885 to 1917, 50.6% of all staff had no doctorate when they accepted positions as professor, and 55.8% of junior teachers were *privatdozent* without a master's or doctoral degree (cf. Rostovtsev et al. 2015: 120). Applicants for the master's degree had to pass an exam before they could defend their theses. Master's and doctoral dissertations were published beforehand and then publicly defended at a meeting of faculty. Dissertation disputes, especially in political economy, often attracted wide public attention and were covered by the press. After defending the dissertation, the strategy of a teacher at a Russian university who taught social disciplines presupposed publishing new monographs and giving lectures at scholarly societies, as well as participating in public and political life (e.g. journalism or interacting in informal associations): "It is impossible to imagine so many private salons and coteries in the capital without university teachers present" (Rostovtsev 2009b:

250–251). Specialized journals in the field of political economy were absent at that moment in Russia, and so Russian scholars preferred to publish purely theoretical works abroad. Tugan-Baranovskii wrote of this to A. A. Kaufman in 1904: "… what interest is there to work in Russian with the complete absence of interest in social theory that undoubtedly characterizes our society. Practical questions are in our grasp, but to theory we are always indifferent. So, sad as it may be for me, I am increasingly convinced that I should spiritually leave Russia and go to the Germans" (Shirokorad and Dmitriev 2008: 97–98).

The Development of Economics: Academic and Socio-political Contexts

Economics at St. Petersburg University, which would set the pace for the discipline's development across Russia, did not develop in a vacuum. Other faculties were growing and offering instruction in economics as well as their own subjects. The Faculty of Law was the largest faculty in the university, as more than half of the students studied there (2391 of 4508 in the 1904–1905 academic year). According to the rules, 400 people were admitted to the first year of the Faculty of Law, and later the number of students increased, spiking in 1906–1907 (8090 students at the university, 4338 in law), when teaching resumed after 1905 Revolution (*Otchet* 1905: 42, 1907: 52; *Pravila* 1904: 1). Study was for four years. While the faculty in 1882–1907 did not have divisions or provide narrow specializations, until 1882 there were legal and administrative divisions, and in 1907 branches of legal, economic, and state sciences were founded. The basis for the curriculum was legal disciplines: History of Roman Law, History of Russian Law, Encyclopedia of Law, State, Civil Police and Criminal Law, the Dogma of Roman Law, and so on. Of the required economic disciplines, only courses in political economy and statistics were offered.

Economics was clearly a theoretical endeavor, and Russian economics developed in a broader field of theoretical discourse into which Russia entered later in the game. In the field of theory, German influence traditionally prevailed in economic teaching and research in St. Petersburg and other Russian universities. Russian scholars oriented primarily to

German scholars and to a lesser extent to the French and English: the standard for being educated at that time presupposed knowledge of German and French languages. Researchers underwent training in Germany as "preparation for professorship." As L. D. Shirokorad noted, "The process of forming political economy … in Russia in the eighteenth and nineteenth centuries was under the enormous impact of the more mature German science. Strengthening ties with German universities over many decades contributed to accelerating the development of Russian economics and to gradual creation of prerequisites for its transition to independent development and at the same time for its deeper integration with European economic science" (Shirokorad 2004: 12).

However, between German and Russian universities there was an important difference that arose in the middle of the nineteenth century: "While after the revolution of 1848–1849, universities in Germany 'agreed' with the state … Russian universities turned into arenas of political discussions and forums of an oppositional public" (Kusber 2013: 205). In political economy, this difference was manifested in the special role of Karl Marx in Russian scholarship in the late nineteenth and early twentieth centuries. In Russia his work was studied and taught at the university level, where in 1872 the first foreign edition of *Kapital* was published. The young I. I. Kaufman (1848–1916),[7] who became a famous economist and specialist on the circulation of money, finance, and credit, published a review about it (Kaufman 1872); Marx thought so highly of this review that he cited it in the Afterword to the second edition. Marx himself noted the strange unity of scholars and revolutionaries in Russia: "In Russia, where *Kapital* is read and appreciated more than anywhere else, our success is even more significant. We have critics on the one hand (mostly young university professors, some of whom I personally know, as well as journalists) and, on the other hand, a central committee of terrorists" (Marx and Engels 1964: 380). Unlike in Russia, "until 1918, there were not a single Social-Democrat among the officially approved professors and adjunct professors at German universities" (Ringer 2008: 172–173).

The Russian *intelligentsia*'s special adherence to socialist ideas was noted by many academics and observers. M. I. Tugan-Baranovskii (1996: 52) wrote: "Socialist sympathies of the Russian *intelligentsia* constitute one of its most characteristically distinctive features." The close connection between Russian scholars and "progressive" political movements and their striving for socially oriented analyses lent a high degree of ideological character to political economy—although it was near impossible to separate academic and socio-political components of disputes in the legal Russian press about "destinies of capitalism." In the 1880s, *intelligentsia* ideology found its expression in economics in the form of legal or reformist populism. On the one hand, the leading representatives of the Populist school of economics (Narodniki), V. P. Vorontsov and N. F. Danielson, proceeded from Marx's economic analysis and criticism of the capitalist mode of production. In fact, Danielson translated all three volumes of *Kapital* into Russian, and for many years, he was in correspondence with Marx and Engels. On the other hand, the Narodniks "were convinced that reliance on capitalism does not correspond to Russian historical traditions and in fact contradicts available objective and subjective conditions by which communities and co-operatives organize the main part of social production, and collectivism in mass consciousness and underdevelopment of the institution of private ownership of land" (Ryazanov 1998: 191). In the 1890s, a new generation of Russian Marxists joined the debate with the populists: P. B. Struve, M. I. Tugan-Baranovskii, S. N. Bulgakov, V. I. Lenin, and others defended the possibility of Russia's unique capitalist development and its progressive nature.

By the beginning of the twentieth century, most scholars in Russia accepted the Marxian theory of value and capital and tried to integrate it with ideas of German *Kathedersozialismus* and Russian populism. In particular, textbooks at St. Petersburg University taking this approach were published by A. A. Isaev and L. V. Khodskii. Isaev was a *privatdozent* in the Department of Political Economy in 1889–1899; his *Principles of Political Economy* (Isaev 1908) enjoyed great popularity in pre-Revolutionary Russia and went through seven editions. Khodskii was *privatdozent* after 1892 and from 1895 on a professor in the Department of Financial Law. He wrote *Political Economy in Connection with Finance* (Khodskii 1908), which also was reprinted several times. Both of these instructors were also engaged in political activities, and they were persecuted and dismissed from the university by order of the Ministry of Public Education as "unreliable": Isaev in 1899, after student unrest, and Khodskii in 1905. (Khodskii returned to the university in 1910 as a *privatdozent.*) At the same time, in the first years of the twentieth century, Marx began to lose his position in the academic sphere, and the Austrian school began to gain popularity. Many Russian economists, beginning with Tugan-Baranovskii, tried to synthesize the labor theory of value and the theory of marginal utility.

The First Russian Revolution and the University

Just as the field of economics seemed to be settling down somewhat—which, as earlier chapters in this volume point out, was no easy task, given the tensions between the profession and the state and the logics of seeking Truth versus serving the state—other tensions exploded into revolution. Students played the initial role in the 1905 Revolution and were the most active in St. Petersburg University itself, making the university one center of revolutionary activism. Russian students traditionally held oppositionist views and sentiments, but after 1899, when an All-Russian student strike took place at St. Petersburg University, their rhetoric acquired the character of a mass political movement, and their practices increasingly included regular meetings and strikes. As well, lectures by unpopular professors were obstructed for political reasons. Both conservative and liberal professors might find themselves subject to rude attacks and insults: the *privatdozent* Isaev noted that he "did not agree with tactics of student obstruction, and on the next day a student leaflet labeled him a 'scoundrel'" (Shchetinina 1995: 191). All this led V. I. Vernadskii to assert that "from 1898 to 1906, there was no proper teaching in higher education; the school was in a state of anarchy, against which Russian professors fought vigorously, since the authorities had no moral authority" (Vernadskii 1912: 4).

In February 1905, with the outbreak of revolution, university classes were again interrupted by a student strike, and they would resume normally only in autumn 1906. Among the students were representatives of different political convictions. Lenin singled out six political groups—reactionaries, the indifferent, academics, liberals, social revolutionaries, and social democrats (Lenin 1967: 343)—but the most active students held the most radical sentiments. This can be shown by party lists and results of elections on October 12, 1905, to the Starosta Council (student "elders" or representatives), formed in the course of the revolution by the student government. From the Faculty of Law, 677 students were present, while the social democrats received 478 votes, Socialist Revolutionaries received 85 votes, and Constitutional Democrats (Kadets) received 105. The Bolshevik A. N. Zamiatin became the first chairman of the University Council of Starosta (Diankov 1907: 64–65; Arshavskii 1982). Under these conditions, professors had to maneuver between students and bureaucrats, while also trying to defend their own interests (Rostovtsev 2009c). One professor managed some success in this attempt, and on August 27, 1905, an imperial decree was issued to the Senate, which put into effect the "Provisional Rules for the Management of Institutions of Higher Education." These rules gave the University Council the right to select the Rector, while the faculty could select their deans, and the Council's obligations and responsibilities included "caring for the proper course of academic life at the university."[8] Regarding political convictions, most professors were moderately oppositional and sympathetic to the Kadets. In 1905, using the "Provisional Rules," professors elected I. Borgman, a Kadet, as Rector; in 1906, professors elected L. I. Petrazhitskii, a member of the Kadet Central Committee and of the first State Duma, to be dean of the Faculty of Law.

Despite these new policies, student unrest failed to dissipate, and the openness of the university was used for subversive revolutionary propaganda and agitation: "In the fall of 1905 this was one of the headquarters of the Petersburg Committee of the RSDWP [Russian Social Democratic Workers' Party]. In some classrooms and committees, a committee was sitting and appointing party workers, while in others Bolsheviks were studying, and rallies were constantly held in assembly halls and the university yard; many lecture halls were turned into places of rallies and meetings of the masses" (Bondarevskaia 1979: 74). By October 1905, the state closed the university for the entire academic year. Student organizations actively continued to influence university life even after the resumption of classes in 1906: "Throughout 1906, the entire inner life of the University was in fact determined by students' decisions and the activities of the Starosta Council. The council of professors could not influence the course of events, nor change them" (Pavlitskaia 1948: 153).

Teaching Political Economy After the 1905 Revolution

One positive outcome of the 1905 Revolution was a weakening of restrictions imposed by the tsarist state on teaching. Some scholars previously banned from teaching were able to return—one of the most important for political economy being Tugan-Baranovskii. I. M. Kulisher, who would become famous as an historian of economics, found his academic career originally hampered because of his Judaism, but now he became a *privatdozent* in the Department of Political Economy. Not unimportantly, xenophobia and similar negative attitudes to Jews were relatively lower at St. Petersburg University (especially in the Department of Political Economy and Statistics) than at other Russian universities and in Europe generally. In 1889, I. I. Kaufman, an ethnic Jew, began to work at the university, but he had to accept Christianity. (Probably being indifferent to religion, Kaufman chose a rare Russian denomination, Anglicanism.) In 1909 Kaufman began teaching at the university, when he already had made his name as an important statistician.

When teaching resumed after the 1905 Revolution (the 1906–1907 academic year), political economy in the Faculty of Law was taught by several people: ordinary professor P. I. Georgievskii, *privatdozent* V. G. Iarotskii, *privatdozent* V. V. Sviatlovskii, and *privatdozent* Tugan-Baranovskii (*Obozrenie prepodavaniia nauk na Iuridicheskom fakultete* 1906).

Pavel Ivanovich Georgievskii stood out among the Faculty of Law for his conservatism and his loyalty to the state bureaucracy (Dmitriev and Chebanenko 2012). In 1905, he was among those "reactionary professors" whose dismissal students demanded. Georgievskii began teaching at the university in 1883 (by 1888 he was a professor). In 1885 he defended his master's thesis on international trade and in 1887 his doctoral dissertation on financial relations between the state and private railway firms. The economics, statistics, and organization of the railway were for many years his main scholarly interest, and after 1883 he headed the railway statistics department for the Statistics Division of the Ministry of Railways. Georgievskii's theoretical views and content of his course can be judged from his textbook *Political Economy*, first published in 1890 and reissued four times. The structure of his course was original and consisted of sections on production of value, circulation of value, distribution of value, and use and destruction of value. The originality of this structure lies in the fact that, following the classical schema of reproduction—which at the beginning of the twentieth century continued to include the political economy of J. S. Mill (*Obozrenie prepodavaniia nauk na Iuridicheskom fakultete* 1906: 66)—Georgievskii emphasized the destruction and use of value, giving himself credit: "The recognition of the importance of this division of economic science, which is not only not inferior in content to other divisions, but even in our opinion is superior to other divisions in its significance, is put forward by us, almost for the first time in economic science" (Georgievskii 1904: 282).

"Use" of value for Georgievskii—he insisted on this term, which he felt was more appropriate than "consumption"—had significance in and of itself and not as the basis for a theory of value. While devoting much space to criticizing Marx, Georgievskii renounced the labor theory of value without automatically accepting the theory of marginal utility, with which he was perfectly familiar. Georgievskii was the first in Russia to introduce Menger's table in a popular course on science (Budovich 2004: 31–32). In addition, being a convinced opponent of socialism, Georgievskii published a Russian translation of Böhm-Bawerk's (1897) criticism of Marx. Georgievskii abandoned the theory of value and replaced it with a theory of production costs and put forward his own theory of "public."[9] His own conceptualization of "public" was vague, by his own admission: "the public does not represent anything definitely definable, that is to say tangible, it is impossible to specify its exactly defined boundaries, but nevertheless, for everyone it is clear that it exists, because only in it does man live and develop" (Georgievskii 1904 [vol. 2]: 244). The "public" for Georgievskii was a fourth factor of production, manifested in the legal order, knowledge, and public cooperation. Proceeding from this, Georgievskii sought to prove the public was the source of "surplus" (pure product) "that should go to support the public or, what is the same, to meet its needs as they have developed" (Georgievskii 1904 [vol. 2]: 257). Georgievskii intuitively grasped what economics in the twentieth century would call externalities and public goods, but he was unable to articulate this insight clearly, and so his theory did not attract the attention of contemporaries, despite his efforts to publish his ideas in French and German and, once he was an honorary professor, to offer a special course on the subject. In 1924, in exile, he gave a lecture at the annual meeting of the Russian Faculty of Law in Prague entitled "The Public in the National Economy" (cf. Dmitriev and Chebanenko 2012: 123–126).

Another important economist, Vasilii Gavrilovich Iarotskii, graduated from St. Petersburg University in 1880 (Dmitriev 2009a). In 1888 he defended his thesis on entrepreneurs' economic responsibilities and in 1896 defended his doctoral dissertation on workers' insurance and the responsibilities of business. The beginning of Iarotskii's academic career was marked by conflict with Georgievskii. In 1887–1888 Iarotskii appeared in the press and in a doctoral dispute criticizing Georgievskii's book on the state railway finances, while Georgievskii published critical comments on Iarotskii's work on entrepreneurial responsibility. Because of this conflict, which might have been due, as Dmitriev suggested, to competition for a professorship in the Department of Political Economy and Statistics Georgievskii occupied in 1888 (Dmitriev 2009a: 484), Iarotskii had to defend his master's thesis and doctoral dissertation at Moscow University. Despite this, in 1888 he was admitted to St. Petersburg University as *privatdozent*, but he worked mainly at the privileged Imperial Alexander Lyceum until the end of his life. He also taught at the Aleksandrovskaia Military Law Academy and other institutions of higher education.

Iarotskii's theoretical views in political economy were reflected in his courses, the lithographs of which were published by graduates of the Imperial Lyceum. In 1916 Iarotskii published a textbook on the first part of his lectures (Iarotskii 1916). A. I. Bukovetskii, who graduated from St. Petersburg University in 1904, noted in his memoirs that Iarotskii's positions were to the left and that he was "a typical representative of the Russian liberal *intelligentsia* of the 80s" (Bukovetskii 1997: 21). Indeed, in the course of his lectures on political economy, Iarotskii revealed himself to be a Narodnik Marxist and adherent of the labor theory of value. Iarotskii emphasized the immutability of his theoretical views: "I did not need, like some of my colleagues, to significantly change my views on the main issues of science, or adapt to fluctuations of temporary fashion trends" (Iarotskii 1916: iii). By "temporary fashions," Iarotskii implied, first of all, the Austrian school and Tugan-Baranovskii: "Current work by economists in this direction should be recognized, if not as completely fruitless, then as of very little use, except for self-esteem and mutual adoration of representatives of such pedantic theorizing" (Iarotskii 1916: 37).

From his youth, Vladimir Vladimirovich Sviatlovskii (cf. Chebanenko 2009) participated in the labor movement and in 1889–1892 was a member of the "Brusnev Group," one of the first social democratic organizations in Russia associated with Plekhanov's Swiss group "Emancipation of Labor." In 1892, Sviatlovskii, fearing police persecution, moved to Germany, where he studied at the University of Munich and in 1897 defended his doctoral dissertation on the economic history of *Rus*. He became a *privatdozent* at St. Petersburg University in 1902, passing his master's exam. In the revolutionary period, he was a member of the bureau and treasurer of the Union of Unions (an alliance of various professional and political unions) and was elected member of the Central Bureau of Trade Unions and of the second executive committee of the St. Petersburg Soviet of Workers' Deputies. After the 1905 Revolution, he left political activity and turned his energies to the history and theory of the trade union movement. In addition, Sviatlovskii published a number of works on a wide range of problems: from the history of economic thought and money circulation to agrarian and housing issues.

His popular "Lectures on Political Economy," published in 1910, attempted to present all the existing theoretical approaches. In particular, he described the following value theories: the theory of supply and demand, the theory of production costs, labor or objective theory, and subjective or Austrian theory. His own theoretical preferences were obvious. About Karl Marx, he wrote that he "provides a strictly logical doctrine, often rising to a high abstract chain inevitably flowing from one and another mathematical formula" (Sviatlovskii 1910: 146). He characterized the Austrian school as follows: "Recently, many general principled and private critical observations have been put forward against the subjective theory of value, which in their totality indicate the precariousness of the methodological and philosophical foundations of the Austrian theory of value" (Sviatlovskii 1910: 165).

Recognized as an outstanding economist in Russia and the West, Mikhail Ivanovich Tugan-Baranovskii needs no special introduction (cf. Shirokorad 2005). In his youth, as a student in the Faculty of Physics and Mathematics at St. Petersburg University, Tugan-Baranovskii participated in the radical student movement, joining the revolutionary circle of his classmate A. I. Ulianov. In 1886, he was expelled from the university and the capital, and he graduated instead from Kharkov University, where he received diplomas simultaneously from two faculties, physics and mathematics, and law. His first serious scholarship, his master's thesis on industrial crises in England and that he defended at Moscow University in 1894, brought him worldwide fame for creating the first endogenous theory of economic cycles. In 1895, Tugan-Baranovskii became a *privat-dozent* at St. Petersburg University, where he would lecture with occasional interruptions until 1895. During this period, Tugan-Baranovskii became one of the leaders of legal Marxism, and his doctoral thesis on Russian factories, defended again at the Moscow University, made an important contribution to discussions about the fate of Russian capitalism. In 1901, he was arrested for participating in a student demonstration and was subject to administrative expulsion from St. Petersburg. He settled on his estate in the Poltava province, where he reassessed his life's goals and values and abandoned active participation in politics. In 1905 Tugan-Baranovskii stayed away from revolutionary activity.

As he began teaching, Tugan-Baranovskii prepared his famous textbook, *Principles of Political Economy* (1909), which went through five editions in his lifetime and was republished in the post-Soviet era. His course was grounded in the classics, although he noted: "I would like to show the possibility of a third direction for economic theory, which does not coincide either with Marxism or the school of marginal utility, but is, to some extent, a synthesis of both of them" (Tugan-Baranovskii 1909: vi). In an effort "to pay tribute to the ideological trends to which it owes its origin" (Tugan-Baranovskii 1909: viii), Tugan-Baranovskii dedicated the book to Quesnay, Gossen, and Marx. Another important distinguishing feature of his course was his desire to introduce elements of social analysis into political economy; the social distribution theory developed later is outlined in this textbook. He also quickly became popular among students, who were attracted not only by his lectures but also student organizations he organized. His student circle was created immediately after his resumption of teaching activity, in 1906. Later, the activity of the circle grew and it began publishing the journal *Voprosy obshchestvovedeniia*. The author of a survey of the activity of the circle told the journal, "There were 181 members of the circle in the academic year 1906–7, 206 in the 1907–8 academic year; at meetings of the circle, which were always open, there were usually several hundred people."[10] The circle took on its finished form in 1908–1909, when it was divided into three parts: an economic pro-seminar, an economic seminary, and an economic circle. The proseminar was a school for beginners, playing a supporting role in relation to the course on political economy; these students also studied Marx's *Kapital*. Seminary work was devoted to theory and methodology of political economy. The circle studied "not only questions of political economy, but also questions of social and economic sciences in general."[11]

Political Reaction and Administrative Pressure

After Stolypin's "coup" of June 1907, Russia entered a phase of political reaction. In higher education, this was associated with the activities of Lev Kasso, Minister of Education. At Moscow University, Kasso's offensive behavior against university autonomy led to the resignation of many instructors. In the capital, events developed differently: "the Council of Professors, rallying around the Rector, chose a different tactic—passive resistance to pressure of the ministry" (Rostovtsev 2017: 624). In response, the ministry tried to overcome the opposition by an unprecedented practice of appointing and "transferring conservative professors from provincial higher educational institutions to St. Petersburg University and transferring St. Petersburg liberals to the provinces" (Rostovtsev 2017: 628). This policy was directed to no small extent against the Faculty of Law, that is, representatives of the social sciences. Ultimately, eleven professors were appointed to the Faculty of Law: "There is no doubt that the 'left' majority was still supervised by the Council of Professors, despite the Kasso's personnel policy, especially in the years of 'reaction' from 1911 to 1913. The Council was enlarged by liberal professors (for example, the Byzantinist V. N. Beneshevich and the Sinologist A. I. Ivanov)" (Rostovtsev 2017: 640).

During this period, Georgievskii, in accordance with established procedures and having served 30 years, left his full-time university post. The question arose as to who would occupy the professor's chair, and the most obvious candidate, in age and in merit, was Tugan-Baranovskii. He was elected, first by the meeting of the Faculty of Law on April 30, 1912 (13 votes in favor, 1 against), and then at a meeting of the University Council on May 21, 1912 (29 votes in favor, 4 against) (*Protokoly* 1914: 77–80). The only vote "against" was Georgievskii's. In a review presented to the Faculty of Law on Tugan-Baranovskii as candidate for professor in the Department of Political Economy, Georgievskii accused him of extreme fascination and praise of socialism and communism. Georgievskii pointed out that even in "enlightened Germany" university chairs were not allowed to engage in social democracy and warned against the danger of "replacing the department of the SCIENCE of political economy with the department of the COMMUNITY of Socialism" (Dmitriev 2008: 100). Georgievskii concluded his gloomy prophecy thus: "The poisoning of students with socialist fantasies, brought to them under the guise of positive science in the university department, can have the most unfortunate consequences for these young people, and in the person of several generations—for the whole state, I consider it my moral duty to warn about this" (Dmitriev 2008: 100). Georgievskii's accusations did not and could not impress the faculty and University Council. Many Russian intellectuals, including colleagues at the university, sympathized with socialist ideas and occupied perhaps more leftist positions than Tugan-Baranovskii. Indeed, Tugan-Baranovskii was known for his work on socialism (and Georgievskii tended to cherry-pick quotes from these works), but he saw socialism as an ethical ideal for the future society of free people. In practical politics, Tugan-Baranovskii actually supported the bourgeois Kadets.

Georgievskii's personal motives for criticizing Tugan-Baranovskii were likely not unimportant. In his response, Georgievskii not only described the scholarly and political views of the candidate but also hinted that he enjoyed profit from capital and land rent and therefore did not adhere to requirements imposed by the socialist ideal (Dmitriev 2008: 98). Relations between Georgievskii and Tugan-Baranovskii were also clear in published correspondence of university colleagues with V. I. Bortkiewicz, an extraordinary professor at the University of Berlin. Kaufman wrote about Georgievskii's hatred for Tugan-Baranovskii, "but in this case the work was so sewn with white threads, that, of course, Georgievskii could not find any support" (Dmitriev 2009b: 101). From a letter to M. V. Ptukha we learn that by 1912 Kaufman, the second professor of the Department of Political Economy and Statistics, defended the candidacy of Tugan-Baranovskii for five years, and "Georgievskii, not wanting to replace the department with M. Tugan-Baranovskii, suggested that Professor Manuilov nominate his candidacy, but Manuilov categorically refused to be a candidate for P. I. Georgievskii. In view of this, as they say, he put forward your candidacy, without first obtaining your consent" (Dmitriev 2010: 116). (Bortkiewicz did not take advantage of this proposal.)

Thus, Georgievskii, intriguing against Tugan-Baranovskii, also strove to ensure that the Department of Political Economy at St. Petersburg University was occupied by a prominent scholar. It is interesting to note that the evolution of the theoretical views of the researchers he proposed developed in the same direction as Tugan-Baranovskii. A. N. Manuilov (1861–1929), an active politician who was a member of the Central Committee of the Kadet Party, served as Rector of Moscow University from 1905 to 1911 and resigned in connection with the "Kasso Affair." Manuilov moved from the Marxist labor theory of value to the Austrian school. A much larger theorist was Bortkiewicz, who after 1901 worked at the University of Berlin but could not get a permanent post there, as he was a Russian subject. Bortkiewicz gained fame as an expert in mathematical statistics and probability theory, and his most important achievement is the application of the cost of production theory to an analysis of the Marxist theory of value and prices.

After the election, Tugan-Baranovskii was supposed to be approved as professor by the Ministry of Education. However, by then the ministry was less interested in the candidate's qualities as a scholar—in the event of a confrontation with the Council of Professors, a loyal candidate was preferred. Such a person was found in Odessa—I. I. Chistiakov, professor of police law at Novorossiiskii University, was transferred to the department of political economy at St. Petersburg University. In addition, the next step of Kasso's Ministry in the offensive against academic freedom was to prohibit *privatdozent* instructors from teaching general courses simultaneously with professors. Tugan-Baranovskii was then able to teach only the course in practical training, and so he left the university and transferred to the Political Institute of Peter the Great.

Ivan Ivanovich Chistiakov graduated from the Faculty of Law of Moscow University in 1898 and in 1901–1910 was a *privatdozent* there; in 1910–1912 he was an "extraordinary professor" at Novorossiiskii University. In 1910 he defended his thesis for a Master of Police Law on education in France, and in 1912 he defended his doctoral dissertation on worker's insurance. Chistiakov's doctoral dissertation had an abstract character and represented an overview of various projects and proposals for workers' insurance. Chistiakov wrote in the Foreword about theoretical insolvency:

> The author's initial intention was to consider theoretically insurance for workers in Russia... But this idea, almost at the very beginning of the endeavor, he had to leave aside as premature, for the complete absence of summary works of a historical nature on the subject of research... Having thus stopped on the history of the subject to be studied, the author also encountered on this path a number of external obstacles relating to the establishment of the volume and the finding of the corresponding material. (Chistiakov 1912: ix)

The speed (two years after the defense of his master's thesis) with which the doctoral thesis was prepared and defended leads to the suspicion that Chistiakov enjoyed some administrative support, in view of the intended appointment.

As a specialist in political economy, Chistiakov did not have time to prove himself. He could and did improve the teaching of economic theory, and he was the first at the university to include the works of Alfred Marshall on the list of recommended readings (*Obozrenie prepodavaniia nauk* 1913: 18). However, Chistiakov served as a professor at St. Petersburg University for only four years. Immediately after the February Revolution, all professors appointed by the ministry were removed from office. Chistiakov left for Tomsk, where on March 1, 1918, he was elected professor of the department of political economy and statistics of the local university.[12] During the Civil War, Chistiakov dropped out of history.

Conclusion

In 1917, unlike in 1905, the radicals did not need the university and its autonomy for organizing revolutionary activity: "after February, the significance of university extraterritoriality no longer played any role for the revolutionary movement" (Rostovtsev 2017: 733). After the February Revolution, for a short time it seemed to liberal faculty that the time had come to realize all their desires. Many received a well-deserved reward. V. G. Iarotskii was awarded the rank of senator. Tugan-Baranovskii was again elected university professor, but rather than return to teaching he cooperated with Ukrainian separatists and died on January 8, 1919, in a train near Odessa, heading to Paris at the head of the Ukrainian delegation (Tugan-Baranovskii 1997: 63). Soon the illusions dissipated, and the new Soviet government quickly managed to demonstrate that university autonomy made sense only when it was protected and supported by the state. Many university instructors were forced to emigrate; according to Rostovtsev's (2017: 755) estimates, the country was left by 37.5% of the law faculty and 23.3% of the total number of those who worked at the university left the country in 1917. The share of economists who departed was less than the faculty and included P. I. Georgievskii, S. O. Zagorskii, M. A. Kurchinskii, P. P. Migulin, and, in 1925, V. V. Leontiev. Many economists from St. Petersburg University who survived Revolution and Civil War continued teaching and researching and were in demand by the Soviet regime and, at the same time, were persecuted and repressed, the last act of which was the Leningrad Affair.

The support of the imperial state created the conditions in which Russian *economist* could approach a global level of quality, while the immaturity of the *academic field* in the social sciences placed them in a difficult situation in which they constantly had to be protected from external influences. (It is no coincidence that the Faculty of Law became the main center of the revolutionary movement and administrative pressure.) Maneuvering between revolutionary students and the state bureaucracy, the professor, in an effort to achieve university autonomy, expanded their powers, acquired the right to elect rectors and deans, but lost autonomy of the academic field. The University Council became a site for political struggle, and departments (including the department of political economy) were occupied by unremarkable researchers.

Notes

1. S. S. Uvarov (1786–1855) was the Minister of Education between 1833 and 1849, and he contributed the concept of an "official nationality."
2. "Ustav Imperatorskikh Rossiiskogo Universitetov 23 avgusta 1884 goda." *Sbornik postanovlenii po Ministerstvy Narodnogo Prosveshcheniia, T. IX. Tsarstvovanie Imperatora Aleksandra III. 1884 god.* St. Petersburg: Tipografiia V. S. Balasheva, 1893: 980–1047; "O vvedenii v deistvit vremennykh pravil ob upravlenii vysshimi uchebnymi zavedeniiami vedomstva Ministerstva Narodnogo Prosveshcheniia. 28 avgusta 1905 goda." *Polnoe sobranie zakonov Rossiiskoi imperii*, vol. 25. St. Petersburg: Gosudarstvennaia tipografiia, 1908): 658–659.
3. Russian universities, then as now, were organized into "Faculties" (*fakultet*) and "departments" (*kafedra*), with the latter subdivisions of the former: thus, the Faculty of Economics contained, say, a Department of Economic Theory, Department of Global Economy, and so on. In the nineteenth century and early twentieth century, a "department" could actually be one professor and perhaps one or two assistant professors, or two professors. The department relevant to this chapter was two professors, one in political economy and one in statistics.
4. This was inspired by the same German practice of the *privatdozent* as an independent teacher who gave lectures for fees, but otherwise did not receive a constant salary, was not a professor, and did not have administrative duties.
5. "Ustav Imperatorskikh Rossiiskogo Universitetov 23 avgusta 1884 goda": 73.
6. Sorokin left for the United States and became a famous sociologist. From 1910 to 1914, he was a student in the law faculty at St. Petersburg University.
7. Kaufman taught at St. Petersburg University from 1889 to 1914 and was a professor of statistics after 1893.
8. "O vvedenii v deistvit vremennykh pravil…": §2.
9. His theory of *obshchestvennost*, a "public" actively engaged with politics and liberating itself from the shadows of the state, mirrors the general emergence of the concept of a "public" in Russia, as manifested in notions of private property, independent courts, a civil society, and so on (cf. Pravilova 2014).
10. "Kruzhok i seminar politicheskoi ekonomii M. I. Tugan-Baranovskogo." *Voprosy ekonomiki* #1 (1908): 311.

11. "Nauchnye organizatsii po politicheskoi ekonomii pod rukovodstvom M. I. Tugan-Baranovskogo." *Voprosy obshchestvovedeniia*, #2 (1910): 352.
12. "Chistiakov Ivan Ivanovich." *Elektronnaia entsiklopediia Tomskogo universiteta* (http://wiki.tsu.ru).

Works Cited

Arshavskii, K.G. 1982. Put k Oktiabriu. In *Leningradskii universitet v vospominaniiakh sovremennikov*, ed. V.V. Mavrodin, vol. 2, 161–163. Leningrad: Izdatelstvo Leningradskogo universiteta.

Blaug, M. 1986. *Great Economists before Keynes: An Introduction to the Lives and Works of One Hundred Great Economists of the Past*. Brighton: Wheatsheaf.

Böhm-Bawerk, E. 1897. *Teoriia Karla Marksa i ee kritika*. St. Petersburg: Tipografiia P. P. Soikina.

Bondarevskaia, T.P. 1979. Bolshevistskaia organizatsiia universiteta v revoliutsii 1905–1907 gg. In *Peterburgskii univeresitet i revoliutsionnoe dvizhenie v Rossii*, ed. E.S. Vostokova, 67–80. Leningrad: Izdatelstvo Leningradskogo universiteta.

Bourdieu, P. 2001. Klinicheskaia sotsiologiia polia nauki. In *Sotsioanaliz Piera Burde. Almanakh Rossiisko-frantsuzskogo tsentra sotsiologii i filosofii Instituta sotsiologii Rossiiskoi Akademii nauk*, ed. N.A. Shmatko, 19–35. St. Petersburg: Aleteiia.

Budovich, Iu.I. 2004. *Kurs politicheskoi ekonomii v Rossii na rubezhe XIX–XX vv. i sovremennost*. Moscow: Maks Press.

Bukovetskii, A.I. 1997. Kratkii obzor prepodavaniia finansovoi nauki i finansovogo prava v Peterburgskom (Petrogradskom) universitete v XIX-pervoi chetverti XX veka. In *Istoriia izucheniie obshchestvennykh fiannsov v Sankt-Peterburge*, ed. I.N. Baranov and A.L. Dmitriev, 9–31. St. Petersburg: Izdatelstvo SPbGUEF.

Chebanenko, E. N. 2009. Vladimir Vladimirovich Sviatlovskii. In *Ocherki po istorii finansovoi nauki: Sankt-Peterburgskii universitet*, 350–355. Moscow: Prospekt.

Chistiakov, I. I. 1912. *Strakhovanie rabochikh v Rossii. Opyt strakhovaniia rabochikh, v sviazi s nekotorymi drugimi merami ikh obespecheniia*. Moscow: Pechatnia A. I. Snegirevoi.

Diankov, A. 1907. *1905 i 1906 god v Peterburgskom universitete*. St. Petersburg: I. Bashalova.

Dmitriev, A. L. 2008. V Iuridicheskii fakultet S_Peterburgskogo universiteta. *Vestnik SPbGU, Seriia 5. Ekonomika* 1: 97–100.

———. 2009a. Vasilii Gavrilovich Iarotskii. In *Ocherki po istorii finansovoi nauki: Sankt-Peterburgskii universitet*, 479–489. Moscow: Prospekt.

———. 2009b. Pisma A. A. Kaufmana k V. I. Bortkevichu. *Vestnik SPbGU, Seriia 5. Ekonomika* 4: 94–113.

———. 2010. Pisma A. A. Kaufmana k V. I. Bortkevichu. *Vestnik SPbGU, Seriia 5. Ekonomika* 4: 113–128.

Dmitriev, A.L., and E.N. Chebanenko. 2012. Pavel Ivanovich Georgievskii. In *Finansovaia nauka v Sankt-Peterburge: istoricheskie ocherki*, ed. V.V. Kovalev, 105–126. Moscow: Prospekt.

Fourcade, M. 2009. *Economists and Societies: Discipline and Profession in the United States, Britain, and France, 1890s to 1990s*. Princeton: Princeton University Press.

Georgievskii, P.I. 1904. *Politicheskaia ekonomiia*. 4th ed., 2 vols. St. Petersburg: Kirshbaum typography.

———. 2016. *Prezrenie bednykh i blagotvoritelnosti. O reforme universitetov v Rossii*. Moscow: Lenand.

Iarotskii, V.G. 1916. *Kurs lektsii politicheskoi ekonomii*. Petrograd: Tipografiia M. A. Aleksandrova.

Isaev, A.A. 1908. *Nachala politicheskoi ekonomii*, 7th ed. St. Petersburg: A. F. Tsinzerlinga.

Kassow, S. 1989. *Students, Professors, and the State in Tsarist Russia*. Berkeley: University of California Press.

Kaufman, I. I. 1872. *Tochka zreniia politiko-ekonomicheskoi kritiki u Karla Marksa*. Vestnik Evropy, vol. III, Book 5.

Khodskii, L. B. 1908. *Politicheskaia ekonomiia v sviazi s financami*, 4th ed. St. Petersburg: Pravda.

Kusber, Ia. 2013. Transfer i sravnenie: universitetskie soobshchestva Rossii i Germanii. In *Soslovie russkikh professorov. Sozdateli statusov i smyslov*, ed. E.A. Vishlenkova and I.M. Saveleva, 191–211. Moscow: Higher School of Economics.

Lenin, V.I. 1967. *Polnoe sobranie sochinenii*. Vol. 7. Moscow: Politizdat.

Marx, Karl, and Friedrich Engels. 1964. *Sochineniia*. 2nd ed. Moscow: Politicheskaia literatura.

Obozrenie prepodavaniia nauk na Iuridicheskom fakultete Imperatorskogo S-Peterburgskogo uiversiteta. 1906. St. Petersburg: Tipografiia B. M. Volfa.

———. 1913. St. Petersburg: Tipografiia B. M. Volfa.

Otchet o sostoianii i deiatelnosti Imperatorskogo S-Peterburgskogo universiteta za 1904 g. 1905. St. Petersburg: Tipografiia B. M. Volfa.
Otchet o sostoianii i deiatelnosti Imperatorskogo S-Peterburgskogo universiteta za 1906 g. 1907. St. Petersburg: Tipografiia B. M. Volfa.
Pavlitskaia, N.I. 1948. Peterburgskii universitet v revoliutsii 1905–1907 gg. *Vestnik LGU* 11: 138–156.
Pravila o prieme v student S-Peterburgskogo universiteta i o plate za slushanie lektsii. 1904. St. Petersburg: Tipografiia B. M. Volfa.
Pravilova, Ekaterina. 2014. *A Public Empire: Property and the Quest for the Common Good in Imperial Russia.* Princeton: Princeton University Press.
Protokoly zasedanii Sovieta Imperatorskogo S-Peterburgskogo universiteta za 1912 god. 1914. St. Petersburg: Tipografiia B. M. Volfa.
Ryazanov, V.T. 1998. *Ekonomicheskoe razvitie Rossii. Reformy i rossiiskoe khoziaistvo v XIX-XX vv.* St. Petersburg: Nauka.
Ringer, F. 2008. *Zakat nemetskikh mandarinov: Akademicheskoe soobshchestvo v Germanii, 1890–1933.* Moscow: Novoe literaturnoe obozrenie.
Rostovtsev, E.A. 2009a. Akademicheskaia korporatsiia Sankt-Peterburgskogo universiteta v nachale XX v.: otnoshenie k vlasti i grazhdanskomu obshchestvu. In *"Byt russkogo po dukhu i evropeitsem po obrazovaniiu": Universitety Rossiiskoi imperii v obrazovatelnom prostranstve Tsentralnoi i Vostochnoi Evropy XVIII-nachala XX v*, ed. Andrei Andreev, 139–156. Moscow: Rosspen.
———. 2009b. Universitet stolichnogo goroda (1905–1917 gody). In *Universitet i gorod v Rossii (nachalo XX veka)*, ed. Trudde Mauer and Aleksandr Dmitriev, 205–370. Moscow: Novoe literaturnoe obozrenie.
———. 2009c. 'Borba za avtonomiiu': korporatsiia stolichnogo universiteta i vlast v 1905–1914 gg. *Journal of Modern Russian History and Historiography* 2: 75–121.
———. 2017. *Stolichnyi uiversitet Rossiiskoi imperii: uchenoe soslovie, obshchestvo i vlast (vtoraia polovina XIX-nachalo XX v.).* Moscow: ROSSPEN.
Rostovtsev, E.A., and D.A. Barinov. 2012. Prepodavatelskaia korporatsiia stolichnogo universiteta 1884–1916 gg.: osnovnye cherty i problemy kollektivnoi biografii. In *Professorsko-prepodavatelskii korpus rossiiskikh universitetov. 1884–1917 gg: Issledovaniia i dokumenty*, ed. M.A. Gribovskii and S.F. Fominykh, 35–50. Tomsk: Izdatelstvo Tomskogo universiteta.
Rostovtsev, E.A., D.A. Barinov, and D.A. Sisnitskii. 2015. Iuridicheskii fakultet Imperatorskogo Sankt-Peterburgskogo universiteta (1819–1917): opyt kollektivnoi biografii. *Vestnik Sankt-Peterburgskogo universiteta*, Series 14 (law), 4: 112–127.

Shchetinina, G.I. 1995. *Ideinaia russkoi intelligentsia: konets XIX-nachalo XX v.* Moscow: Nauka.

Shirokorad, L.D. 2004. Vliianie nemetskoi ekonomicheskoi nauki na formirovanie politicheskoi ekonomii v Rossii v XVIII-pervoi polovine VIV v. *Vestnik Sankt-Peterburgskogo universiteta*, Series 5 (economics), 3: 3–13.

———. 2005. Nauchnaia i pedagogicheskaia deiatelnost M. I. Tugan-Baranovskogo v S-Peterburge. *Vestnik Sankt-Peterburgskogo universiteta*, Series 5 (economics), 1: 48–66.

Shirokorad, L.D., and L.A. Dmitriev. 2008. *Neizvdestnaia M. I. Tugan-Baranovskii.* Nestor-Istoriia: St. Petersburg.

Sorokin, Pitirim. 1991. *Dolgii put: avtobiograficheskii roman.* Syktyvkar: Soiuz zhurnalistov Komi ASSR, MPO Shypas.

Sviatlovskii, V.V. 1910. *Konspekt lektsii po politicheskoi ekonomii.* St. Petersburg: Tipografiia V. Bezobrazova i Ko.

Tugan-Baranovskii, D.M. 1997. M. I. Tugan-Baranovskii. Biograficheskii ocherk. In *Tugan-Baranovskii M. I. Izbrannoe. Russkaia fabrika v proshlom i nastoiashchem.* Moscow: Nauka.

Tugan-Baranovskii, M.I. 1909. *Osnovy politicheskoi ekonomii.* St. Petersburg: Tipografiia AO Slovo.

———. 1996. *K luchshemu budushchemu.* Moscow: Rosspen.

Uvarov, S.S. 1864. *Desiatiletie Ministerstva narodnogo prosveshcheniia: 1833–1843.* St. Petersburg: Tipografiia Imperatorskoi Akademii Nauk.

Vernadskii, V.I. 1912. *1911 god v instorii russkoi umstvennoi kultury.* St. Petersburg: Trud.

Zweinert, Joachim. 2008. *Istoriia ekonomicheskoi mysli v Rossii, 1805–1905.* Moscow: Higher School of Economics.

7

Repressive Fields: Economic Theory in Late Stalinism and the Leningrad Affair

Denis Melnik

Introduction: Soviet Economics and Stalinist Fields of Discourse

"Discussions"[1] in the second half of the 1940s and early 1950s, some regarding economics, had already attracted the attention of contemporary observers (e.g. Barghoorn 1948; Domar 1950). At that time of the dawn of "Sovietology," such public and academic discussions in the USSR were considered indicators of possible trends in Soviet domestic and foreign policy, as well as examples of the impossibility of applying dogmas of Marxism-Leninism. Two decades later, the prominent American sociologist Alvin Gouldner (2003: 510–512) used an example of the discussion in linguistics to substantiate his thesis about the transformation of Soviet Marxism during the Stalin period: abandoning the original liberating potential and turning the theory into a conservative ideology. This focus on ideology was characteristic of the totalitarian

D. Melnik (✉)
Department of Theoretical Economics, National Research University Higher School of Economics, Moscow, Russia

J. K. Hass (ed.), *Re-Examining the History of the Russian Economy*,
https://doi.org/10.1007/978-3-319-75414-7_7

model that dominated Western research until the 1970s and 1980s. Within the framework of this model, society as a whole and its individual institutions and social groups were objects of suppression and control from above on the part of the state apparatus. Ideology was considered to be a central pillar of the Soviet system, where individuals and social groups were left to passive observation and subordination (or, much less often, resistance). In essence, it was this model that informed work on the history of Soviet economics that began to appear in the *perestroika* period. Ideological campaigns in the decade after World War II were presented as a stage in the repression of economics (Manevich 1991), even though the very nature of Soviet economics did not allow for a clear-cut demarcation between spheres of theory and dogmatism. The same approach was followed by special work on the history of the Faculty of Economics at Leningrad State University during the Leningrad Affair (Sidorovskii 1990). It became a valuable source, because it was based on oral histories of eyewitnesses. At the same time, it corresponded to historical journalism prevailing at that time (and understandable in that context) aimed at a moral assessment of the Soviet past.

An implicit moral assessment of the Stalin period, and of the Soviet system as a whole, infused Sovietological research within the framework of the totalitarian model. This is why the rise of the alternative "revisionist model" of the 1970s was accompanied by fierce criticism by older scholars: for example, "revisionists" were accused of seeking to justify Stalin's regime (cf. Fitzpatrick 2008). This accusation was connected with the fact that "revisionism" departed from the consideration of unilateral repressive influence from top to bottom and proposed to take into account forces "from below" to explain Soviet history and politics. Actions related to the desire of actors to realize their own interests were seen as an essential factor in stability and dynamics of the political regime. In the context of the Cold War, this could have seemed like jumping to the enemy's side. However, in retrospect, this new direction resulted in the expansion and complication of the model of Soviet society. Attention to individual groups and classes was expressed in the transition from political history, with its attention to ideology, to social history and its analytic focus on the micro level. This trend was fixed in the course of the "archival revolution" of the 1990s (see Raleigh 2002).

An offspring of this trend was studies in the social history of science.[2] These studies focused mainly on the natural sciences and different branches of the social sciences, touching economics only indirectly. However, they are important for reconstructing the social context of late Stalinism and general conditions of the interaction between scholarly communities and authorities. In addition, archival data made it possible to escape from dependency on memoirs of contemporaries and retrospective evaluations in describing the political system of the late Stalin period and, in particular, the Leningrad Affair (see Gorlizki and Khlevniuk 2004: 79–89).

Another trend emerged in studies of Soviet history in the 1990s, associated with the linguistic and discursive turn (Pereira 2011: 518–519), and which forms and methods of communication are viewed as constituting elements of social reality. Here, in particular, one can note the analysis of ideological campaigns of the late Stalin period as fundamental change in language, especially in ideological language (Yurchak 2006: Chap. 2). Unlike the vertical logic of the totalitarian model, of pressure from above and subordination or resistance from below, this alternative approach embeds ideology in regular everyday communication and, ultimately, even in individual consciousness. This provides an analytical justification for the (originally satirical) concept of *homo sovieticus*, the person who speaks and thinks in a "Soviet way." This makes it possible to generalize micromodels of social history of the Soviet period by introducing interactions under consideration into the general discursive field. In addition to interests, goals, and limitations, describing the conditions of behavior and interaction of actors under special conditions, an additional significant parameter is introduced, that is, postulating the features of actors themselves. However, the significance of this parameter is often substantiated by speculation and by deconstructing individual examples of narratives of the Soviet period. Introducing discursive analysis into Soviet studies presumably allows one to penetrate the consciousness of the Soviet person—although this can also be considered an example of excessive multiplication of variables, and it is not always obvious whether its application adds something new for understanding Soviet history. This said, in considering the history of Soviet social sciences, especially economics, a combination of discursive and field frameworks—with the latter providing some structure to the former—might help us to move from exclusive consideration of economics as a set of institutions and peculiar individual interactions to considerations of a project or "product": theory understood as a discourse and a discursive field.

Judging from the current state of Soviet studies, it seems that the history of Soviet economics cannot be fully analyzed using traditional methods of research in intellectual history. Ideology and political pressure played a too significant of a role in the history of Soviet economics (and social sciences in general), such that it cannot be reduced to distorting the "standard" scholarly process: ideology and politics were not only external forces but also key *endogenous* parts to academic (economics) practices and discourses. They were manifested as censorship and control by colleagues (e.g. "discussions," departmental and party organizations, etc.) and self-censorship. If we accept the concept of discourse, then influences of social and political factors are woven into scholars' own language, especially in the social sciences, including economics. But it is precisely for this reason that the unconditional use of the concept of repressed science, which either explicitly or implicitly proceeds from the possibility of a clear demarcation between theory and purely ideological statements (and that call for focus on the first in the context of the second), and also between "good" and "bad" actors in the scholarly process, is problematic.[3]

This chapter focuses on the process of institutionalization of Soviet economics as a concrete example and for a specific period, and on the behavior and interaction of concrete actors. In the field of intellectual history generally, such a focus (akin to a case study) is often ignored, and instead attention is given to recreating the logic of the development of ideas and theories. Scientists themselves appear to be impartial (ideally) observers; their departure from this normative ideal acts as a factor hampering the progress of scientific knowledge. In research on the social history of science, the focus is on institutions and social contexts; behavior and interaction of actors is often considered to be a function of social conditions and normative (ideological) attitudes. This presumption of determinism of scholarly activity corresponds to both the official Soviet doctrine of partisanship and the totalitarian model of Sovietology. However, the evolution of research in recent decades has complicated this picture. If we are to take seriously not only ideology, discourse, and institutions but also actors and interactions, we need a framework that appreciates contexts and actors, structures and practices. One potentially fruitful approach draws on the field framework of Pierre Bourdieu and Neil Fligstein. While both scholars' models of "fields" have differences, they also share these significant elements:

1. Interests are not ignored but are embedded in field positions. The notion of one's interest as a basic motive has influenced political philosophy at least since Machiavelli and Hobbes and is at the heart of modern economic thinking. However, with reference to the history of societies dominated by communism, the influence of ideology was traditionally considered a factor strong enough to eliminate or suppress this motif (through fear or indoctrination).[4] Yet interest and ideology can coexist without one being reduced to the other—rather, both are shaped by one's experiences of field rules and positions.
2. The concept of the field as a social order at the meso level can also be interpreted as a site in between the space of official ideology and normative guidelines imposed "from above" and individual consciousness "from below": an area within which actors (individually or in groups) pursue their interests and seek to increase their capital, whether following the rules of the game or seeking to change them.
3. Vectors structuring fields cannot be specified arbitrarily or a priori. They are composed of the interaction of individual actors within a field, via horizontal interactions of collective actors and vertical hierarchical links. Stresses arising along all these relations shape the dynamics of the entire social system and its individual elements.
4. Power structures (dominance-subordination) arise and are supported in separate fields rather than in the framework of the social system as a whole: their sources at the micro level are related to *habitus* (internalized knowledge and dispositions) and *doxa* (field rules). The political field itself appears as arena of interaction and competition between different fields.
5. The difference between settled and unsettled fields is essential to characterizing equilibrium conditions in a social system.

Drawing on these elements of a field framework, this chapter advances the hypothesis that Soviet political processes in the second half of the 1940s and early 1950s that affected economics were a result of the desire of the country's political leadership to undermine the autonomy of fields that had developed in the leading scientific and cultural centers, and to fix its position as supreme arbiter of knowledge through a series of conflicts initiated from above. At the heart of these conflicts lay explicit and implicit signals from above to individual fields—addressed primarily to newcomers—about the possibility of drastic change in the rules of the game in their (newcomers') favor. At the same time, the ultimate goal was the final centralization of power and the establishment of complete control over discursive space.

Soviet Political Economy by the 1950s: An Unsettled Field

Soviet economics (or political economy, as it was known in the Soviet period) occupied a special position in the hierarchy of Marxist-Leninist social studies. The official genealogy was grounded in the legacy of Marx and through him to classical political economy. The October Revolution and subsequent experience of building socialism provided seemingly empirical evidence of the correctness of the classics and proof of progress both in the history of society and in the history of proletarian science. In this country of the victorious proletarian revolution, political economy should indicate the shortest way to building a communist future for both Soviet Russia and the world. This scheme, however, possessed one essential flaw carefully concealed behind the rhetoric: political economy as a science was absent from the Bolshevik arsenal.

This does not mean the absence of Marxist economics as such, but there was no theoretical unity among its followers at the end of the nineteenth and beginning of the twentieth centuries. Such was Russian Marxism. Bolshevism, as one offshoot of its framework, was for the most part not an academic phenomenon. Even those Bolsheviks who wielded theory never set before themselves primarily theoretical problems. They used their pen to fight opponents, to respond to the evil of the day, to propagate their ideas, and to constantly assert their own exclusive right to the truth in their interpretation of "doctrine." In the texts of Bolshevik authors, rhetoric of suppression came to the fore (easily seen in the texts of Karl Marx himself). They approached cardinal changes in views and approaches easily, while putting solutions to practical tasks at the forefront rather than developing a coherent analytical system. (A vivid example here is the work of Lenin.) Consider the Marxist theory of economic development. It provided an explanation for dynamics of capitalism and the inevitability of its collapse. These processes were assumed to be so inevitable and automatic that the Bolsheviks could concentrate on political struggle. After they came to power in a backward agrarian country and survived the Civil War, it became obvious that expectations of the rapid collapse of global capitalism and victory of the global proletariat were not justified, and they faced their most important theoretical challenge: the lack of a clear concept of post-revolutionary economic development to guide policy. Attempts to respond to this challenge in the 1920s reflected the struggle for power within the Bolshevik leadership. Despite the prevalence of the rhetoric of suppression, in the first post-revolutionary decade valuable theoretical studies did appear, although no theoretical consensus was reached. The final confirmation of Joseph Stalin's power put an end to the theoretical debates of the 1920s. The lack of a theoretical understanding of the Soviet economy was hidden behind the ideological façade of Stalinism.

Soviet political economy should have played a significant role in the creation and maintenance of this façade. It had to construct a reality for new generations of the people of a huge country; it had to teach, convince, and confirm. For this, its arsenal had to include not only complex "dialectical" constructions but also clear didactic schemes. However, in addition to unresolved analytical problems, institutional factors also hampered this. Successful production and dissemination of knowledge required a system of educational institutions and availability of qualified teachers. In the field of natural sciences, the Soviet authorities learned early enough to involve researchers from the pre-Revolutionary system. In the social sciences, which were supposed to be subject to strict ideological requirements, the demonstration of personal loyalty was not enough. But in the first post-revolutionary decade, attempts to establish ideological control ran into a shortage of teaching "cadres." The main role for teaching economic disciplines was played by teachers from the pre-Revolutionary era. In these conditions, the authorities tried to exercise party control from below, by politically "conscious" students, while periodically conducting campaigns of "proletarianization" of higher education. However, success in implementing this strategy led to failures in the educational area (Andreev 2008). "Staff hunger" was an important factor preventing the establishment of strict control from above. Low pay did not attract young and ambitious people and forced existing teachers to lead the life of "freelancers" (e.g. non-tenured or temporary staff) who taught in several educational institutions simultaneously—which also made it difficult to control them. The universal system for "internalizing" ideological control (with the exception of sporadic attempts at pressure from various party bodies) did not yet exist in the 1920s. And by the end of the decade, during the beginning of rapid industrialization, policies were adopted for the rapid development of technical education. Attempts to develop theoretical training in economics were actually curtailed.

In the 1930s, measures were taken to strengthen administrative control over higher education. After 1930, a policy to "consolidate" teachers for a single place of work was adopted; the system for payment was unified by setting the annual teaching load for each position (equivalents of assistant professor, associate professor, full professor) and monthly salary was set in accordance with position held. After 1933, a unified hierarchy of academic advancement and centralized control system were introduced, initially only for institutes of technical educational. However, as in the previous decade, the desire to strengthen the administration of higher education faced a shortage of qualified teachers and constant reorganization of higher education: organizational schemas appeared, were united, were separated, and were liquidated. Mergers and liquidations of educational institutions primarily involved the transfer of students or of tangible assets, while teachers themselves, because of their lack and relevance, remained unlinked to any single main place of work. By the end of the 1930s, the failure to train theoretical specialists in the social sciences (including political economy) had become glaring. This was seen as proof of a dangerous "bias" and led to the subsequent identification of leaders of higher education responsible for this situation as "wreckers" and "enemies of the people." Administrative measures followed. The All-Union Committee on Higher Education (VKVSh), established in 1936, launched activities to develop curricula common to all educational institutions in the country, including those in the social sciences. At the beginning of 1938, VKVSh issued a decree establishing as of September 1938 a unified number of hours for teaching courses in dialectical and historical materialism and political economy for all institutes of higher education for the entire country, regardless of the fields in which those institutions specialized.

By 1936 decisions were taken to prepare a unified textbook on political economy. However, a theoretical consensus that could ensure a seamless transfer of scientific knowledge into pedagogy did not develop, and during the mass repressions of that time, a significant number of Bolshevik economists were killed. In those circumstances, survivors clearly avoided anything that could be interpreted as a claim to an authoritative statement—and the official textbook obviously had to provide an authoritative interpretation of the main economic problems. They sought to replace the need to formulate theoretical propositions with copious quotes from Stalin, Lenin, Marx, and Engels. The unsatisfactory nature of early drafts of this textbook drew the attention of the only person in the whole country who, by the end of the 1930s, had any right to make authoritative judgments on any issues. In January 1941, Stalin personally met with a group of economists and provided them with his comments and instructions. The outbreak of war distracted his attention from this issue. The next meetings of Stalin with economists, to discuss new versions of the textbook, took place only in 1950. Finally, in 1951, with the participation of a significant number of participants, a discussion was organized toward drafting the textbook. Stalin did not take part personally in this, but he closely monitored its progress.[5] In 1952, he published a pamphlet (Stalin 1952) in which he not only expressed his remarks on some ideas voiced during these discussions but also formulated the main provisions to be set forth in the textbook (and, correspondingly, to form the theoretical foundation of Soviet political economy). The final revision was entrusted to a group of economists from participants in those meetings who had worked on previous versions.

The result was published in 1954, after Stalin's death, and 18 years after the original order for preparing the textbook. During de-Stalinization, the provisions formulated in the textbook were criticized; for purposes of this chapter, it is important to note that up to the early 1950s, there was no authoritative and systematic exposition of the foundations of Soviet economic theory.

There was also no mass system for training theoretical economists. Despite the centralization of academic management, any large-scale administrative control measures for teaching political economy could not be effective because of the very small number of qualified specialists: at the end of the 1940s, there were only about 100 doctors of economic sciences in the whole of the USSR—more than half worked in Moscow and Leningrad—and there were only three or four defenses of doctoral dissertations per year. Only those who held a doctorate could formally apply for the position of professor, according to the centralized system of academic promotion introduced in the early 1930s. At that time, the authorities tried to find an institutional solution to the problem of teacher shortages, and leading universities began to open economic faculties. The first of these was the Faculty of Economics at Leningrad State University (LGU), which began its work in September 1940. In the first academic year, 200 students and 15 graduate students enrolled. The Faculty of Economics at the main university of the country, Moscow State University, began work during the war, in December 1941. The number of students in the first academic year there was 28 (Kolesov 2011: 6).

The Voznesenskiis: Early Newcomers

It was not by chance that the Faculty of Economics at LGU was a pioneering project for training cadres of Soviet economists. First, its creator and first dean, Aleksandr Voznesenskii (1898–1950), was a person of high ambitions and organizational abilities. He belonged to the first generation of Soviet economists who began their academic careers after the Revolution. In 1921, he was sent to study at the Faculty of Social Sciences (FSS) at

LGU, and after his graduation in 1923, he remained at the university as a teacher of political economy. Faculties of social sciences were organized in leading universities shortly after the Revolution, and economics departments were opened. The initial intention for creating these faculties was to place the teachers from the pre-Revolutionary era in new conditions and providing them with proper ideological control. A large number of teachers who remained in Petrograd after the Revolution from several pre-Revolutionary centers of economic education gathered at FSS. From an educational standpoint, this project was quite successful. From 1921 to 1924, future Nobel laureate Vasilii Leontiev trained at FSS, where his father, Vasilii Leontiev the elder, taught. However, the degree of ideological control was not considered sufficient; effective control measures were absent because of the lack of necessary personnel, and efforts to prepare theoretical economists under those conditions were an excessive luxury. As early as 1924, its activities were curtailed, and in 1925 this Faculty was closed. Training of teachers in economics virtually ceased, but teaching of those disciplines did not. Teachers at the Faculty of Social Sciences formed the basis for the teaching staff of the Faculty of Economics at Leningrad State University, which opened in 1940. In creating the Faculty of Economics, Aleksandr Voznesenskii showed "initiative from below," highly valued in Soviet social practice. However, this initiative was based on substantial support "from above."

Voznesenskii's younger brother Nikolai belonged to the generation (part of which would later be called the "Brezhnev generation") that were teenagers during the 1917 Revolution. Approval of the new political regime striving to realize the grandiose project of building a new society provided enormous opportunities for them and generated ever greater ambitions. Their training and subsequent social progress was rarely measured and consistent, but the speed of their rise up the ranks was unprecedented. In addition to the explosive growth of new Soviet institutions in the early 1930s, this speed was supported by a specific Stalinist mechanism of social rotation: repression. One of the main targets of repression was representatives of older generations of the Soviet elite. While repression was launched from above, representatives of this generation, voluntarily or unwittingly, were their main beneficiaries. The career of Nikolai

Voznesenskii serves as a vivid illustration. Like his elder brother, in 1921, he was sent by his local Komsomol organization for training, but unlike his older brother, Nikolai was sent to Moscow. Then, after several years of party work in the Donbass, he returned in 1928 to the economics department of the Institute of Red Professors in Moscow, created to train Bolshevik scholars and experts. Like his brother, he remained a teacher after completing his studies. In 1935 he received a doctorate in economics and being in the capital opened up opportunities for a parallel party career. He was transferred to a leading position in Leningrad, where a large-scale purge was initiated after the murder of city boss Sergei Kirov in December 1934. The subsequent wave of mass repression freed up leading positions at the national level, and at the end of 1937 (16 years after his arrival as a student) Nikolai returned to Moscow to the post of deputy chairman of the Gosplan of the USSR. In early 1938, after the arrest of his immediate superior, he headed Gosplan. After 1939, he also gained the position of deputy chairman of the government of the USSR (Stalin occupied the post of chairman after May 1941). After 1947, he was a member of the Politburo and was considered a possible successor to the aging Stalin.

We should not automatically assume that Nikolai Voznesenskii was directly involved in the opening and subsequent work of the Faculty of Economics, but having the name of one from the country's leadership on the faculty was undoubtedly a significant asset. However, Aleksandr Voznesenskii's ambitions and abilities went beyond the framework he created. In the 1920s, he chose the unpromising career path of a university economics teacher. With increasing attention of the country's leadership turned to the development of Soviet political economy in the late 1930s, this choice began to pay dividends. He did not lead the Faculty of Economics for long, and at the beginning of World War II, he was appointed Rector of the university, which was evacuated deep into the hinterland and continued its work at Saratov State University (where Voznesenskii was named Rector as well).

Leaders of the Stalin period had a particular management style based on rudeness, rigidity, and cruelty. Apparently, Aleksandr Voznesenskii as a leader fully possessed these skills (and his brother had them to an even greater extent), but he was able to combine them with an understanding of principles of a normal academic environment and to a large extent contributed to their maintenance and development. He also had charisma that attracted many to him (and also repelled many, apparently). His influence in Leningrad went beyond the walls of the university. As one eyewitness noted, due to the influence of Voznesenskii's personality, after returning from evacuation in 1944, "the university as a whole and its humanitarian faculties in particular were especially privileged" (Ganelin 2004: 46). In the summer of 1948, he moved to Moscow, to the post of Minister of Education. By 1949, the Voznesenskii family could serve as an ideal example of a Soviet success story and demonstrate the opportunities the government provided to ambitious and capable people. In addition to the two brothers, two sisters successfully developed their careers. One also chose economics and, by 1949, headed the Department of Political Economy at one of the leading higher educational institutions in Leningrad; the other held a leading position in Leningrad's Party organization.

Leningrad, like the whole country, began to heal after the terrible war. Outlines of a new post-war normality were emerging. However, the internal logic of the political regime remained unchanged. In conditions of normality, the all-powerful "party" Stalin personified would inevitably become only one of the centers of decision-making. The system provided opportunities for winning (the basis of Stalinist social dynamics), but the stakes remained high. Of the four Voznesenskii brothers and sisters, only one of the sisters survived the year 1950.

The Leningrad Affair

The history of the Faculty of Economics turned out closely interwoven with the fate of the Voznesenskii family. Their fall was the result of the disturbance of the shaky balance of the country's leadership in the post-war years and the beginning of the final redistribution of forces in the closest circle around Stalin. Even in the pre-war years, Stalin set out to weaken the positions of the "old guard," including his faithful ally from the early 1920s, Viacheslav Molotov. The process of elite rotation (which in the Stalinist system almost inevitably meant another wave of repression) stopped in the war years, although the positions of those who entered the higher echelon in the second half of the 1930s increased: Georgii Malenkov, Andrei Zhdanov, Lavrentii Beria, and Nikita Khrushchev. Stalin clearly favored Nikolai Voznesenskii, who was from this cohort. In 1946 Malenkov's position weakened. Khrushchev was running the Ukraine, and Beria concentrated his efforts on leading the Soviet nuclear project. Zhdanov and Voznesenskii saw their statuses increase.

The informal structure of the Soviet ruling stratum was characterized by clans: leaders advanced upward in the Party hierarchy and "elevated" their former subordinates from those regions or administrative structures they managed earlier. In 1946, the rapid rise of "Leningraders" (a group of Zhdanov's former subordinates during his tenure as the head of the Leningrad party organization from 1934 to 1945) began. It was widely believed that the "Leningraders" were under the patronage of Zhdanov. Nikolai Voznesenskii was not a "Leningrader"—his rise was not directly related to Zhdanov's influence, and by that time he was quite self-sufficient in the struggle for power and influence. In particular, he attempted to consolidate his position by referring to the academic sources of his career: in 1943 he ascended to the highest level of the Soviet academic hierarchy, becoming a member of the Academy of Sciences of the USSR, and in 1947 he published *War Economy of the USSR in the Period of the Patriotic War*. The book itself did not contain any significant theoretical advances, but it became the first book since the 1920s with a significant statement by a representative of the country's leadership about a socialist economy. It was quickly translated into English (Voznesenskii 1948) and in 1948 was awarded the highest distinction of that period, the Stalin Prize. This seemed an unambiguous sign, and a campaign began in print claiming the book provided a significant theoretical advance.

For opposing clans and their informal bosses, the rise of the Leningraders and Voznesenskii in the first post-war years was a threat. Stalin himself might have seen a threat from a change in the "war of all against all" among his closest associates. In July 1948, he weakened Zhdanov's position, and he transferred some of his powers to Malenkov, who had returned from disgrace by then. By decision of the Politburo, Zhdanov was sent on leave for health reasons. It is possible Stalin only wanted to stir things up for the moment, and Zhdanov could have regained lost influence, as Malenkov had managed to do, but in August 1948 he died. Initially, this did not affect positions of other Leningraders, but the balance of power in the top leadership turned against them, and opposing clans struck back. In mid-February 1949, Politburo member Aleksei Kuznetsov, chairman of the government for the Russian Soviet Federated Socialist Republic Mikhail Rodionov, and head of Leningrad Pëtr Popkov were removed from their posts. In Leningrad itself, repression began at the end of February 1949, when, under Malenkov's direction, a plenum of the regional Party organization was held. Nikolai Voznesenskii managed to take the blows for a while, but in March he was also removed from all his positions. In August 1949, arrests of high-ranking Leningraders, including Aleksandr Voznesenskii, began. Nikolai Voznesenskii was arrested in late October. A year later, the main defendants of the Leningrad Affair were sentenced to be shot.

In Leningrad, large-scale purges continued after the verdict came down on the Leningraders themselves. From June 1950 to June 1952, 18 rectors and 29 heads of departments in social and economic sciences were expelled from the city's institutions of higher education; around 300 people were dismissed from Leningrad State University alone (Demidov and Kutuzov 1990: 120). The purges continued of their own inertia after Stalin's death in April 1953: the Leningrad branch of the Institute of History of the USSR Academy of Sciences was closed, for example (Paneiakh 1993). Among educational institutions in the city, Leningrad State University, and especially the Faculty of Economics, was affected to the greatest extent because of the proximity of the Voznesenskiis. Almost all teachers from the first rank and some younger teachers and graduate students in 1949–1950 were arrested. Voznesenskii's successor as dean of the faculty, Viktor Reichardt, and head of the statistics department Likarion Nekrash died during "interrogations." In the 1950s, the Faculty of Economics had only seven assistant professors and not a single full professor for three departments (cf. Peshekhonov 1998); in 1947–1948, there had been 347 students at the faculty (*Leningradskii universitet* 1948: 288).

The Leningrad Affair was not widely covered in the press, and instead news of the new terror disseminated among average people via rumors and meager informational notices in newspapers. For the elite, however, this was a fundamental shock, and many representatives of the future "Brezhnev generation" (which could have been known as the "Voznesenskii generation"), who were the main beneficiaries of the Great Terror of the 1930s, realized that the main blow of repression could now fall on them.[6] The immediate consequence of the Leningrad Affair at the top of the power pyramid was the strengthening of the Malenkov-Beria alliance and Khrushchev (who was reported to have joined them, given that his positions began to strengthen). One of the witnesses of events characterizes this period thus: "Then Stalin gradually began to move away from management of current affairs, and the solution to operational issues, and even management of the country, was in the hands of Malenkov and Beria… After the Leningrad Affair, remaining members of the country's leadership feared them so that no one dared to object when they actually took over the reins of government" (Zhirnov 2011).

One of the characteristic features of historical interpretations within the totalitarian model is the concentration on the personality of the leader, including his physical and mental health, as one driving force of policy. This was typical of interpretations of late Stalin period, where old-age infirmity and aggravations of paranoia were considered key reasons for these unfolding processes. However, an interpretation using field theory (based on historical studies of recent decades) suggests that any social system is characterized by the presence of a multiplicity of players and the intersection of many fields; vectors of forces shaping political dynamics consist of the collision of a set of largely autonomous interests. One of the generally accepted characteristics of a totalitarian political regime is the desire to absolutize the leader's personal power. Within the framework of field theory, the realization of this aspiration means not turning the leader into a single significant player (by destroying competitors) but moving him from the position of player to the position of arbiter. The power of the referee rests on the possibility of removing from the field (e.g. by destruction) any player or their group.

However, the simultaneous removal of too many players from the game without replacing them reduces possibilities for "refereeing" and returns the referee to the field. In this sense, the basis for leader's total control was creating constant competition at the top of the bureaucratic apparatus, supported by periodic circulation of elites via promotion of representatives of younger generations. This youth created important pressure from below. In these conditions, the clan nature of the political system was inevitable, with clans as particular networks of actors anchored into particular fields or areas within fields. However, such clans were situational alliances. The emergence of cohesive teams on the basis of personal dependence would mean increasing the autonomy of individual players. Perhaps it was the excessive cohesion of the Leningrad clan, characterized by pragmatic collective leadership in Leningrad during the 872-day Blockade, that signaled the potentially dangerous degree of their autonomy.

In this way, political rules of the Stalinist game—generated not by institutions themselves, but by actors and networks in institutional fields—generated general paranoia, reflecting uncertainty as to who would be removed from the field and the obvious or implicit awareness that anyone could be condemned at any time as an "enemy of the people." The very irrationality of periodic repressive campaigns was an element that allowed the arbiter to retain the right at any time to change the rules of the game as rationally calculated by individual players and his exclusive right to be a source of uncertainty (to shake up *doxa*). But the logic of the entire system, not only of its higher administrative levels, was completely subordinated to instrumental rationality of retaining power. It was this instrumental rationality of Stalinism—perhaps clouded by paranoia but certainly containing rational elements—that is probably its most appalling characteristic. After Stalin's death and until 1985, the USSR was ruled by Stalinist cadres, "promoters" of the Stalin era. Despite differences between them, they refused a forceful rotation via mass repression after Stalin's death. This was the post-Stalin elite consensus and undoubtedly reflected personal experiences of living in a system of uncertainty that provided significant opportunities for gain but also was associated with enormous personal risks. Khrushchev's shift was the final chord in fixing the gains acquired by representatives of that generation of the Soviet elite, which was obliged to promote Stalinism and retained its positions right up to the period of *perestroika*.

Without the mechanism of repression, trends toward autonomy began to increase and soon embraced the entire system. Individual players were given the opportunity to accumulate symbolic capital and to shift the rules of the game in areas they controlled, to their advantage. Without mechanisms of mass repression, the authorities substantially reduced opportunities for nullifying capital accumulated by individual players. The system underwent a certain transformation, allowing individual players to build and implement their own strategies. At the same time, a significant reduction in uncertainty by the "authorities" in the post-Stalin period did not undermine the legitimacy of the system. Dissident sentiments for the vast majority of Soviet citizens did not go beyond what was permitted, remaining within boundaries of expanded private spheres of the post-Stalin period. However, in public spaces, this same majority reproduced official discourse and rigorously followed "Soviet" rules (cf. Yurchak 2006). The absence of mass repression was compensated for by internalization of control at the individual level.

The question of factors of formation and the principles of the functioning of the late Soviet social system (and of its different elements) caused and will cause significant disagreements between scholars. Remaining within a field framework, it is possible to put forward a hypothesis that reproduction of the established order in the social interactions of Soviet economists was ensured by consolidation of the rules of the game at the level of *habitus*. This consolidation made it possible to provide a "doxic situation" in Soviet economic science. Its institutionalization began in the late Stalin period, and at the same time, rules of behavior were assimilated that were reproduced in the future. The instrument of indoctrination became ideological "discussions."

Ideological "Discussions" in the Stalinist System

Ideological "discussions" in the late Stalin era are often viewed in historical reconstructions as stages in the successive suppression of all dissent, inextricably linked to the work of the repressive machine. They did indeed follow this general model, as was demonstrated during "discussions" widely covered in the press and copied at all levels of the academic and cultural hierarchy. A typical "discussion" would start after a signal from above, usually organized around public debates on the content of published texts.[7] The model presupposed criticism of the author or authors (identifying "mistakes" and "errors") and their repenting speeches in response ("self-criticism"). The Party was present at such events in the form of a "moderator" or their group; the status of Party observers determined the value of the discussion itself. The moderator, as a rule, did not interfere directly in the course of the event itself but only summed up its results. However, despite the obvious staging, most participants were invited to guess the roles, and the final distribution of roles was fixed only in the process of summing up results. Thus, everyone could try on the role of "beater," trying to guess the final decision of the "Party" at the choice of the victim and in advance to predict the moderator's decision. But no one could be completely sure of the correctness of the choice and in the degree of punishment, while all speeches and remarks were carefully stenographed. Uncertainty was also supported by the fact that consolidation of the role of the accused did not necessarily mean the subsequent inclusion of punitive organs into the process or complete loss of status.

The position of moderator was also uncertain. A number of model discussions included: a series of discussions of the Party resolution of August 1946 about journals *Zvezda* and *Leningrad*, condemning their editorial policy and the publication of "harmful" works by Mikhail Zoshchenko, Anna Akhmatova, and other Leningrad authors, and a session of the Academy of Agricultural Sciences (the highest level in the Soviet hierarchy of agrarian research) in August 1948. The first case was a blow to the relative autonomy of the Leningrad cultural space fixed in the years of the Blockade. The curator of this process on the part of the country's leadership was Zhdanov, who remained in the memory of subsequent generations of the Soviet *intelligentsia* an irreconcilable persecutor of art. However, the fulfillment of this function implicitly meant the need to strike at the Party organization in Leningrad, which he headed for more than ten years, and positions of the Leningraders closely associated with his own—in fact, a blow to his own positions. In the second case, according to several studies (e.g. Esakov 1994; Kojevnikov 1998), this "discussion" was originally the idea of Iurii Zhdanov,[8] at that time influential as the head of the science department of the Central Committee of the Communist Party, as the final blow to the positions of Academician Trofim Lysenko. Zhdanov's idea was to support representatives of classical genetics in Soviet biology. However, Lysenko succeeded in delivering a preemptive strike, drawing attention of Stalin himself and skillfully grouping his supporters. As a result, the session became a terrible blow to Soviet genetics and natural science in general.

"Discussions" followed a complex script that was not an invention of the late Stalin period; they were a common occurrence in Party struggles since pre-Revolutionary times. What was new was expanding the reach of this method beyond the boundaries of the Party, bringing in new sets of participants and spectators. Evaluating this process in retrospect, one eyewitness noted that the first sign of "cooling" in society was the decree on two Leningrad journals, which became "a warning to all humanitarian workers, although not all understood it at once" (Ganelin 2004: 51). However, the real campaign to hold discussions began to unfold in 1947, especially after the so-called philosophical discussion took place under the supervision of Andrei Zhdanov in the summer of that year.[9] However, "cold snap" is hardly worth linking with the fight against dissent: there were no places left for it in public discourse (and precisely published texts were discussed) after two decades of single-handed Stalinist rule and mass repression. The choice of this or that object for discussion was the result of several factors, and the very content of the texts only provided material for seeking "evidence" for subsequent criticism. Almost all the texts discussed not only passed through all stages of censorship, without which their publication was impossible; they were also marked by awards in the same system that now fell on them. Not the content of the texts, but the fact of publication in the first post-war years was the real source of threat.[10]

"Cooling" was associated with a sharp increase in control in the system of production of meanings and symbols. "Discussions" in a concentrated form reflected the increase in uncertainty. Participants were required to guess the rules of the game in extreme conditions, and any previous successes were not guarantee against expulsion from the field. At the same time, the growth of uncertainty was produced by a developed management system. The desire for total control and concentrating virtually all that control threads in the hands of one person turned out to be completely unpredictable: "The system reacted to random stimuli, but with excessive application of force:

not particularly significant 'signals from below' could provoke an avalanche of disproportionately large-scale political consequences" (Ganelin 2004: 14–15).

The first public campaign in the field of economics was the discussion of Evgenii Varga's book, *Changes in the Economy of Capitalism as a Result of the Second World War* (1946) in May 1947. That discussion concerned a special area of the international economy and was directed primarily against Varga himself and his research institute. As noted above, economists as a professional community were honored with a special discussion only in 1951; this particular discussion was outside the general model. In this case, Stalin himself acted as "arbiter," despite the fact that formally the discussion connected the draft of a textbook, the participants were charged not with accusing the authors, but preparing draft versions suitable for later editing and processing into a new text. After summarizing the discussion in a brochure Stalin had written on the subject, and just a few months before his death, a campaign began in the professional community of economists to "discuss" the outcome of the discussion and hastily to reorganize the discourse in the light of Stalin's utterances. This campaign, of course, affected the Faculty of Economics at Leningrad State University. By that time, the forced transformation of its internal structure was finished and implanted in the memories of shocked participants and in a large number of sources that later appeared in archives.

That forced transformation of the Faculty of Economics began with the discussion of the book *Essays on the Development of Russian Social and Economic Thought of the Nineteenth and Twentieth Centuries* (1948) by Professor Viktor Stein in June 1948. Shortly before this, the book had received a university award, and the discussion followed the standard model. Over 200 participants were gathered to participate in this discussion, not only from LGU but also from other educational institutions in Leningrad. As per the standard model, Stein's behavior was attacked, as he had not demonstrated sufficient "self-criticism" and was obviously trying to put forward arguments of a purely scientific nature. Immediately after this, no measures were taken against the author or against the editor responsible for publishing the book, Viktor Reichardt (still the dean of the Faculty of Economics). Summer holidays began—one symbol of the time is a photo of Reichardt peacefully resting at the university boarding house, preserved in one editions of the university newspaper for summer 1948. However, by September, he was forced to vacate his post as dean, and a few months later he left the Faculty. The formal reason Reichardt was let go was that an audit of the use of library funds for the Faculty revealed that the library had retained Soviet journals from the 1920s containing articles written by "enemies of the people" who had been repressed in the 1930s. Arrests of teachers, including Reichardt and Stein, began as part of the "Leningrad Affair" in 1949.

Two powerful external factors exerted significant influence on the dynamics of events at Leningrad State University in 1948–1950: the atmosphere of ideological campaigns that embraced all scientific and cultural spheres and the "Leningrad Affair." Without the latter, the impact of the former likely would not have been so tragic. However, an important factor in the purge was in implicit conflict of interests between various layers of faculty, mostly between "founding fathers" and postgraduates and young teachers of the first generation of students; "newcomers," largely comprised students, postgraduates, and young teachers who joined the faculty since 1944–1945. Many "founding fathers," the core of the teaching staff, knew each other since the beginning of the 1920s. In the first years of work, the Faculty was compact, and the experience of World War II and the Blockade of Leningrad (including evacuation and return) created almost familial relations among students and teachers (cf. Eliashova 1998). Some of the wartime students after the university's return from evacuation completed their training under the guidance of their teachers and began their careers as junior faculty. However, faculties of economics were created for mass production of economics teachers, and the war had temporarily suspended the growth in the number of teachers until 1944–1945. By 1948–1950, students of the first post-war cohort (many of whom began their studies immediately after demobilization from the Red Army) completed their education or already entering graduate school of the ranks of junior faculty. Between "new" and "old," there was an alienation of sorts, caused by different experiences, reinforced by institutional forces.

In early 1946, a long process of unifying academic promotion and centralization of its control gained a financial dimension: the official salaries of university teachers were sharply increased. Associate professors and, in particular, full professors got substantial raises, which promoted them to a well-paid social stratum. Hence, acquiring scholarly degrees ceased to be a matter of sheer prestige, as they were the preconditions for academic promotion. On the other hand, one of the main tools of financial management in a centralized system became an amount of "slots" allotted to each university. "Slots" could be redistributed on intra-faculty or intra-university levels, but a general increase in their amount was subject to bureaucratic procedures and scrutiny at all levels of university management system. As the amount of "slots" was the basic indicator for financing, the requests for increase were not always approved. So, a person could not be admitted to the faculty or promoted to a higher academic position if there was no vacant "slot" for that position. A junior faculty member of that period later recalled, "Graduates returned to the Faculty [from military service], but all teaching positions were filled." Questions were raised publicly and privately: "Where do we work? It is time to drive out these bourgeois professors" (*Leningradskoe delo*: 394). Actually, archival materials and memoirs of contemporaries reveal that the real problem was a shortage of teaching staff. But some younger scholars saw in the status quo of the Voznesenskii era an obstacle to rapid advancement of their careers. This provided fertile ground for manipulation from above. As the same source recalled, "And then the secretary of the [Vasileostrovskii] district committee ... and the head of the department of science for the city Party committee ... are constantly giving instructions" (*Leningradskoe delo*: 394). This combination made it possible to launch a self-sustaining process obeying the logic of a power struggle. According to an eyewitness, "There were people who, at that time, pursued purely personal interests and showed the worst features of their characters" (Berezhnoi 2005: 36).

In the specific conditions of the system that had developed by that time, scholars as individuals or in groups often tried to involve higher party institutions as "moderators" to resolve internal conflicts and contradictions. However, any signal from above was sensitively picked up, giving rise to personal and professional struggles for power and influence in hierarchies and ideological discourses. However, at Leningrad State University during the Leningrad Affair, one can see the impact of one more dimension of institutional transformation. In the first post-war years, the role of local Party and university structures sharply increased. They were entrusted with functions of management and control at the micro level, and they became the direct embodiment of power for Soviet teachers in the coming decades. In the first years of the Faculty's existence, most teachers were not members of the Party. A Party bureau had existed at the university, but its role was minimal. Mass membership in the Party intensified during the war, and in the post-war years, Party membership became indispensable to a successful career. Simultaneously, the role of the local Party organization strengthened. In 1946, Serafim Ilyin became the head of the rapidly growing Party organization at the Faculty. He was one of the "aliens" who ended up among the Faculty in 1944. His field of expertise was the agrarian economics, but surviving evidence characterizes him primarily as a tough and skillful manager of the Stalinist blend. In the following decades, he occupied administrative positions in various educational institutions; in 1948–1950 he was dean of the Faculty of Economics and in later years was university pro-rector.

The head of the local Party organization had another lever of power: he was entrusted with functions of collecting personal data on Party members and preparing reports on personnel to higher Party authorities. Thus, in 1947, Ilyin signed an overall positive report on Reichardt that still mentioned his origins among the nobility and that in 1936 he was expelled from the ranks of Party candidates (not yet a full-fledged member) for showing too little political vigilance.[11] Later, Ilyin pointed out, "The Party organization of the Faculty since October 1946 [when Ilyin became the secretary of the Party cell for the Faculty] was continuously working towards a radical restructuring of scientific and educational work of the Faculty, to expose and expel alien elements from the Faculty."[12] To this end, higher-level organizations "were repeatedly informed of the state of affairs at the Faculty," but it was only after "exposing the anti-Party grouping of Popkov and others" that it was "completely cleansed" with the support of higher Party organizations," that is, after January–February 1949, the beginning of the Leningrad Affair. This undoubtedly included no small elements of self-promotion, but it was equally clear that any party secretary could act as an important source of information on the ground. A report on Ilyin himself, dated 1950, noted that he is "an active participant in the Party struggle in the

Faculty and the University with cosmopolitanism, bourgeois nationalism, and formalism in science."[13]

Along with the strengthening of Party organization, the role of the departments also grew. Departments and their chairs were assigned the function of holding discussions at this local professional level regarding great ideological discussions and other forms of control. For example, in October 1948, an extended meeting of the department of statistics was devoted to revising the programs of theoretical statistics courses in connection with a decision by the Agricultural Academy on the situation in biological sciences. In December 1948, at a meeting of the department, department chair Nekrash gave a report on shortcomings in department work as noted by authorities at the faculty and university levels, especially about professor Nekrash not attending sufficient lectures by department instructors.[14] The miscalculation was all the more serious because 38 people (i.e. the whole staff of the department) were "enrolled" by the dean and his deputy at the department of political economy to attend and observe lectures and other activities, and a discussion of transcripts of classes made it possible to identify "the non-Marxist character of Bukovetskii's lectures" (a prominent economist whose career began before the Revolution, an instructor in the Faculty of Social Sciences, and a member of the first cohort teaching at the Faculty of Economics—who was also arrested).[15]

The institutionalization of new forms of management and control at Leningrad State University took place during the worst of the Leningrad Affair in which it was embedded, but it also reflected the processes common to all Soviet science and culture beyond this particular city. Ideological campaigns in different contexts spilled over into each other until 1953. This resulted in the creation of an all-pervasive surveillance system in which practically all statements were brought into the public domain—a further "nationalization of the private sphere," as it were, and a shift in the rules and relations of different fields (state, Party, professions, etc.). In addition to lectures and speeches, instructors continuously moved from meetings of departments, to meetings of Party and trade union organizations, to scholarly and dissertation councils. "Discussions" completed the picture. All statements were carefully recorded in writing, leaving behind a paper trail. After Stalin's death, the most odious forms of control (e.g. stenographies of lectures and visits by colleagues and administration representatives) began to lose their significance or acquired a formal character without much real bite. Yet basic elements of the institutional system remained unchanged. The results of the seemingly chaotic and irrational ideological campaigns of the late 1940s and early 1950s led to the internalization of control at the level of individual behavior and practices.

Conclusion

In the first decades of the twentieth century, the processes of embedding production and dissemination of scientific knowledge into mass industrial society began. Knowledge production became the object of administration; the goals of such administration were specified "useful" results and the production of "personnel" necessary and "useful" for certain sectors of the economy. Scientific disciplines and subdisciplines were separated not only theoretically but also administratively. Research and teaching were structured by a complex internal hierarchy of universities, research centers, professional associations, and regulatory bodies. One of the most important (and often the only) source of funding was the state. Due to the conditions of the emergence of the Soviet planning system, many tendencies of the institutionalization of scientific research were presented to observers in a concentrated form. In general, institutionalization of scientific knowledge in the Soviet period was not entirely divorced from the rest of the world (which seems to be an assumption of comparative studies).

The institutionalization of Soviet economics as an academic discipline took place in the three decades after the 1917 revolution. This process coincided with the launch and in the framework of ideological campaigns in the second half of the 1940s. They took place in the form of numerous "discussions" affecting various areas of science and the humanities. The ultimate goal of these campaigns was to undermine the autonomy of professional communities that had developed by that time and to establish complete control over the production of discourse.

The reproduction of social systems is structured by power, but the source of that power is not external to individual institutions, nor does it not stem solely from the state's capacity for violence and coercion. Probably the most terrible discovery in the history of totalitarian societies of the twentieth century was that the field of power is not formed exclusively in political space, subordinating society from above, but relies on spaces of subordination and control from below, at the level of individual social institutions and groups. In the second half of the 1940s, Soviet power faced the formation of new fields, in particular, in fields of science and art. In theory, individual collectives had to act as conductors of the "party line." In fact, especially in the presence of significant influence of leaders (as was the case in LGU and its Faculty of Economics prior to 1948), they acquired a significant degree of independence.

Under late Stalinism, strategies to combat the complexity and autonomy of professional communities in fields of science and art were not mass repression as in the 1930s—the Leningrad Affair was an exception—but a permanent shock of existing fields and the nullification of accumulated symbolic capital. At the same time, unfolding processes moved not only from above as pressure but also through stresses within fields themselves, primarily between established beneficiaries and newcomers. Simultaneously, the administrative system was transformed. Internalization of control over discourse and institutionalization of supervision and management at the micro level was the legacy of Stalinism, which ensured the maintenance of legitimacy of the late Soviet system without reliance on mass repression until the 1980s.

Notes

1. "Discussions" were held at regular or ad hoc gatherings as opportunities for particular speakers to criticize certain pieces of writing or to indoctrinate listeners in latest instructions from the highest authorities (e.g. the Central Committee of the Communist Party). Later, we shall consider them as a form of social control.
2. A representative sample to these studies is presented at the Internet project of the S.I.Vavilov Institute for the History of Science and Technology of the Russian Academy of Sciences: http://www.ihst.ru/projects/sohist/.

3. This tends to be the stereotypical view of the history of Lysenko and genetics: an opportunist using Stalinism to "warp" proper biological studies.
4. This shift in the understanding of interests and behavior (albeit with different conclusions) was characteristic of official Soviet official, unofficial Soviet dissident discourse, and Western Sovietology.
5. According to Evgenii Primakov, who evidently received information from one of the participants in the event, "Stalin did not participate in heated discussions, but sat in his office and listened to the speakers through his headphones" (Primakov 2015: 19). Primakov's access to this kind of knowledge is not in doubt—he headed Soviet and Russian foreign intelligence—but the reliability of many statements pertaining to Stalin cannot be verified. This statement, rather, reflects the general atmosphere of that period. In any case, it is hard to believe that any of the participants in the discussion could observe Stalin in his office. Undoubtedly, Stalin carefully studied the materials and transcripts of the meetings.
6. There was, apparently, an unspoken elite consensus that the Leningrad Affair was unjust. An indication of this is the fact that the surviving victims of the trials began to return from the camps shortly after Stalin's death, while the process of their rehabilitation started as early as 1954. We should note that the post-Stalin practice of returning victims of repression from the Gulag did not mean their automatic rehabilitation, that is, the right to have legal judgments against them lifted. The process of mass rehabilitations of victims of the "Great Terror" of the 1930s began only at the end of the 1980s.
7. According to the sphere of debates, published texts as the subject for "discussions" could be replaced with pieces of art or music. However, the latter required certain authoritative interpretations, while the former directly provided materials for criticism in the forms of quotations (often out of context) aimed to vilify "mistakes" and "deviations" on the part of the author. Sometimes subjects for "discussion" were provided by publications in leading Party or professional media, directives of the Central Committee, interventions of Stalin, and so on. In such cases, debates lacked well-defined targets for criticism, and participants were required to perform acts of "self-criticism" in light of "valuable instructions" received from above.

8. Andrei Zhdanov's son and, for some time, Stalin's son-in-law through his short-lived (1949–1952) marriage to Stalin's daughter Svetlana.
9. This was formally connected with the discussion of a book by an important party functionary, Georgii Aleksandrov, *The History of Western European Philosophy*, which had been awarded the Stalin Prize. The administrative result of the discussion was that Aleksandrov lost his position as head of the propaganda and agitation division of the Central Committee. He was replaced by Mikhail Suslov, a future influential member of the Politburo and main Soviet ideologist of the Brezhnev era, who happened to be present at this discussion.
10. This applied not only to texts but also to other forms of utterances in the space of public discourse. For example, in February 1948, the Central Committee of the Communist Party issued a resolution on the opera "The Great Friendship," which gave rise to a campaign to combat "formalism" in Soviet music.
11. OA SPbSU [institutional archive, St. Petersburg State University], fond 1, opis—personal files of employees dismissed in 1942–1949, sviazka 32, delo 1496, list 51.
12. TsGA SPb [Central State archive, St. Petersburg], fond 7240, opis 14, delo 1614, list 2.
13. OA SPbSU, fond 1, opis—personal files of employees dismissed in 1955, sviazka 6, delo 156, list 66.
14. TsGA SPb, fond 7240, opis 14, delo 1493, list 3.
15. TsGA SPb, fond 7240, opis 21, delo 58, list 3.

Works Cited

Andreev, D.A. 2008. "Krasniy student" i politika proletarizatsii visshey shkoli. *NLO* 90. http://magazines.russ.ru/nlo/2008/90/an5.html.

Barghoorn, Frederick C. 1948. The Varga Discussion and Its Significance. *American Slavic and East European Review* 7 (3): 214–236.

Berezhnoi, A.F. 2005. *V pervye poslevoennye gody: 1946–1954. K istorii Leningradskogo-Sankt-Peterburgskogo universiteta*. St. Petersburg: St. Petersburg University Press.

Demidov, V., and V. Kutuzov. 1990. *"Leningradskoe delo." Sbornik*. Lenizdat: Leningrad.

Domar, Evsey D. 1950. The Varga Controversy. *American Economic Review* 40 (1): 132–151.

Eliashova, L.L. 1998. "Papa" Voznesenskii. *Neva* 10: 147–159.

Esakov, V.D. 1994. Novoe o sessii VASKhNIL 1948 goda. In *Repressirovannia nauka*, vyp. 2, 57–75. St. Petersburg: Nauka. http://www.ihst.ru/projects/sohist/books/os2/57-75.htm. Accessed November 5, 2017.

Fitzpatrick, Sheila. 2008. Revisionism in Retrospect: A Personal View. *Slavic Review* 67 (3): 682–704.

Ganelin, R.Sh. 2004. *Sovietskie istoriki: o chem oni govorili mezhdu soboi*. St. Petersburg: Nestor-Istoriia.

Gorlizki, Y., and O. Khlevniuk. 2004. *Cold Peace: Stalin and the Soviet Ruling Circle, 1945–1953*. New York: Oxford University Press.

Gouldner, Alvin. 2003. *Nastupaiushchii krizis zapadnoi sotsiologii*, The Coming Crisis of Western Sociology. St. Petersburg: Nauka.

Kolesov, V.P. 2011. *Ekonomicheskomu fakultety 70 let*. Moscow: TEIS.

Kojevnikov, A.B. 1998. Rituals of Stalinist Culture at Work: Science and the Games of Intraparty Democracy circa 1948. *The Russian Review* 57: 25–52.

Leningradskii universitet za sovietskie gody, 1917–1947. 1948. Leningrad: Izdatelstvo LGU.

Manevich, V.E. 1991. Stalinizm i politicheskaia ekonomiia. In *Repressirovannaia nauka*, ed. M.G. Iaroshevskii, 181–198. Leningrad: Nauka.

Paneiakh, V.M. 1993. Uprazdnenie Leningradskogo otdeleniia Instituta istorii AN SSSR v 1953 g. *Voprosy istorii* 10: 19–27.

Pereira, N.G.O. 2011. Post-Soviet North American Historiography of Russia. *Canadian Slavonic Papers/Revue Canadienne des Slavistes* 53 (2/4): 513–525.

Peshekhonov, V.A. 1998. Alma mater: stranitsy istorii. *Vestnik Sankt-Peterburgskogo universiteta, Seriia 5* 3: 89–95.

Primakov, E.M. 2015. *Vstrechi na perekrestkakh*. Moscow: Tsentpoligraf.

Raleigh, Donald. 2002. Doing Soviet History: The Impact of the Archival Revolution. *Russian Review* 61 (1): 16–24.

Sidorovskii, L. 1990. Gde zhe sovest, professor? In *Leningradskoe delo*, ed. V.I. Demidov and V.A. Kutuzov, 386–399. Leningrad: Lenizdat.

Stalin, I.V. 1952. *Ekonomicheskie problemy sotsializma v SSSR*. Moscow: Politizdat.

Voznesenskii, N.A. 1948. *The War Economy of the USSR in the Period of the Patriotic War*. Moscow: Foreign Languages Publishing House.

Yurchak, A. 2006. *Everything Was Forever, Until It Was No More: The Last Soviet Generation*. Princeton, NJ: Princeton University Press.

Zhirnov, E. 2011. Posle rasstrela Berii Malenkov vso vremia ulybalsia. *Kommersant Vlast*, 33. August 22. http://www.kommersant.ru/doc/1752481. Accessed November 4, 2017.

Part II

Fields, Economic Policies, and Economic Practice

Introduction to Part II

From the story of economics as *theory* in Part I, we now turn in Part II to economic phenomena and policies. While the attention to social relations as well as actors and institutions remains, as well as a holistic account of discourses and practices, the narratives and analyses here are also relatively closer to the typical application of a field framework. The theme of problems with innovation and development, versus *status quo*, is not new to Russia's economic history, and *Rumiantsev* takes us back into the past to seek roots of logics of this conundrum in various field and institutional logics. Fields of practice were constituted by three institutions: those of the state, of the Orthodox Church, and of the peasant commune. One common logic to all three was a moral economy of reciprocity and redistribution: this was grounded in survival for peasants, in the theology of the Church, and as a foundation of legitimate authority for the state. As such, the logics of all three subsumed economic activity under relations of cooperation as well as competition, survival as well as profit. To the extent a capitalist spirit was going to emerge in Russian economic fields, it would have to contend with counterbalancing forces and logics from these other institutional fields. In fact, the state would have to exert some effort not only to speed up Russia's lagging industrialization to catch up with the West; it would have to confront these very non-capitalist logics from which it had benefited. *Ryazanov* follows this logic further, exploring how markets were not entirely a state creation, but owed their existence to state policies as much as anything else. In Russia more than in England, Karl Polanyi's dictum rings true—market economies require market societies, a task to which states are best suited. Alas, even state elites did have their eye on innovation and reform; elites and actors of other fields and institutions could stand in the way, not only with opposing interests but also with opposing practices.

That the politics and political economy of post-Soviet Russia's economy are tied up with field dynamics becomes clearer in our last chapters, which also bring contentious politics into the open—addressing one problematic feature of current field theory (i.e. privileging reproduction over conflict). *Rumiantseva and Musaeva* attempt to bring together a field framework and economic theories linked to growth and innovation, including evolutionary economics. One could argue that field theory tends to privilege reproduction over change, and while a field framework can accommodate change, scholars usually invoke it to examine reproduction. These two authors draw on Fligstein's insights that exogenous shocks—whether economic downturns or sudden state intervention—can force field leaders and members to seek new field rules and conceptions of control. These might not always favor innovation, however, as rent-seeking or simply maintaining stability can outweigh perceived opportunities and gains from innovation. Further, field rules in contemporary Russia are not so kind to innovation, as too many (but not all) field leaders in the state and business prefer stability, often enough for rent-seeking. To the extent there might be some hope, however, it is with professionalization, especially in arbitration courts.

Protasov discusses a conflict theory of inflation, i.e. that one driver of inflation is distributional conflicts, and he locates roots in conflict partly in the different rules and interests of fields. Not content to focus on Russian inflation as a phenomenon *sui generis*, he uses the Russian case to dive into the thorny issue of what causes inflation to begin with. Inefficiencies and rent-seeking, among others practices, are built into existing Russian economic fields, such that competition for rents rather than returns on innovation contributes to inflationary pressures (although

Protasov is careful not to exclude other usual macroeconomic variables). While Protasov's argument has more than a passing resemblance to that of Mancur Olson, he does not postulate that distributional conflicts are a function of the quantity of economic associations alone, nor does he pin the blame on unions; rather, he points to fields that have crystallized into informal cartels. Continuing along the lines of money and finance, *Bartenev* provides an extended commentary on how contemporary Russian fields of finance have become intertwined both with the state and with a new bifurcated ruling class of state and business elites. The elites of state and business fields use financial institutions and organizations for rent-seeking gain, rather than as a vehicle for profit from serving economic growth and the general population. Bartenev also suggests that the political and media fields have been co-opted by state and business elites and drawn into the orbit of their fields: both media and democracy have created a quiescent, tamed population that, if not supporting the elite and their wealth, at least does not have means or expectations to resist the new political economy.

Turning to the first decade of Russian post-socialism, *Hass* explores how institutional shocks in the last years of Soviet socialism opened up the environment of institutional fields to reconstruction. This opened the door for confusion and contention over what the post-Soviet economy should look like, that is, what the operative logics of organizational strategies and structures, the nature of property, and the basis of authority should be. On the one hand, key actors' Soviet-era experiences and knowledge (embedded in *habitus*) shaped how different actors perceived risks and a normal economy; this provided some path dependency. However, the Soviet experience was not uniform, and from that past came three competing "new classes": Soviet-era managers (Red Directors), financial entrepreneurs (paradoxically), and state cadres, especially from the massive security apparatus. As these classes set about organizing and defending field rules and boundaries, they came into conflict over whose conceptions of normality would predominate. That conflict was at the heart of the drama of the 1990s and the rise of Vladimir Putin. Finally, *Shevelev* suggests that Soviet-era habits and worldviews have persisted, but they have also been reshaped after the 1990s. Inequality, gain, and an almost cynical approach to economic relations and practices seem to have emerged out of the "wild 1990s." Put differently, the existence of fields and field rules might create a stable equilibrium of economic strategies and practices, but these need not automatically translate into innovation or growth. Rather, they can reproduce rent-seeking or predation and latent conflict. This suggests that the potential for real reform and development faces not only entrenched interests but also entrenched practices and identities—a problem that plagued the USSR in its last decade as well.

8

Empire, Orthodoxy, and Economy: The Influence of Russian Orthodoxy and Empire on Economic Fields in Pre-Revolutionary Russia

Mikhail Rumiantsev

Economic Fields, Empires, and Religion

Neil Fligstein's (1990, 2001) theory of markets and fields allows us to expand the social context of economic research, as it turns our attention to how actors' economic practices are embedded in structured networks of socio-cultural interactions and plexuses—fields—which then "allows us to create and maintain social worlds through cooperation with other actors" and "encourage others to cooperate" (Radaev 2013: 8). Yet existing applications of field theory to economics are still at an early stage, and many other institutional and cultural contexts, and other components of field cultures, remain to be explored. In particular, can Fligstein's political-cultural approach to fields and economies, derived primarily to explore market-state structures in industrial societies, generate insights and explanations

M. Rumiantsev (✉)
Faculty of Economics, Department of Economic Theory,
St. Petersburg State University, St. Petersburg, Russia

J. K. Hass (ed.), *Re-Examining the History of the Russian Economy*,
https://doi.org/10.1007/978-3-319-75414-7_8

when applied to historical and economic experiences of pre-industrial societies? I suggest field theory not only can provide insights about the economy of an ideocratic, traditional society, such as the Russian Empire of the nineteenth century; this case can also expand how field theory is used. While nineteenth-century Russia had not reached the same level of "modernity" as other European countries, cognitive prerequisites and expectations of actors—common meanings and cultural understandings—that prompt actors to seek and find stable social conditions for interaction in the market were far from irrelevant.

This chapter is a first step to explore the nature of rules and practices that influenced earlier Russian economics, beyond the usual actors of the state, agricultural elites living off of peasant labor, and occasional nobles and foreign entrepreneurs investing in fledgling industry. One source of rules for economic actors in pre-Revolutionary Russia was the related fields of *Russian Orthodox ideology* and a long-standing *moral economy* originally derived from peasant culture. In Russia's traditional ideocratic society, economic fields were largely non-market-oriented and determined by the influence of Orthodoxy and the sacral power of the tsar. As such, economic fields were grounded in relationships of redistribution and reciprocity, including rituals of gift exchange. Unlike market transactions, governed by a one-stage transaction using money as a means of exchange, gift exchange is based on a time-consuming exchange of services between non-anonymous social actors based on the principle of "deferred reciprocity." Like other forms of economic practice, this earlier non-market economy had a "supportive mode of communication between people"—markets for exchange, a center for redistribution, and symmetrical groups of people capable of reciprocity (Polanyi 2010: 47–81).

A second source of rules stemmed from politics and ideology of Russia as *empire*. While the Russian Empire existed formally from 1721 to 1917, essential attributes of empire (territorial expansion, authority of a metropole, integration of heterogeneous territories) emerged earlier, during the time of the Duchy of Moscow at the end of the fifteenth century. It was Marx who drew attention to Russia's transformation into an empire: "The astonished Europe, at the beginning of the reign of Ivan [Ionna III], hardly aware of the existence of Muscovy squeezed between Tatars and Lithuanians, was stunned by the sudden appearance on its eastern borders of a huge empire, and Sultan Bayezid himself, before whom Europe trembled, first heard the arrogant speech of Moskovit" (Marx 1989: 3). Typically, "empire" is defined through relationships of control of the center (core) over peripheral-dependent political communities (Go 2011: 7; Lachmann 2016: 96), for example, the capacity of the core to use subordinate territory for taxes and cheap imports (Hanneman et al. 2015: 409). Note that I am not going to address the imperialist aspect of empire and fields—seizing land and exploiting people for gain. Rather, I explore possible influences of social and economic structures of empire on fields and how non-capitalist principles of Orthodoxy and power could integrate heterogeneous segments of the economy into the multidimensional socioeconomic structure of empire.

"Empire" was once thought of as a centripetal force of connected systems of spiritual, political, and economic power that regulated the life of peoples and integrated local worlds with diverse logics of practices into a relatively self-sufficient socioeconomic organism. Despite varieties of strategies for economic behavior and the heterogeneity of economic structures, the highest regulatory authority in empire is the core state. Mulgan (2007: 271) suggested that such a state had three sources of power: violence, money, and trust. Of those three sources of power, the most important for sovereignty was shaping expectations to generate trust; in contrast, violence is costly and possibly negative, and money can be used only via issuance and withdrawal. I suggest that religious meanings and empire's cultural hegemony—in Mulgan's words, "power over thoughts"—created social conditions for interaction in the economy of the Russian Empire. In particular, I explore how three dynamics might have some significance in shaping pre-Revolutionary Russian economic fields: inter-consecrated gift exchange, taking root in the era of Muscovy; fields of cooperation originating in peasant communities; and the role of railway construction that connected the empire and contributed to capitalist development but also led to a deformation of traditional economic fields. During the epoch of Muscovy, religious fields took shape, and they later contributed to institutional integration through the sacralization of tsarist imperial power. Peasant agriculture, united in communes, represented the prevailing way of life in pre-Revolutionary Russia. Large-scale construction of railways was a driver for the development of capitalism, and at the same time led to an increase in social conflicts and economic imbalances.

In the Beginning: Fields of Religion and Power and Muscovite Institutions

How can we characterize principles of institutional integration at this point in Russian economic history? Richard Pipes (1997) considered pre-Petrine *Rus* a deformation of European feudalism, generated by patrimonial ideology and institutions. Pipes proceeded from the premise that the absence of positive law and of restrictions on the sovereign's rights proves Russian tsars used unlimited power as *votchina*, the monarch's fiefdom, which eclipsed or negated other types of property and private ownership. In the early twentieth century, historian N. P. Pavlov-Silvanskii (1988) attempted to substantiate the feudal system of pre-Petrine *Rus*, and other historians of that time, V. O. Kliuchevskii (2002) and N. I. Kareev (2015), agreed with his arguments about the existence of certain elements of feudalism (the nature of land ownership, the relationship between power with landowning, state fragmentation) from the thirteenth century to the middle of the fifteenth century. Later, Soviet Marxist historians imputed the theory of Russian feudalism to the era of Muscovy (from the second half of the fifteenth to the end of the seventeenth centuries).

Yet this interpretation is not without disputes. Historical work has shown that Russian institutions in the era of Muscovy did not have the same hierarchical relations between subjects of feudal vassalage in stereotypical feudalism. The basis of land ownership was local land tenure. The *pomeste*, a form of estate, was a land plot given as a personal possession to a nobleman under condition of his service to the state. Possession of the estate was temporary but could be lifelong. The *pomeste* differed from the *votchina*, which was complete and hereditary property of its owner, a boyar. Peasant property until the second half of the eighteenth century was based on elements of personal freedom, expressed in the "right of exit" (transition from one landlord to another) and laws of local self-government. Enslavement of the peasantry in the second half of the seventeenth century was based on abolition of the "right to exit" and recognition of the heredity power of landlords over peasant progeny. The Russian patrimonial boyar and noble landlord qualitatively differed from European feudal lords by the fact that they did not have independent sovereignty. If in Europe suzerain and feudal lord could interact as gentlemen, in Russia boyars and nobles received power and property only with the sanction of the tsar or Grand Duke. Feudal property law in Europe was grounded in the belief that land belonged to lord and vassal, the latter having the right of immunity (limited sovereignty) in relation to lord and king. The seigneur had no right to interfere in management of the feudal estate and its population, subordinate only to its baron, but not to the lord and king. In Russia, there was nothing like a "feudal ladder," in which any landowner or boyar did not even have limited sovereignty with respect to tsarist power, and legitimacy of their rights was established by the tsar's authority. Also, if in Western Europe rights for land ownership were inherited from father to eldest son, in Russia property rights were passed from father to son only when inheriting compulsory military service to the state (Gorskaia et al. 2001; Zimin 1960; Liubavskii 2002; Skrynnikov 1997; Fursov 2001).

The existence in Russia of landlords and peasant serf labor is not a strong argument that a feudal system was dominant in Russia. Reducing the social system to norms of positive law or a materialist understanding of history, to productive or paternalistic relations, elides potential roles of religious and ethical content of the era. As a rule, the economic system is determined not only by production factors but also by norms and meaning systems, symbolic capital, and religion and political ideology that can shape how those institutions operate in practice. One could make a case for an "Asiatic mode of production," based on centralized distribution of resources and communal ownership of land. Even then, one can find traces of this mode of production elsewhere (North Africa, the Middle East, Central Asia, South-East Asia)—making this model of political economy not much more useful than reference to generic "feudalism."

Taking a more holistic approach to economic history that takes into account material and cultural forces—sometimes called a "civilizational view" of history (Erasov 2002)—might have utility, and a field framework does this without being trapped in limits of "feudalism" or "Asiatic mode of production." Historically, Russia borrowed and interpreted Orthodoxy and related institutions from the Byzantine Empire. Many everyday habits in Russia in this early time were grounded in a theocentric worldview, coupled with religious and extra-social dimensions.

According to A. S. Panarin, institutional organization of Russian society was not considered an autonomous site of natural rights. The Eastern Christian archetype of power was based on imitating the Heavenly King: "My kingdom is not of this world … the tsar is the supreme ruler and is not considered to be earthly and of social hierarchies, but like Christ is heavenly" (Panarin 2002: 138). Tsarist power was formally unlimited in relation to property and individual rights, although in reality it was limited by the state system, public institutions, and religion. Orthodoxy and social conformism were not established by the tsar's will. Imperial power depended on these and sought to maintain conformism in fields of religious and social institutions. Power was limited by tradition, bureaucratic clans, and religious canon (Zazykin 1924). According to St. Joseph of Volotskii in the second half of the fifteenth century, "The unrighteous king is not God's servant, but that of the devil" (Florovskii 1991: 18). The Byzantine theory of cooperation and complementarity between state and church was also widespread (Bishop Nicodemus [Milash] 1897). "The power of the tsar was considered to flow from God through the mediation of the Orthodox Church" (Vernadskii 1993: 271). Russia's ruling class was not a sovereign estate but an estate in the service of the tsar, who was a bearer of divine power and the symbol of the Christian state.

Russian Orthodox clergy as a class was not as sharply opposed to laity as in Catholicism. Priests and laity communicated equally and had equal access to Holy Scripture and Tradition. Worship was conducted in Slavonic, and the laity invoked the right to choose priests. A particularly close connection between laity and clergy was at the parish level: churches were built "by the whole world," and the parish decided land set aside for clergy. Landholdings of the Orthodox Church never acquired the same status as in feudal Europe, and ecclesiastical principalities or monastic orders did not become independent estates. The lower clergy had no real right to tithe (as in Catholicism) and lived off offerings of parishioners and their own labor (Dmitriev 1990). A Papal nuncio to Rome in 1622 is instructive: Orthodox "priests are so poor that to earn a piece of bread, they must cultivate the land with their own hands" (Dmitriev 1990: 43).

A result of this historical development was that classical and hierarchical divisions of people were weakly expressed, and corporate and social relations vis-à-vis property were blurred and underdeveloped. A man in pre-Petrine Russia was much less attached to his class than to the state. Fundamental social strata co-existed in this framework of a uniform patriarchal way of life, without great differences in attitudes and outlooks. The status of a person was determined by his place in the order of estate service to the tsar, manifested in duties and social functions. During this earlier period of Muscovy and the rise of Orthodoxy, a religious field formed that ensured institutional integration of poorly structured Russian society via sacralization of royal power.

As a result, the social universe of Muscovy was grounded in strategies of nationwide gift exchange, sanctioned by religion and power: inter-class and intergenerational transfers to ensure stability of society. The landowner was "to be with the people, on his horse, and armed" to serve the Tsar's sovereignty and to ensure protection of the Orthodox kingdom from external threats. The peasant, in turn, would "feed" the nobleman during his long-term military service that could risk his life. Nobles before the 1762 decree, which expanded their liberties, were obliged to serve in the army for 40 years—although in practice, service was less. Russian society was reproduced exactly as a servile state, in which each estate had its own taxes to bear, fulfilling its economic or social function in the state and church system (Liubavskii 2002). The usual Muscovite did not see in the "estate" an exclusive right to private property; rather, this was material provision for the obligation to serve the tsar. As K. S. Aksakov (1998: 157) noted, "there is no private property." In Muscovy, the legitimacy of property was determined by the will of the sovereign and personal duty of each carrying his duties. This was one foundation for the fragility and instability of property rights in Russia. Inter-class exchange in the performance of mutual obligations required a necessary level of trust between classes and the stability of economic fields.

Gift exchange between estates as a guiding principle in Russian history was thoroughly undermined in the era of the Petersburg Empire, after Peter the Great. Territorial expansion required new, more severe forms of social control over vast territories. Russian emperors and empresses strove to realize power through the nobility and made a series of fateful concessions. Decrees by Peter III and Catherine the Great liberated the nobles from obligations, such as compulsory military service to the state. In her "Manifesto," Catherine the Great stressed, "We intend to preserve landowners with their possessions in an unbroken manner, and to keep peasants in due obedience to them" (Liubavskii 2002: 463). As a result, landed estates became embryonic private property, and peasants were enserfed to noble landowners. The nobility as a new dominant group exploited serfs and provided the state with control over vast territories. The reverse side of strengthening social control was a growing distance between top and bottom of the empire and a deformation of social justice. At one time, even a partial attachment of peasants to the land, legalized by the Council of Aleksei Mikhailovich's Code of 1648, was perceived in popular consciousness as one form of universal tax (service to the state) that all estates had to bear. "While there was obligatory service of the nobility, serfdom was justified in the eyes of the state and the nobility, and even of peasants themselves" (Liubavskii 2002: 466). The "revolution of the Nobility" in the eighteenth century weakened legitimacy of landlord power over peasants, who, unlike the nobles, continued to pay their compulsory tax in the landlord economy. The strategy of inter-consecutive reciprocation lost its quality as a systemic regulator. Reciprocity began to be sectoral in nature and manifested in certain areas of economic life and above all in peasant communities.

Fields of Gift Exchange in Peasant Communities

In the economic life of Russian peasants, *pomoch* played a significant role as a form of gift exchange embedded in peasant communes. *Pomoch* is an umbrella term for various forms of exchange, the main feature of which is reproduction of human life by reproduction of the group. Gift exchange participants were not unaware of long-term interests of the community and the coincidence of these interests with their goals, that is, the typical moral economy. From this emerged such characteristics of gift exchange fields as length in time, asymmetries in exchange of goods and services on the basis of reciprocity, and mandatory personification (rather than anonymity). Moss (2011), one founder of social anthropology, saw three duties running through this practice: the obligation to give, the duty to accept, and the obligation to compensate. The last was the most interesting, as it provided the functional equivalent of legal and economic norms compelling a backward or archaic type of society to respond with a gift for a gift. Various gift strategies were driven not so much by economic as by social logic in its institutional, communicative, religious, and ethical modalities. According to economic anthropologist Marshall Sahlins, any exchange embodies a certain coefficient of friendliness that cannot be understood apart from its social dimension (Sahlins 1999: 169). This idea correlates with Polanyi's claim that economies provide livelihoods, and so are embedded in values, motives, and policies (Polanyi 2010: 42).

The practice of gift exchange (*pomoch*) in the peasant communities of the nineteenth century was driven not only by strategic considerations of subsistence survival (traditional moral economy) but also by cultural fields in which the peasant ethic of cooperation and Christian moral values had some force. Collective help to others in the commune occupied an important place in village life and had a mass nature. "It happened that the world sent healthy people to fire the stoves, cook food, and take care of children in those courtyards where all working family members were sick. To widows and orphans the community provided assistance by labor of the commune: during sowing, harvesting, mowing" (Gromyko and Buganov 2000: 263). In the province of Pskov, according to data for 1879, the commune provided labor assistance to a peasant who suffered from fire, crop failure, and cattle death (Gromyko and Buganov 2000: 265). Widespread help in the form of labor and money was provided by the commune to those who had suffered from fires. "According to data from the Cherepovets *uezd* [county], if a fire hit the whole village, the commune of another village would help with rebuilding" (Gromyko and Buganov 2000: 264). The help was of a lump-sum nature and was not given to just any impoverished peasant: it was believed that the peasant himself was to blame if things went badly for him.

The informal institution of *pomoch*, defined by custom, morality, and peasant "public opinion" was normative in practice. According to reports from the Ryazan province for 1877, "peasants certainly do not consider *pomoch* as something for themselves, but they realize *pomoch* as a moral obligation so deeply that there is almost no refusal to help." According to data from the Tula province in 1879, "in those rare cases when someone from one village, under the pretext of lack of horses, refuses to participate in aid, the commune does not take any punitive measures," but general opinion would condemn such behavior, "and rarely does anyone go against the commune" (Gromyko and Buganov 2000: 265). *Pomoch* was both *gratis* and for a price. Gratuitous help was provided to a commune member in an extremely unfavorable situation (fire, illness, widowhood, being orphaned). Reimbursed aid ended in refreshments and, according to peasants themselves, at times "hiring help would be more expensive." Reimbursement could also be monetary: victims of a fire got back on their feet and might repay the commune.

Two sources of the logic of cooperation in fields of peasant agriculture were the state and the peasant moral economy. Russian economist A. V. Chaianov suggested a core set of principles made up the rural moral economy. Goals of the peasant economy lie beyond the limits of market motives and were expressed in the "labor-saving balance," which established a balance between severity of labor and family consumption (Chaianov 1989). James C. Scott (1992) noted a peasant "ethic of survival," mutual assistance related to the need to support the less well-off strata and to insure against ever-present risks (hunger, climate change). L. V. Milov (1998) claimed the northern continental climate led to arrhythmic work and mobilization in the short summer, and nature and climate conspired to keep agricultural productivity low. By the end of the eighteenth century, average yields of cereals in European Russia were two to three collected grains per sown crop; in Western Europe of the sixteenth century, grain yields reached five harvested grains per sown (Milov 1998: 189). This gap was not due to developmental delays in agricultural technology: the net yield of plant biomass in Russia was 2–2.5 times lower than in Western Europe (Milov 1995: 15). Under such conditions, collective insurance of risks of uncertainty and ensuring stable long-term reproduction of peasant life (intergenerational transfer) was a necessary condition for survival. Life in peasant communities demanded relatively equal distribution of land and high fertility and closely related and structurally complex households that united three generations" (Hoch 1993: 86). The main driver of participation in *pomoch* routines was peasant confidence in the fact that neighbors would provide aid. Participants in *pomoch* received an insurance of sorts against future risks and guarantees of social transfers from neighbors. The main principle of the gift exchange strategy was visible in mutual obligations of cooperation, which might seem nonequivalent but which ensured stability and long-term reproduction of the peasant world.

The normative nature of gift exchange via labor and services was sanctioned by the religious field of Orthodoxy. Studies of Orthodox economic ethics note several main components to these ethics: the priority of spiritual and moral values over economic goals; the secondary nature of labor in relation to prayer; a negative attitude to wealth and property in and of themselves; the value of collectivism; and orientation to an equitable distribution of benefits rather than efficiency of production (Koval 1994; Rumiantsev 2005; Simonov 2005; Tiugashev 1998; Rumiantsev and Raskov 2008). The "non-market" nature of Orthodox ethics and orientation to collective solidarity contributed to the formation and consolidation of redistributive norms in peasant communes. The Gospel call that "whosoever shall compel thee to go one mile, go with him two. To him that asketh of thee, and from the one who wants to take from you do not turn away" (Matthew 5: 41–42) was perceived as the guiding religious principle of peasant cooperation.

The state as well played a role in propagating similar norms of economic practice, including redistribution, which was connected with uncertainty of conditions of economic activity. Risks to agriculture meant the need for regular collection of taxes to create insurance funds that could be used in case of sudden unfortunate events (crop failure, famine, fire). All this required a strong state with authoritarian undertones generated by fields of religion and culture. In the economic context of the fields of culture, resources not expended on production can be invested in acquiring intangible symbolic capital. The social function of such expenditures from the redistributive Center is to unite society and limit aggressive competition for the sake of nationwide cooperation, thereby strengthening the authority of that Center. According to Fernando-Aramesto (2009: 261), "Eastern despots" that made their people build irrigation systems knew this could unite people. The despot follows an ethical code of the benevolent person who interferes in others' affairs and organizes neighbors for demonstrations in defense of rights. Societies joined by common festivities, and leaders who use manifestations of generosity, find ways to use intensive farming and food reserves. Such "symbolic expenses" might seem irrational in today's economic theory, but such rituals had their own rationality for long-term economic perspectives.

Severe natural, geographical, and climatic conditions of economic management in the Russian empire generated labor and consumer austerity for the majority of the population. There were more than 90 days of fasting in the Orthodox year. "Our people are worn on weekdays," as V. I. Dahl once said sadly. The lack of rest under such conditions led to physical and mental exhaustion. Holidays as a symbolic waste of the aggregate labor allowed people to escape routine briefly, participate in a solemn renewal of life, "touch" the transcendent, and restore labor potential. In the traditional peasant and urban worlds, ceremonial rituals were not a conditionality but an experience of the cycle of life—a continuing tradition involving ancestors, contemporaries, and descendants. For an ideocratic state such as Russia's, a significant part of holidays was transferred to the Church and had a church-state character. The law of June 2, 1897, "On the duration and distribution of working hours in factory institutions and the mining industry," established obligatory "non-residential" holidays (according to the Julian calendar), most of which were Orthodox except for royal days and the New Year (Aleksandrov et al. 2005: 616–646). Workers of "heterodox" religions were allowed not to include in their calendars holidays not honored by their religions, but it was possible to add other holidays to the schedule of non-residential days in accordance with the law of their faith. Sundays, Orthodox holidays (the beheading of John the Baptist, the icon of the Kazan Mother of God, etc.) were considered days off, as were other holidays for those with particular positions on the Table of Ranks, as well as days declared by the tsar's authority. Orthodox Christians were supposed to prepare themselves for holidays by prayer and fasting, to visit the church and take communion, and to help the needy.

In the peasant world, church and state holidays supported "faith in the Tsar" and a symbolic field of imperial power. Peasant ideas about power stemmed from religious belief in its sacral origin: "Any power from God," "Obedience to the authorities is obedience to God," and "there is in the Scriptures that the authorities are to be obeyed" (Dahl 2000: 154, 151–152).

The symbol of state administration and of the whole order in the state was the tsar's figure: "It is awe-inspiring, terrible, but without the tsar is impossible" (Dahl 2000: 152). It is from the principle of the sacredness of tsarist power that peasant ideas about the legitimacy of Russia's institutional system of power as instruments of the tsar's will flow. If the tsar enjoyed absolute confidence in the peasant world, then distrust of the state bureaucracy, especially petty bureaucrats whom peasants faced in everyday life, was a traditional feature of their worldview. The scholar of pre-Revolutionary life N. M. Astyrev noted that in the peasant environment, there was a centuries-old belief that all bureaucrats had nothing in common with the peasant people. In fact, two lay categories, *surtuchnik* (official) and *lapotnik* (peasant) were interrelated elements with no common interests, according to the peasant. If the bureaucrat is at least the most scrutinizing boss, like a clerk of the police department, every sober peasant would try to conceal his thoughts, desires, and hopes from the eyes of this lowly representative of officialdom when possible (Astyrev 1886: 30–31).

With the alienation of peasants from the bureaucratic environment, the sacralization of the tsar ensured legitimacy of power and some integration of society in the Russian empire, and a constant symbolic relation between people and tsar was carried out through the field of religious culture. During the Liturgy every Sunday, obligatory for every believer, the name of the reigning monarch was pronounced five times, and the paternalistic power of the Russian monarch was claimed to have historical grounds as the sole parental authority for the extended peasant family. Because all levels of power in the Russian empire were considered legitimate due to their submission to the authority of the Emperor, the strength of the entire system of state redistribution was determined by public confidence in the highest, royal authority of the government.

For insight into peasants' economic motives, let us turn to testimonies of researchers who studied peasant farms directly in Russian villages. Peasant S.T. Semenov, who became a writer and Pushkin Prize laureate, noted egoism in peasant behavior: "Peasants in public life are far from philanthropists. Everywhere they pursue benefits and hardly allow anyone to use the public good for free" (Semenov 1915: 75). Landowner A. N. Engelhardt observed village life and wrote of peasants' individualism: "Envy, distrust of each other, subjugation of one under another, humiliation of the weak before the strong, worship of wealth—all this is strongly developed in the peasant environment ... Each peasant, if circumstances favor, will in the most excellent way exploit anyone else, whether peasant or master, he will squeeze out the juice from it, exploit its need" (Engelhardt 1987: 521). Engelhardt also stated that "while [the peasant] is working the land and is still not a real *kulak*,[1] he does not think of taking everything for himself, does not think how good it would be if everyone else were poor or in need, he does not act in this way." The peasant, he continued, "rejoices at his creations, his livestock, his hemp, his bread. And it is not only because it gives him so many rubles. He expands his farming not only for profit, he works until he is tired, he does not sleep, he does not eat." The author drew attention to peasants' Christian humanism toward the disabled, beggars, and criminals (Engelhardt 1987: 521–522).

We see a combination of different motives in peasant economic behavior. Of course, they correlated benefits and costs and their behavior can be called rational. Yet peasant goal-setting cannot be called entirely market-oriented, as it was aimed not at maximizing income but instead at long-term reproduction of large families. Further, peasant motives contained individualism and an attempt to benefit from exploiting others. At the same time, daily life was based on help—the relationship of gift exchange within the community. In their attitude to unprotected strata of the population, peasants were guided by principles of Christian ethics. How to connect such a variety of motives? Let us turn once again to Engelhardt, who summed up his long-term observations of peasant economic behavior: "according to the concept of the peasant, everyone thinks for himself, about his personal usefulness, each person is selfish, only the commune and the tsar think about everyone, only the commune and the tsar are not selfish. The tsar wants everyone to be equal, because he loves everyone the same way, everyone feels equally sorry for him. The function of the tsar is to equalize all" (Engelhardt 1987: 540–541). Peasant economic behavior seems to resemble Max Weber's value-rational social action within the community, determined by conscious reproduction of cooperation and belief in social justice of the institution of the commune. Outside the commune, peasants focused on an individualistic model of behavior aimed at maximizing their benefit. The peasantry as a whole had equal social ideals, which, in their opinion, the Orthodox tsar should have realized in the state. Thus, economic fields of peasants were heterogeneous and depended on the type and scale of social interactions.

After the liberation of peasants from serfdom in 1861, and after Stolypin's agrarian reform (1906–1917), capitalist relations penetrated the agrarian sector, leading to an increase in social differentiation and escalation of peasant revolts against landowners and kulaks (Mironov 2003; Osipova 1974; Tomsinskii 1932). The theme of agrarian reforms and social change in pre-revolutionary Russia requires special consideration. We are interested in interaction of traditional peasant economic fields with bourgeois and socialist paradigms. Equalizing social ideals and norms of gift exchange in conditions of agrarian capitalization in the beginning of the twentieth century led to the fact that "apostates"—peasants who stood out from the commune and became independent rural masters—were killed or expelled from villages and their houses burned. Newspapers at the time wrote that "reprisals because of land allotments occurred all over Russia" (Semenov 1915: 127–129). Yet it is problematic to speak of peasant communism. Let us turn to the testimony of I. F. Nazhivin, a famous writer, publicist, and public figure whose views were close to socialism and who followed the teachings of Tolstoy. In 1918 he traveled through Russian villages and observed peasants' reactions to the Revolution. He drew attention to negative attitudes toward socialization of the land—that is, to the free transfer of all landed estates to peasant ownership. The peasants said, "The land is for ransom and nothing else. By robbery, for nothing … If you take, then for money, so that it's legal and irrevocable." The farmer wanted to obtain land for ransom and wanted firm state power (Nazhivin 2016: 25).

Peasants refused to consider merchants, traders, and "bourgeoisie" to be enemies, as the private trader (in contrast to socialist cooperatives) "knows his business" and sells his goods cheaper and without delay (Nazhivin 2016: 47). Nazhivin noted with regret that when trading with urban clients, peasants could be greedy speculators. Claims that workers in Russian cities were dying of hunger led only to new speculative rises in prices for agricultural goods. In Nazhivin's opinion, during the war years and the first year of the Revolution, peasants speculating in trade with cities "hid in the ground" billions of rubles (Nazhivin 2016: 89). Peasants were indifferent and disinterested in problems of fair land ownership in other regions, and they sought only to increase land holdings for their communes. For them, "there was no Russia—there is only the parish" or commune (Nazhivin 2016: 77). At the same time, peasants' attitudes to tsarist power in a Russia engulfed by revolution did not fundamentally change. The typical peasant in 1918 wanted "to choose the tsar as quicker and wiser, so as to live again in order, in the old way, in a good way" (Nazhivin 2016: 93). The author noted that in the ancient village of Bogoliubovo, near the city of Vladimir, armed peasants did not allow revolutionary soldiers to demolish a monument to Aleksandr II (Nazhivin 2016: 67). Empire in the form of power of the Orthodox tsar remained a factor of social integration in archaic fields of Russian peasants.

Empire and Capitalism: The Impact of Large Projects on Economic Fields

Religious fields in Russia had their own peculiarities. The Puritan idea of a successful entrepreneur as God's chosen was not typical in the contemplation of traditional Russian merchants and industrialists. In Russia, wealth was perceived through other moral and social lenses. If one knew that a person was wealthy, then that person might face social prejudice and suspicion, even if there was no doubt about the legality of how that wealth was acquired. As one Russian proverb notes, "from the labors of the righteous you cannot make palaces of stone." The rich themselves felt the most acute sense of illegitimacy of their wealth. In their midst, an existential sense of responsibility for property and increased religiousness, and a penchant for repentance were cultivated (Rumiantsev 2008). One historian of Russian entrepreneurship who was a native of the merchant class, P. Buryshkin, recalled: "We were told about riches that God gave to use and about which He will demand a report" (Buryshkin 1991: 113). One means to alleviate any psychological stress of wealth and to justify one's wealth before society was to contribute to charity. Charity of such important entrepreneurs as Tretiakov, Tsvetaev, Mamontov, and Morozov is well-known in Russia.

Ulianova (2014) showed that the overwhelming majority of less well-known entrepreneurs also contributed to the construction of churches, hospitals, schools, theaters, and museums. Yet much of this was due to the fact that many of these entrepreneurs were, as a rule, Old Believers living in Moscow and the central regions of Russia.

The question of peasant entrepreneurs, especially those who lost their place in the first half of the nineteenth century, is of interest. An ability to conduct economic affairs allowed many serfs to begin their own endeavors and even new industries. According to modern estimates, most enterprises of the Russian textile and cotton industry were founded by peasants of the Moscow and Vladimir *gubernii* from the 1800s to the 1830s (Raskov 2012: 56–57). Fields of peasant entrepreneurs were based on common religious meanings: as a rule, they represented a persecuted religious group of Old Believers. Rich merchants helped persecuted communities, and in turn members of Old Believer communities often worked at enterprises of their co-religionists, which ensured the loyalty of workers and high labor discipline. Entrepreneurs "were in the center of the spiritual and material life of the Old Believer communities." Over time, new generations of entrepreneurs turned to official Orthodoxy in the interests of business, and others "abandoned profitable enterprises, preserving the traditional and measured way of life, of which business is a part" (Raskov 2012: 126). We should be cautious when talking of a "spirit of capitalism" and formal rationality with regard to the genesis of peasant entrepreneurship among Old Believers. In this case, it is appropriate to emphasize the existence of common religious meanings and cultural understanding based on a community outlook.

In the late nineteenth and early twentieth centuries, capitalist development followed a "Petersburg business style" of state procurement and speculation, railway construction, and close links with vertical power (Raskov 2012: 260–261). According to a contemporary, "Russian banks ... operated almost entirely from the means of the State Bank. The administrators of these banks under a fiction of electivity were essentially officials of the Ministry of Finance" (Kolyshko 2009: 132). Intertwining of state and private interests bred "distributive capitalism" of the distribution of state orders, subsidies, concessions, preferential loans, and so on. Strategies of large private businesses were oriented to affiliation with the state and obtaining preferences. This was not new: the Russian state always was a "distributive state" that "distributed" resources to economic agents, replacing market coordination with redistribution (Bessonova 2006). The large-scale redistribution of the aggregate social product made it possible to realize long-term social and geopolitical strategies of the imperial state. In return, there was a long dependence of economic agents on the state and an increase in risks of opportunistic behavior and corruption.

Among the different facets in the development of Russian capitalism, let us consider the railway megaproject for the Russian empire because of the significance of "Great Projects" in the economy of empires, as Arrighi (2010) noted with his separation of "capitalism" and "territorialism" as opposing modes of governance. Unlike capitalist states, "territorial rulers identify their power with the extent and population of their possessions and regard wealth/capital as a means or by-product of the desire for territorial expansion" (Arrighi 2010: 75) Following this, we can assume that, from an economic point of view, the imperial logic of territorial expansion due to redistribution of resources forms a special value or benefit: large infrastructure projects for the development of territories. This imperial logic of territorial expansion from center to periphery was embodied in major infrastructural projects for economic development of territories. Among such projects, the construction of dams and irrigation communications in Eastern despotism, pipelines and roads of the Roman Empire, the Great Wall of China, and the unique system of postal and transport communications in the Mongol empire were of such importance. Such projects involved investment of available resources in exchange for benefits of future development. Due to redistribution of resources from the periphery to the center and taxation of the entire population, sectors with potential for long-term social and economic development were most likely to receive financial support. As a result, the economic system of empires produces special benefits: infrastructure projects, to which the standard criterion for dividing economic benefits into goods created by the market and the benefits provided by the public sector of the economy is not applicable (Rumiantsev 2005: 256–264). Of particular importance is the scale effect from the unification of diverse economic structures into a single economic complex of empire. For the economy of the greater spaces of Continental empires, an effective system of transport communications, linking local markets and accelerating economic turnover, was of key importance.

In the second half of the nineteenth century, the dominant direction for development of the Russian economy, which would alter its entire appearance, was railroad construction. For 1861–1881, the length of railways in Russia grew from 1600 versts to 23,100 versts. This would grow to 53,200 versts by 1900, after the purchase of about two-thirds of the railways by the treasury in the early 1880s. By the 1880s, investment in railway construction reached 1.8 billion rubles, of which 1.4 billion rubles was from state funds; by 1890 investments in the railway network amounted to 4.27 billion rubles, of which foreign loans accounted for 3.2 billion rubles (Soloveva 1975). In 1860–1870, the railways were built by the private sector by distributing state subsidies to entrepreneurs, as well as gratuitous loans and guarantees of profitability—if the railway did not bring profit, the state paid the owner an agreed premium. After the treasury purchased the rail line, construction and operation were financed directly from the imperial budget. The next megaproject from 1891 to 1916, the Trans-Siberian Railway was financed through state funds and foreign loans. Expenses for the Siberian railway were about 1.5 billion rubles—approximately equal to internal revenues of the empire from taxes for previous 15 years of construction and the amount of all foreign loans during railway construction (Gilko 2012: 92–97).

All this investment brought a system of transport arteries that became the driver of industrial and social development. The Trans-Siberian Railway connected European Russia with the Urals, Siberia, and the Far East, and from there to the northern Korean peninsula, China, and Mongolia. The railway facilitated transportation to western ports central for trade with European markets. In the second half of the nineteenth century, Siberia was a backward outpost with a small population, but railway construction opened up opportunities for agrarian and industrial development. Thanks to transcontinental transport beyond the Urals, new cities, towns, and industrial centers appeared, and the railway fleet of locomotives and wagons, stations, warehouses, and depots was a fast-growing industry for the Siberian economy (Mogilevkin 2005: 165–174). From the railroad's stimulus came social change. Agricultural regions could export products, contributing to an increase in marketability of farms and increased income for peasants-turned-farmers. The rural population began to migrate to cities, providing labor for capitalist industry.

Yet the unavoidable shadow companion of such projects is the aggravation of distributional conflicts. The redistribution of resources to priority industries through the transfer of resources from other industries created imbalances in the movement of material and financial flows, such that economic development could seem close to a zero-sum game. Railway construction in the second half of the nineteenth century contributed to this contradiction between beneficiaries and donors of this project. Internal savings could not become the main source for financing railways, given a shortage of necessary capital and the inadequacy of farmers' incomes for this task. As Minister of Finance, Sergei Witte suggested, issuing paper money ("Siberian" rubles) would help pay for railway construction in Siberia, but more cautious financiers dissuaded him from this step (Gilko 2012: 92–97). Perhaps cheap money could increase business activity in the country, accelerate economic growth, and improve incomes and consumer demand—in the spirit of Keynesianism, it would cause an increase in domestic savings and investment. But a different path was chosen. To finance the construction of the railroad, the state used taxes, many introduced or increased from the late 1870s to the early 1890s. In addition, in the first ten years of the construction of the Trans-Siberian Railway, an apartment tax was introduced; taxes on fishing and indirect taxes on sugar, tobacco, many imported goods, kerosene, and matches were strengthened, and stamp duty was increased. The growth of the tax burden on the peasantry—the main tax-paying estate of the empire—was such by 1903 that Witte was forced to admit that direct and indirect taxes had reached their limit, and so the state turned to foreign loans. In 1893, there were 2.1 billion rubles of foreign capital, of which 70.6% were state railway loans (Gilko 2012). Thus, railway construction was based on a centralized redistribution of budgetary resources from the agricultural sector to the industrial sector due to the growth of taxes and government foreign loans—that is, the transfer of the state's debt to future generations.

The construction of railroads was made possible by "donations" of peasant farmers and landlords, as well as the next generation that would have to pay back foreign loans. Who benefited from the Russian railway project? Speculators. "Private railway enterprises operated at the expense of the treasury. In such conditions, when the state guaranteed profits and prevented losses, when the favor of the powerful could replace millionaire capital, favoritism, and corruption blossomed in a magnificent color" (Ananich and Ganelin 2000: 55). A common railway scam ran as follows. The state announced a tender for construction of a segment of line. Immediately a joint stock company consisting of Russian and foreign entrepreneurs was created. High-level imperial dignitaries were offered bribes, such as including them among shareholders or simply paying a large sum of money. The dignitaries, in turn, ensured that company would win the tender. A little-known entrepreneur, S. S. Poliakov, paid a record bribe of 7 million rubles to obtain the concession for building the Transcaucasian railway (Laverychev 1974: 55–56). After receiving the concession, the company issued and placed shares on the stock exchange. Since profitability of shares was guaranteed by the state, these were in high demand and brought considerable profit to the founders. In the first year, the company received guaranteed payments and did not build anything and, under the pretext of difficulties, sought a large subsidy from the state so that construction could finally start. The difference between the size of the subsidy from the state and the cost of construction was appropriated by "effective owners." The entrepreneur's calculation, an eyewitness noted, was to build cheaply and save as much money as possible. The state paid 5% of the capital (more was difficult to obtain), even if the venture was unprofitable (Chicherin 1934: 89). After the purchase of two-thirds of the railways by the treasury in the early 1880s, the situation with corruption and misappropriation of budgetary funds improved.

Another distributional conflict was connected with tariff policy. The cost of transporting agricultural products was differentiated and calculated inversely with distance. The economies of Central Russia were clearly at a disadvantage. One landlord from the Ryazan province noted the consequences of this policy for commodity farms: "The scheme of differentiated tariffs was such that the greater the distance, is cheaper the cost of transportation ... Under certain conditions, payment for a pood[2] of cargo transported to the Baltic and Black Sea ports from local places was more expensive than transportation of similar goods from more distant locales" (Kolyshko 2009: 116). Due to the decline of productive farms of the Central Gubernias, economic growth in Siberia intensified and made the territory a major exporter of agrarian and raw materials.

Gains available to railway swindlers were not available to peasants, landowners, or other entrepreneurs. Imperial redistribution in the course of railway construction gave birth to a new stratum of capitalists and millionaires, whose wealth was immoral and parasitic to these other strata more firmly embedded in other fields of meaning. On the impartial assessment of Prince V. P. Meshcherskii, in 1873, "There was an abyss of merchants and financiers whose existence until now still one could not suspect, and they did not even think of preparing for such an activity" (Carpi 2012: 111). While a new industrial class and corrupt officials became wealthy, peasants, landlords, and entrepreneurs (mainly from Central Russia) had to bear the new tax burden. While the railway project was the driver of industrial capitalism and facilitated the flow of resources from agriculture to industry, it also contributed to the deformation of economic fields. The redistributive conflict between capitalists and elite beneficiaries of the railway project and the population at large intensified. Imperial authorities used the power of the state for industrial modernization, but traditional entrepreneurial ethics were undermined by the spirit of profit and speculation cultivated among capitalists and the administrative elite of the empire. The social contradictions of the Russian Empire significantly increased.

Conclusion

This excursion into religion, empire, and economic fields only begins to scratch the surface of this important topic. A few conclusions reached in this discussion, which could be possible launching points for further work, are as follows. I draw these out by summarizing this history of institutional integration of Russia, which had three stages.

In the first stage (the second half of the fifteenth century to the beginning of the eighteenth century), the principle of integrating actors formed through inter-class gift exchanges. The significance of gift exchange was associated with seeking stable conditions for social interaction under conditions of growth of the Muscovite kingdom. In conditions of territorial expansion and wars requiring mobilization, market and fields of state power could not provide necessary stability in Great Spaces of Russia. This chapter considered the negative impact of natural conditions on labor productivity in agriculture. It follows that in Russia there was a small amount of surplus product, and opportunities for creating a significant bureaucratic apparatus were limited. Russia was a "non-bureaucratized state." Further, the vast expanse reduced the speed of market turnover and limited the influence of market fields on coordination of actors. Non-market relations of gift exchange went beyond boundaries of separate groups and created conditions for actors to interact in a growing empire. Moreover, relations of solidarity and interaction on an ethical basis were supported by the Orthodox Church. Muscovy's economic fields were normative relations of gift exchange between estates. Legitimacy of ownership was determined by the state as material provision of a "tax" as compulsory service to the tsar. The power of the tsar depended on established fields of religion and culture and sought to maintain conformism in fields of religious and social institutions. The hegemony of Orthodoxy ensured enough institutional integration of a poorly structured Russian society through sacralization of tsarist power. Fields of religion normatively enshrined the tsar's sacral power and supported nationwide inter-social transfers.

The sacred power of the tsar in a centralized state required a balance of interests of the main actors (peasantry, boyars, nobility, merchants). The social world received ethical restrictions and cultural understanding on the religious basis of Orthodoxy. The Church sought to use its authority to equitably distribute the burden of public service among estates, condemning inequality and the growth of private wealth. For example, in *The Ruler* (*Pravitelnitsa*), the monk Ermolai-Erasmus, a well-known church writer of the sixteenth century, developed a system of a just class state, calling for restrictions of land ownership by landlords and the reduction of peasants' obligations (Rezvanov 2016). The fields of royal power and the field of Orthodoxy did not coincide with fields of individual economic actors and social groups: they were autonomous, national in character, and ensured long-term social interactions of the main actors.

The second stage covered the period from the beginning of the eighteenth century until 1861, the beginning of bourgeois reforms in Russia.

This stage was characterized by nationalization of Russian life under the reforms of Peter the Great, especially strengthening of state control over social and economic fields (Anisimov 1989). The religious field underwent "state secularism" and the clergy became a "frightened estate" (Florovskii 1991: 89). It is significant that in 1721, when Peter officially proclaimed Russia an empire, the title of Patriarch, as elected by the council of church elders, was liquidated, and direction of church affairs was transferred to the bureaucratic institution of the Holy Synod. Territorial expansion required tough forms of control. The empire exercised power by distributing privileges to nobles in return for maintaining control over vast territories. Nobles were freed from compulsory service to the state, their land ownership became unconditionally private, and peasants became serfs dependent on landlords. Fields of state power lost their autonomy and a symbiosis with the fields of the nobles developed.

Peasant communes were reproduced through fields of cooperation. A particularly important practice was *pomoch*, asymmetric exchanges based on reciprocity. Assistance arose as a specific form of moral economy in connection with risks in natural conditions. The normality of peasant reciprocity was sanctioned by the religious field, in which solidarity, collectivism, and orientation to justice, rather than efficiency, were cultivated. The legitimacy of redistribution was supported by church-state institutions. In conditions of alienation of the peasants from the state bureaucracy, the integration of rural economic fields and communities into a broader institutional system of the empire was a result of peasants' acceptance of the sacred nature of tsarist power.

In the third stage (1861–1917), capitalism expanded. The ruling elite pursued accelerated industrialization on capitalist principles (Ananich and Ganelin 2000). Industrialization was carried out at the expense of agriculture, which was taxed 3–3.5 times higher than industry. This limited the domestic market and reduced consumer demand by the population (Kitanina 2011: 49–50). Peasants and landlords suffered losses, but incomes of new actors—speculative banks and industrial enterprises—grew. At the same time, new capitalist actors, such as bankers, were created artificially and were dissociated from the norms and logics of earlier economic traditions (Epshtein 2011). A class of elite capitalists emerged thanks to state subsidies and preferential credits—a new kind of gift exchange, perhaps. The state tried to incorporate industrial capitalists and bankers into political and cultural fields by conferring honorary status on them, for example, noble ranks and orders (Ananich 1991: 134–137).

In sum, imperial policy was ambivalent. The ruling elite tried to preserve cultural identity and power of the Empire. Under the last emperor, Nicholas II, much was done to maintain the cultural hegemony of Orthodoxy. In his reign, 30,300 temples were built at the expense of the state—more than by any other Russian ruler. The number of saints grew more quickly than any time since the eighteenth century, and church institutions, even if subordinate to the state, were employed actively (Tsypin 2012). On the other hand, the government distributed gifts to new bourgeois actors and created conditions for accelerated development of capitalist institutions. This undermined cultural understandings between actors, as norms of Orthodox ethics and gift exchange were not relevant to the new "spirit of capitalism." As a result, imbalances developed between capitalist and traditional actors. In economic fields, chaotic trends began to increase, which, in particular, manifested themselves in an escalation of peasant riots after the bourgeois reforms of 1861. In sum, the influence of empire on Russian capitalism was embedded significantly in the project of railway construction. This is not unique: transport communications played a special role in economies and politics of Continental empires. Further, such politics of empire contributed to the creation of a transport infrastructure of industrial capitalism, growth of industry, and acceleration of the movement of private capital. At the same time, economic fields were deformed due to social and economic imbalances between beneficiaries and donors of railway construction, which further aggravated conflicts in Russian society.

In terms of methodology, this discussion was grounded in a synthesis of Fligstein's field theory and Polanyi's political economy of exchange, with provisions from economic anthropology and moral economics. Such an extension of research is necessary when studying the influence of religion and power on economic fields in pre-capitalist societies, in which actors' relations are often non-market. Of course, in order to fully cover the influence of Empire and Orthodoxy on economic fields in pre-Revolutionary Russia, more extensive research is required, going beyond the scope of a single chapter to cover all economic fields and stages in Russian history.

Notes

1. *Kulak* (fist) was a term for wealthier entrepreneurial peasants.
2. A pood was a pre-Revolutionary Russian unit of weight. One pood was approximately 16.38 kilograms.

Works Cited

Aksakov, K.S. 1998. Khomiakov A. S. *Khomiakovskii sbornik. T. 1*. Tomsk: Volodei.

Aleksandrov, V.A., I.V. Vlasov, and N.S. Polishchuk. 2005. *Russkie*. Moscow: Nauka.

Ananich, B.V. 1991. *Bankirovskie doma v Rossii 1860–1914 gg. Ocherki istorii chastnogo predprinimatelstva*. Leningrad: Nauka.

Ananich, B.V., and R.Sh. Ganelin. 2000. *Sergei Iulevich Witte i ego vremia*. St. Petersburg: Dmitrii Bulanin.

Anisimov, E.V. 1989. *Vremia petrovskikh reform*. Leningrad: Lenizdat.

Arrighi, G. 2010. *Dolgii dvadtsatyi vek: Dengi, vlast i nashego vremeni*. Moscow: Territoriia budushchego.

Astyrev, N.M. 1886. *V volostnykh pisariakh. Ocherki krestianskogo samoupravleniia*. Moscow: Tipografiia V. V. Isleneva.

Bessonova, O.E. 2006. *Razdatochnaia ekonomika Rossii: evoliutsiia cherez transformatsii*. Moscow: Rossiiskaia politicheskaia entsiklopediia.

Bishop Nicodemus [Milash]. 1897. *Pravoslavnoe tserkovnoe parvo*. St. Petersburg: V. V. Komarov.

Buryshkin, P.N. 1991. *Moskva kupecheskaia*. Moscow: Byschaia shkola.

Carpi, G. 2012. *Dostoevskii—economist. Ocherki sotsiologii literatury*. Moscow: Falanster.

Chaianov, A.A. 1989. *Krestianskoe khoziaistvo*. Moscow: Ekonomika.

Chicherin, B.N. 1934. *Vospominaniia. Zemstvo i Moskovskaia duma*. Duma: Sever.

Dahl, V.I. 2000. *Poslovitsy russkogo naroda*. Moscow: Eksmo-Press.

Dmitriev, M.V. 1990. *Pravoslavie i Reformatsiia: reformatsionnye dvizheniia v vostochnoslavianskikh zemliakh Rechi Pospolitoi vo vtoroi polovine XVI v*. Moscow: MGU.

Engelhardt, A.N. 1987. *Iz derevni. 12 pisem. 1872–1887*. Moscow: Mysl.

Epshtein, E.M. 2011. *Rossiiskie kommercheskie banki*. Moscow: Politicheskaia entsiklopediia.

Erasov, B.S. 2002. *Tsivilizatsii: Universalii i samobytnost*. Moscow: Nauka.

Fernando-Aramesto, F. 2009. *Tsivilizatsii*. Moscow: ACT MOSKVA.

Fligstein, Neil. 1990. *The Transformation of Corporate Control*. Cambridge: Harvard University Press.

———. 2001. *The Architecture of Markets*. Princeton: Princeton University Press.

Florovskii, G. 1991. *Puti russkogo bogosloviia*. Vilnius: Vilniuskoe pravoslavnoe eparkhialnoe upravlenie.

Fursov, A.I. 2001. Russkaia vlast, Rossiia i Evraziia: Velikaia Mongolskaia derzhava, samoderzhavie i kommunizm v bolshikh tsiklakh istorii. *Russkii istoricheskii zhurnal* 4 (1–4): 15–114.

Gilko, M.A. 2012. Transsibirskaia magistral: dlia kogo i za chei schet. *"Obshchestvennye nauki v sovremennom mire: sotsio9logiia, politologiia, filosofiia i istoriia": materialy mezhdunarodnoi zaochnoi nauchno-prakticheskoi konferentsii (23 aprelia 2012 g.)*. Novosibirsk: Sibirskaia assotsiatsiia konsultantov.

Go, J. 2011. *Patterns of Empire: The British and American Empire, 1688 to the Present*. New York: Cambridge University Press.

Gorskaia, N.A., L.V. Danilova, et al. 2001. *Sobstvennost v Rossii: srednevekove i rannee novoe vremia*. Moscow: Nauka.

Gromyko, M.M., and A.V. Buganov. 2000. *O vozzreniiakh russkogo naroda*. Moscow: Palomnik.

Hanneman, R., R. Collins, and G. Mordt. 2015. Kak modelirovanie kompaktnoi teorii mozhet vosproizvesti zaputannye puti istorii. In *Makroistoriia: Ocherki sotsiologii Bolshoi dlitelnosti*, 398–424. Moscow: URSS; LENAND.

Hoch, S. 1993. *Krepostnoe parvo i sotsialnyi control v Rossii*. Moscow: Progress-Akademiia.

Kareev, N. 2015. *V kakom smysle mozhno govorit o sushestvovanii feodalizma v Rossii? Po povodu teorii Pavlova-Silvanskogo*. St. Petersburg: LENAND.

Kitanina, T.M. 2011. *Khlebnaia torgovlia v Rossii v kontse XIX-nachale XX veka. Strategii vizhivaniia, modenizatsionnye protsessy, pravitelstvennaia politika*. St. Petersburg: Dmitrii Bulanin.

Kliuchevskii, V.O. 2002. *Russkaia istoriia. Polnyi kurs lektsii. Tom 1. Lektsiia 20*. Moscow: ACT Kharvest.

Kolyshko, I.I. 2009. *Velikii raspad*. St. Petersburg: Nestor-Istoriia.

Koval, T. 1994. Etika truda pravoslaviia. *Obshchestvennye nauki i sovremennost* 6: 55–70.

Lachmann, Richard. 2016. *Chto takoe istoricheskaia sotsiologiia?* Moscow: Delo.

Laverychev, V. Ia. 1974. *Krupnaia burzhyaziia v poreformennoi Rossii (1861–1900)*. Moscow: Mysl.

Liubavskii, M.K. 2002. *Rosskaia istoriia 17–18 vv.: Uchebnoe posobie*. St. Petersburg: Lan.

Marx, Karl. 1989. Razoblachenie diplomaticheskoi istorii XVIII veka. *Voprosy istorii* 4: 3–14.

Milov, L.V. 1995. Prirodno-klimaticheskii factor i mentalitet russkogo krestianstva. *Obshchestvennye naukii i sovremennost* 1: 76–87.

———. 1998. *Velikorusskii pakhar i osobennost rossiiskogo istoricheskogo protsessa*. Moscow: ROSSPEN.

Mironov, B.N. 2003. *Sotsialnaia istoriia Rossii perioda imperii (XVII-nachalo XX veka)*. St. Petersburg: Dmitrii Bulanin.

Mogilevkin, I.M. 2005. *Transport i kommunikatsii: proshloe, nastoiashchee, budushchee*. Moscow: Nauka.

Moss, M. 2011. *Obshchestva. Obmen. Lichnost. Trudy po sotsialnoi antropologii*. Moscow: KDU.

Mulgan, J. 2007. *Good and Bad Power: The Ideals and Betrayals of Government*. London: Penguin.

Nazhivin, I.F. 2016. *Zapiski o revoliutsii*. Moscow: Kuchkovo pole.

Osipova, T.V. 1974. *Klassovaia borba v derevne v period podgotovki Velikoi Oktiabskoi sotsialisticheskoi revoliutsii*. Moscow: Nauka.

Panarin, A.S. 2002. *Pravoslavnaia tsivilizatsiia v globalnom mire*. Moscow: Algorithm.

Pavlov-Silvanskii, N. 1988. *Feofalizm v Rossii*. Moscow: Nauka.

Pipes, R. 1997. *Rossiia pre starom rezhime*. Moscow: Nezavisimaia gazeta.

Polanyi, Karl. 2010. *Izbrannnye raboty*. Moscow: Territoriia budushchego.

Radaev, V. 2013. Predisloviee nauchnogo redaktora russkogo izdaniia. In *Arkhitektura rynkov: ekonomicheskaia sotsiologiia kapitalisticheskikh obshchestv XXI veka*, ed. Neil Fligstein, 6–16. Moscow: Higher School of Economics.

Raskov, D.E. 2012. *Ekonomicheskie instituty staroobriadchestva*. St. Petersburg: St. Petersburg University Press.

Rezvanov, R.I. 2016. Reformatsionnyi proekt Ermolaiia-Erama: skvoz prizmu kontseptsii vlasti gosudarstva Maikla Manna. *Razvitia territorii* 2: 59–64.

Rumiantsev, M.A. 2005. *Religioznye osnovaniia khoziaistvovaniia*. St. Petersburg: NPK ROST.

———. 2008. Pravoslavie i sotsialnaia otvetstvennost predprinimatelstva. *Problemy sovremennoi ekonomiki* 3: 509–513.

Rumiantsev, M.A., and D.E. Raskov. 2008. *Khristianstvo i islam ob ekonomike*. St. Petersburg: NPK ROST.
Sahlins, M. 1999. *Ekonomika kammenogo veka*. Moscow: OGI.
Scott, James C. 1992. Moralnaia ekonomika krestianstva kak etika vyzhivaniia. In *Velikii neznakomets: krestiane i fermery v sovremennom mire*, ed. Teodor Shanin, 201–211. Moscow: Progress-Akademiia.
Semenov, S.T. 1915. *Dvadtsat piat let v derevne*. Petrograd: Posrednik.
Simonov, V.V. 2005. *Tserkov-obshchestvo-khoziaistvo*. Moscow: Nauka.
Skrynnikov, R.G. 1997. *Istoriia Rissiiskaia. IX–XVII vv*. Moscow: Ves mir.
Soloveva, A.M. 1975. *Zheleznodorozhnyi transport Rossii vo vtoroi polovine XIX v*. Moscow: Nauka.
Tiugashev, E.A. 1998. Pravoslavnoe otnoshenie k trudu v zerkale nravstvennogo bogosloviia. In *Chelovek. Trud. Zaniatost. Vyp 2*. Novosibirsk, 40–50. Rossiiskaia akademiia nauk, Sibirskoe otdelenie, Institut filosofii i prava.
Tomsinskii, S.G. 1932. *Krestianskoe dvizhenie v feodalno-krestianskoi Rossii*. Moscow: Zhurnalno-gazetnoe obedinenie.
Tsypin, V. 2012. *Istoriia Russkoi Pravoslavnoi Tserkvi: Sinodalnyi i noveishii periody. 1700–2005*. Moscow: Sretenskii monastyr.
Ulianova, G.P. 2014. *Blagotvoritelnost moskovskikh predprinimatelei. 1860–1914*. Moscow: Forum.
Vernadskii, G.V. 1993. Ocherk istorii prava Rossiiskogo gosudarstva XVIII–XIX vv. (Period imperii). In *Russkaia filosofiia sobstvennosti (XVII–XX vv)*, 271–279. St. Petersburg: Ganza.
Zazykin, M.V. 1924. *Tsarskaia vlast i zakon o prestolonasledii v Rossii*. Sophia: Tipografiia Novaia zhizn.
Zimin, A.A. 1960. *Reformy Ivana Groznogo. Ocherki sotsialno-ekonomicheskoi i politicheskoi istorii Rossii serediny XVI veka*. Moscow: Sotsegiz.

9

State, Markets, and Fields in Russian History

Viktor Ryazanov

The formation of a market economy in Russia has historically been characterized by two principal features. The first is the presence of an established, centralized state and its near monopoly of the organization and governance of social processes. The second is the predominance of political reorganization in the peasant-based agrarian economy, with a noticeable role played by non-economic dependency relations by main producers, which led to a lag in creating market relations in the countryside.[1] This could not but affect the formation of markets as economic fields and clashes of interests by main participants, including capitalists and representatives of the state. Such a classical scheme of market formation as a field of interaction (e.g. Fligstein 1990, 2001) provides an important framework for adding to insights regarding an issue usually analyzed using typical political economy. For national economies with unfinished market transformations and persisting problems of social and economic consolidation, the composition of active participants in

V. Ryazanov (✉)
Faculty of Economics, Department of Economic Theory,
St. Petersburg State University, St. Petersburg, Russia

J. K. Hass (ed.), *Re-Examining the History of the Russian Economy*,
https://doi.org/10.1007/978-3-319-75414-7_9

economic fields should be expanded at the expense of other classes. In Russia, these were landowners and peasants, whose role was very significant, given the predominantly agrarian nature of the country's economy until the 1930s. Before the revolutions of 1917, the share of the Russian rural population was about 85%. If we take into account that fields include collective actors trying to create a system of domination in this space using (or resisting) existing institutions (Fligstein 2001), then one of these circumstances determines the severity of confrontations in economic fields in the process of market transformations.

In such a complex social situation, the state had to solve two interrelated tasks simultaneously: to ensure the transition to a predominantly industrial economy and to create necessary effective economic and legal institutions, fixing property rights and exchange rules, and creating working structures to control, manage, and coordinate economic activity (not only labor). Both these tasks had to be solved in a historically short period of time. This meant crafting policies and institutions, but how actors applied those new rules and scripts, and the routines that emerged in intersubjective activity, bred fields of perceptions, positions, and practices. This chapter examines how those tasks were solved in Russia in early historical stages.

The State in Russia: Historical Preconditions for Institutional and Field Construction

Economies are made up of institutions and institutional structures that can generate fields. An institutional structure is an ordered set of institutions that shape diversity and integrity of economic interactions and relations and that determine the nature of economic development. Institutions provide constraints and incentives that affect choices and decisions for coordinating economic activities and shaping appropriate behavior (e.g. North 1990). In general, the existing institutional structure is an important characteristic of the socioeconomic structure and can be viewed as an integral part of the Russian institutional matrix (economy, politics, ideology, culture). This institutional structure is formed in two main ways. First, institutional structures can emerge almost spontaneously, under the influence of a complex of internal factors and constraints correlated with actors' interests and needs. In this case, institutional structure provides a means for adaptation and readaptation to the existing environment. Second, an institutional structure can emerge from conscious and purposeful influences of different actors—a less random politics of institutional design—which, as a rule, is connected with active policy of state elites and often related to reform programs. The first pathway involves questions of natural variation, that is, reflecting factors and needs in a country and economy. In the second, institutional design is a response (if imperfect) of state and societal elites primarily to external challenges (global economic competition, threats of war, etc.) and in which a wide range of techniques and accumulated experience are used, sometimes drawing on foreign practices of governance (i.e. "importing institutions"). These two impulses can exist in two different states: "self-sustainability and sustainability" and "incompleteness and variability."

Considering the vastness of its space, heterogeneity of the natural environment, and the multiethnic composition of its population, Russia's economy cannot but differ from other developed countries in the complexity of its institutional structure. I will highlight the most significant links that shaped historical peculiarities of and sources for the Russian economy, starting with the sixteenth and seventeenth centuries as providing an institutional foundation and then tracing institutional and field evolution up until the early twentieth century (when revolution drastically altered the institutional trajectory, albeit leaving templates for the socialist epoch). During this period, major economic reforms resulted in serious institutional changes. The state played a key role, actively participating in forming and reforming such embedded institutions as serfdom, community, and private property, among others. Fligstein (e.g. 1990) consistently has noted how states shape institutional fields, whether by active construction or by setting legal boundaries for strategies and structures. As he remarked in the preface to the Russian edition of his book, "Regarding the example of Russian history, it seems useless to reflect on an economy that is not rooted in the state or in social relations that exist in the markets. The fact that states and markets mutually generate each other seems self-evident" (Fligstein 2013: 21).[2]

The state is a historically established political form of organizing society based on public authority, tools of centralized governance of society, and a monopoly on the use of force. According to Weber's definition: "the state, as well as political institutions that historically preceded it, is the relation of people's domination over people, based on the means of legitimate (that is, considered legitimate) violence. Thus, in order for it to exist, people under domination must obey the authority claimed by those who now rule" (Weber 1990: 646). While mainstream economics tends to elide states, taking for granted that they either support market institutions and provide basic public goods (and let market actors get on with business) or interfere and distort market signals and relations, the state in reality acts as one center of power influencing economic fields. The universality of the state as an institution is due in part to isomorphic mimicry[3] and that states enjoy comparative advantages performing particular functions due to lower transaction costs of management—the usual nation-state enjoys an economy of scale employing power to defend and pacify a territory (Tilly 1990). Historically, regimes and states have not relied on brute force alone to maintain order and instead have appealed to formulations legitimating the arrangement of rules and power that give the state its primary position. This begets the idea of a contractual facet to state authority, complemented by characteristics associated with a "formula of rule" as a compact between those who rule (government) and subjects (members of the polity) about the exchange of public goods (law and order, security, etc.) for taxes and obedience.

In institutional economics, the state is a corrective for market failure. That not everything in the economy is effectively regulated by the market, violating the Pareto optimality criterion, requires that the welfare level be improved at least for one participant and the welfare level for all the others should not be reduced. In reality, not everywhere and not always it is possible to observe such a condition of economic development—hence market failures where the Pareto principle does not work. This includes cases of monopolies, externalities, incomplete or asymmetric information, and so on. This is when states should and often do step in. In theory, if market failures could be eliminated, then a market economy could function independently of the state influence. However, such an economic utopia has yet to be observed in reality, and rather than invoke market failures to explain the existence or activity of states in economies, it is better to accept that no economy, even a market economy, can develop effectively without the state's participation. States not only defend contract, currencies, and property rights and provide public goods, as economic theory assumes; states as complex institutions not only obey existing rules, like other economic entities (firms, households), but also create and protect them, and its representatives (officials) are involved in relations with other economic agents. And often enough, state elites create rules that serve their political interests, or the unintended consequences of state actions contribute to the forms and functions of institutional fields.

The main feature of the formation and subsequent development of the Russian state was that it always played an active and dominant role vis-à-vis economy and society. Having emerged as a centralized monarchical (autocratic) entity at the end of the fifteenth century, it directly participated in the formation of the economic system with its own special features. It was characterized by two interrelated economic functions: (1) priority of mobilizing resources (labor, material, financial, military, governance) in the interests of ensuring defense and realizing geopolitical and economic interests and (2) reliance on predominantly non-market methods of seizing and using surplus output (a method of redistribution). This led to a model of economic relations and practices in which institutions for mobilizing resources were central, and the state performed functions of governing and regulating the economy, relying on predominantly non-market methods for influencing economic entities.

This is the essential difference between Russia and the countries of the West. In the latter, the fundamental basis in organizing economic activity was the progressive advance of contractual relations within the elite, and then their expansion to other sections of society, ensuring legal regulation of property rights and effective operation of legal norms and laws. In Russia's social and political order, scarce resources could not be distributed among owners and therefore were given exclusively to the highest authorities for disposal. At the same time, this nature of resource management was based on a combination of coercion and violence with a corresponding ideology for their justification. Violence and coercion by themselves could not ensure stability and order in the country over any expected long-term horizon. Therefore, it was necessary to introduce an appropriate state ideology to explain and justify policies of power. Unsurprisingly, the basis of this state ideology was the idea of sacrificial service to the Fatherland, and Russian Orthodoxy became the bearer of this state idea. This also expressed a key difference with Western Christianity. If Catholicism in its context implies a "contractual relationship" with God, Orthodoxy was characterized by love and disinterested service in God as a supreme value—no contractual relations were implied, nor could they be in this formulation of lopsided relations and agency. Through this, tsarist power acquired the significance of divine power. And this contributed to a special cultural code in Russia, which in different historical forms influenced the formation of economic fields, shaping the nature of the interaction between the state and economic agents that were more than calculation and negotiated exchange alone. A power dynamic permeated said calculations and negotiations, and studying them requires a broader set of analytical tools than is usually found in the typical economic methodology.

In explaining the origin of the mobilizational and distributive nature of Russian statehood, historians usually offer different versions of the liberal interpretation of autocracy as a form of eastern despotism with sacred sources of authority that helped the Russian state reach beyond a legal framework and into mystical roots associated with a divine origin of the highest autocratic power. While such a myth of the state's *raison d'être* and formula of rule provided tools for legislating and justifying state power, there were also objective reasons for the rise of the Russian state in its particular historical form, which resulted in particular state-economy relations, institutions, and fields. First, geopolitical dynamics contributed to the Russian state's early form. Early elites, in particular those in Muscovy, had to extend much time and effort to defend their independence and statehood against potential and real external threats, as well as to drive development in a country already lagging behind other geopolitical powers. The original impulse to the centralization of state power arose in the confrontation with the Golden Horde; as was the case in some other countries, such struggles turned the state into the main guarantor of internal stability and overcoming internecine strife.

The very formation of a centralized state—originally Muscovite Rus, and at the end the Russian Empire—continued for several centuries and required tremendous efforts and skills solving complex political and organizational problems. This was accompanied by the simultaneous formation of the Russian nation and the unification of more than two dozen foreign-speaking peoples into one political community. Among the most important incentives were struggles for independence and creating more favorable conditions for normal economic activity. Although the Mongol invasion did not lead to the direct inclusion of the country into the Golden Horde, its dependence was expressed in heavy tribute that people had to pay to the conquerors and in controlling activities of the Russian princes. Only by uniting and overcoming internal disunity could one become free from external dependence. Before Moscow became such a center for organization, a disjointed and often warring principality had to go through a period of strife and invasion, and to unify various lands, Muscovite princes used projections of strength and constructed successful economic relations. Thus, the historical experience of the formation of the Russian state created the conviction in the need for a strong state able to protect the country and provide for life and economic activity.

From the end of the fifteenth to the end of the nineteenth centuries, Russia's territory continuously increased, given the lack of significant natural borders. During this period, Russia made its way to the Atlantic Ocean via the Black and Baltic Seas and become the only European power to extend its territory to the Pacific Ocean. During this time, its territory increased from 550 thousand square kilometers to 22 million square kilometers, with a population increase from 2 million to 125 million people. This process involved more than just mastery of a physical space; there were confrontations with competitors and the need to pacify and incorporate local peoples. Such state building, with unique tasks and scales, demanded colossal effort and enormous costs, objectively limiting the capacity of the state to focus on creating and consolidating market institutions.[4] Such extended borders—at the start of World War I, Russia's border was almost 70,000 kilometers long and incorporated 20,000 square kilometers of land—meant monetary expenditures and institutional development oriented toward security, with an important institutional focus on the army. Even in peacetime, the Muscovite kingdom maintained an army of 200,000 men under arms; this was about 4% of the total population and 8% of the male population. During moments of intense military activity, of which there were more than few, such expenses multiplied many times.[5] For example, during the reign of Peter the Great, about 75% of the state budget was spent on the military. The extremely important role of achieving military and political goals could not but be reflected in weak attention to economic activity. The economy often turned out to be a secondary topic and sphere, the development of which was neglected. For all the importance of economic factors in social development, not all of them became predominant. In certain circumstances, the achievement of political tasks proved to be predominant, as evidenced by the history of Russia and modern practice.

The very need for necessary economic changes was often determined by military and political circumstances. The peasant reform of 1861, the need for which arose long before it was actually enacted, entered the political agenda after Russia's defeat in the Crimean War. This geopolitical setback convincingly confirmed the growing economic backwardness of the country vis-à-vis European countries, which could not but lead to a backlog in the production of more advanced types of weapons. Therefore, one of the main tasks in conducting these reforms was to restore the country's military potential, which required a change in its economic system and intensification of industrialization. In turn, the curtailment of reforms in the 1880s was also conditioned by political circumstances. The refusal to accept a constitution and changes in the political system was due in part to perceived dangers of the collapse of the Empire and the emergence of turmoil in the country.

Another important reason for the rise of an active state role in the early Russian economy was because impulses to economic development from non-state actors were traditionally weak, especially in the agricultural sector, because the usual risks of innovative or entrepreneurial activity could prove fatal. Unpleasant conditions for agricultural activity, in contrast with more accommodating climatic conditions in European countries, increased the significance of a survival orientation. This widespread subsistence agriculture contributed to the creation of a large and powerful centralized state that could perform compensatory and insurance functions in difficult times. According to available estimates, the yield of many crops, perhaps 50% to 70%, were at the mercy of unpredictable weather conditions. The impact of climate turned out to be one of the biggest obstacles to agricultural innovation, to the point that hunger was persistently a threat (cf. Ryazanov 1998: 316, 326).[6] The fact that natural anomalies in the development of agriculture increased political risks is borne out by historical facts. One tragic episode in the country's history was the "Time of Troubles" (1598–1613). At this time, the severity of internal contradictions, the weakness of the state due to a sudden dynastic crisis, and discord and strife actually led to the loss of the country's political independence. At the same time, the dramatic nature of this period was largely due to economic factors. For three consecutive years (1603–1605), there was catastrophic crop failure due to a climatic anomaly, resulting in widespread hunger and even starvation deaths. All this contributed to the emergence of large-scale peasant uprisings and led to the onset of the "great turmoil." As a result, many industrial workshops ceased to exist, resulting in economic decline and a drop in the urban and rural populations.

The formation of Russian statehood also had an autocratic form of government. By the beginning of the eighteenth century, this autocracy had evolved into a bureaucratic set of institutions based on universal service to the state, in the person of the monarch. Such a system of centralized and autocratic state power was entrenched by terminating the convocation of the Zemskii Sobor, and the abolition of the Boyar Duma and the Patriarchate. This presupposed an administrative power vertical headed by a monarch, ensuring subordination of all estates to the state. Military, legislative, administrative, and financial authorities were concentrated through the bureaucracy in the position of the tsar. Main features of Russian statehood acquired a stable character, influencing the subsequent historical evolution of the country. The state became the central link in the institutional structure of the Russian economy, acting as an active economic agent and catalyst for economic processes (Ryazanov 1998: 321). Such an active state role can be useful and even necessary if the state can timely and accurately address objective needs for a country's development and enhancing its potential and its people's well-being. At the same time, the state, bureaucratic in nature and endowed with excessive powers, often proves to be a brake on economic development. This manifests itself both in badly chosen policies and the suppression of economic actors' own initiatives.

The rise of a strong centralized state before the coming of capitalism is not unique to Russia; Japan and France, for example, also had relatively strong states before the development of capitalist institutions (Tilly 1990). The advantage of such a political structure is the capacity for one set of elites (presuming sufficient unity and skills among that elite) to mobilize a broad range of resources and pursue an active economic policy across numerous fronts (cf. Zysman 1982). Its drawback is the risk of diminishing effectiveness of non-market use of resources and risks of rent-seeking and other forms of opportunism, due to mismanagement, corruption, bureaucratization, and so on—that is, "bureaucratic failure." In this regard, Russia's uniqueness was that as general political problems were resolved, the dominant role of the state spread beyond the economy, especially as the gap between Russia and better developed neighbors increased. Unable to rely on more efficient market forms of governance that would facilitate a more intensive transition from an agrarian to industrial economy, the state became the main initiator and promoter of reforms designed to accelerate economic development and overcome those gaps with geopolitical rivals. Thus, the leading role of the state for economic development was fixed as a characteristic feature of Russian statehood in the long-term historical perspective.

At the same time, the duality of the state's position graphically illustrates attitudes to the institution of serfdom, which, before its abolition, shaped the main characteristics of the country's economic structure. During the formation and development of serfdom, the state pursued a decisive policy to consolidate it. This not only corresponded to economic interests of the nobility as the ruling class but also brought benefits to the state itself in the realization of its functions of control. To consolidate this institution, the state made considerable efforts to introduce serfdom into territories where it did not exist. The authorities realized the existence free territories undermined its foundations. However, when preservation of serfdom was fraught with a revolutionary explosion capable of destroying the state and eliminating the monarchy, the authorities turned to abolish it, even if this risked losing the confidence and support of the ruling class. Interests of state were greater than economic interests of landowners in preserving cheap labor.

The Role of the State in Early Market-Building and Industrialization

The role of the state in industrialization is well-known for the Soviet era, and economic historians have also shown how the state played an important role in the second half of the nineteenth century in facilitating the creation of industrial centers in St. Petersburg, Moscow, and the Urals (Rieber 1991). If during the period of Peter's reforms the policy of industrialization, which he actively pursued, was based on setting high customs tariffs and using coercive methods and serf labor in newly created industrial enterprises, then after the abolition of serfdom, industrialization in the country developed more successfully and at a higher rate. The ineffectiveness of serf labor at newly created industrial enterprises was realized even before the abolition of serfdom. Therefore, a law was passed already in 1840 that effectively forbade the use of serf labor in industry. After the abolition of serfdom, there was a tremendous upsurge in railway construction: from 1865 to 1875, the average annual growth of the railway network was 1500 versts, and from 1893 to 1897, it was 2500 versts. If by 1881 the length of railways in the country reached 22,000 versts, then by 1901 this total was 58,000, a threefold increase. A characteristic feature of industrialization at the end of the nineteenth century was the rapid growth of concentration and monopolization of industrial production. About half of all industrial workers were employed in enterprises with more than 500 workers in this period. Only Germany had a comparable figure. The largest enterprise in Russia, the Putilov works in St. Petersburg, had more than 40,000 workers. Only the Krupp factories and Armstrong factories in England were bigger.

How did the Russian state pursue an active economic policy aimed at a transition to a predominantly industrial economy? An important facet of this story is that existing institutions and fields—not only laws and rules, but conventions and practices—were grounded more in logics of redistribution and moral economies at the local level and of rent-seeking by landlords. As such, Russia's story sounds similar to the story of the rise of capitalism in England as noted by Karl Polanyi (1944): the state had to create institutions and practices for a market society before a market economy could emerge, and this entailed eradicating local obstacles grounded in non-market, even non-industrial, foundations. Starting in the eighteenth century, the state engaged in "clearing" economic fields, removing institutional barriers and eliminating obstacles, and wagering on economic incentives from within. For illustration, we can refer to a number of important changes related to removing barriers to the development of regular market exchange in the country. In 1754, internal customs and duties were finally removed. The customs barrier between Russia, Ukraine, and the Don Military Region was also eliminated, and a new customs procedure less inimical to market trade was also established. Customs regulations of 1755 expanded trade rights for all segments of the population. At the same time, foreign merchants were not allowed to engage in retail trade and could not trade with each other. An even more significant innovation was the Decree of Catherine II, adopted in 1775, which authorized "to each and every person" to engage in production. In 1779, the Manufactory College was abolished, which also confirmed economic policy for the development of free enterprise and the lifting of restrictions in the sphere of industrial production (Ryazanov 2013: 289–290). The fundamental importance of such measures was the fact that they undermined the universality of relations of serfdom in the country, which influenced the sphere of formally free labor. A typical example is the actual absence of a hiring in the strict sense of the term. Instead, it was the custom that every person who was legally free and entering labor market found himself in a relationship with the employer who was a "master" rather than "employer."

On the other hand, state innovation relied primarily on use of its own administrative resources and power to create new forms of governance, introducing new routines into established institutional fields in the hope of rapid transformation and adaptation to new tasks. Often enough, this approach was motivated by the need quickly to solve specific and urgent problems. The practice of this administrative and compulsory approach was laid down already during the reign of Peter the Great. Peter compensated the lack of initiative and capital for organizing large-scale industrial production (due to a weak bourgeoisie or entrepreneurial class) by creating state-owned manufacturing, with the transfer of some enterprises to designated industrialists—an early variant of "*nomenklatura* privatization" that would characterize the 1990s. About 180 relatively large manufactories were created, half of which belonged to the state. A characteristic example of privatization of a state-owned firm is from the entrepreneurial activity of Nikita Demidov, who in 1702 received an Urals factory that was supposed to fulfill state needs. This practice of governance at that time was widespread, which led to entrepreneurs working for the state rather than the market. As a result, being rid of competition and having a guaranteed buyer for their output, such "entrepreneurs" did not seek to improve production, which could not but lead to preserving the technical level of the Ural factories and their lagging behind European producers.[7] That is, the rules of industrial fields—or to use macroeconomic jargon, the "equilibrium" in industry at this time—were to orient production to the state, rather than to a competitive market. Innovation brought its own risks and little chance of rewards, given the underdeveloped market—a problem that would be all too familiar to Soviet planners in the twentieth century.

The way in which the state directly tried to stimulate the development of industry is evidenced by the practice of providing personnel for nascent industrial production—and this became one of the worst bottlenecks for industrialization. For a long period, this problem was solved not through the formation of an operational labor market but by assigning state peasants to factories as payment of state taxes and performance of natural duties. The same measure was adopted in 1721, by a law that allowed non-nobles to buy serfs to work in factories. A new category of peasants began to take shape, those who remained serfs but who were linked to a particular factory, regardless of the fact that ownership of the factory might change. As a result, industry developed not in a market environment but on the basis of bonded serf labor.

In subsequent periods, the state actively participated in developing the industrial sector through patronage in various spheres of economic life:

- Creation of state-owned factories (mainly in military production) that were protected from market competition;
- Concentration of most of the railways in the state's hands and the establishment of state control over their activities—in 1910, of 62,400 versts of railways, the state owned 42.5 thousand versts, that is, 68%;
- Development of state regulation of economic activity, such as unified tariffs for railways, regulation of domestic and foreign trade, provision of state loans and purchases, and so on;
- State ownership of a significant part of land—in European Russia, state ownership accounted for about 30% of land ownership;
- The introduction of a state monopoly on trade in alcohol (1894), bringing an annual net income to the treasury of 443–675 million rubles for the period 1903–1913; and
- Regulating the activity of the State Bank.

The policy of active state patronage of industrial development brought positive fruits. However, it became clear that without eliminating obstacles to the formation of full-fledged market relations, it would not be possible to achieve significant advancement in the industrial sphere. The main obstacle here was the preservation of serfdom, which blocked the possibility of widespread deployment of economic incentives for economic activity on the basis of market relations, limiting the possibility of attracting workers under market conditions of employment. Its elimination was determined by the need for economic development and at the same time was necessary to relieve the social tension in society. The difficulty of implementing agrarian reform eliminating serfdom and creating more favorable conditions for industrialization testifies to the complex and contradictory nature of the interrelationship of economic and social tasks to be resolved in state administration. For the state, additional complexity was associated with the need not only to create new institutions and norms but also to take into account specifics of economic incentives and practices of various social groups. Established traditions and conservatism weighed on the decision-making process.

The Role of the State in Removing Barriers to the Formation of a Field for Markets

The establishment of serfdom in Russia took place over a long historical period. A key moment in that evolution was 1649, when the Sobornoe Ulozhenie was adopted under Tsar Aleksei Mikhailovich (Peter the Great's father), which gave landowners authority over peasants who lived off their land and who were attached to that land for an indefinite period. Historians continue to discuss the reasons for approval of the Ulozhenie. However, various details notwithstanding, the role of the state in this process was significant. By attaching peasants to the land, the state pursued a policy of settling and developing a vast territory, forming a tax-paying population and fulfilling various other needs. From the point of view of state interests and the possibility of saving resources, attaching peasants to the land, as well as attaching the urban population to cities, was the cheapest and perhaps the only way to ensure the protection and economic development of new territory.

The rise of serfdom was part of the process of the emergence of the Russian state and not only of a new form of economic relations. The process of enslaving peasants occurred in the context of the emergence and consolidation of new institutions through the operation of a mechanism for selecting economic institutions. When the state emerged, being a "weak state" initially, institutions (norms and rules) initially took the form of "customs" reflecting the effect of social norms, usually in accordance with entrenched religious beliefs or backed by authority among direct participants in economic activities, that is, landowners and peasants. As the state grew in strength, it was transformed from a set of elite networks and customs to a complex institution in which previous customs were fixed as legally binding norms. At the same time, established legal norms (in Muscovite Rus tsarist decrees and *sudebniki*) were gradually transformed under the influence of changes in the economic sphere (Ryazanov 2013: 275).

Serfdom was a continuation and expansion of the nature of land ownership. At the turn of the end of the fifteenth century, the Muscovite state established a *pomeste*, under which the Grand Duke transferred land, of which he was owner, to an individual in return for the duty of carrying out military or civil service. The owner of the estate rented the land under contract to peasants in return for quitrent and performing other duties. Therefore, the state as the owner in the person of representatives of the highest authority was directly interested in the general ordering of property rights and their precise regulation. In the years that followed, the autocratic state acquired necessary power and economic leverage to dictate its rules, which expressed and defended the interests of this ruling feudal landlord class; ultimately, this was a symbiotic relationship. This was manifested in the establishment of the nobility's monopoly over the ownership of serfs. It was natural that during the next 150 years, serfdom was consistently toughened, with state coercion its ultimate foundation. This was economically reinforced by gradual transformation of the landlord economy from the natural into a market economy. This also caused active resistance by peasants, as evidenced by large peasant uprisings.

In general, as long as the nobility was obliged to provide military and public service to the state—a duty was initially abolished in 1762 (Peter III) and confirmed in 1785 (Catherine the Great)—serfdom acted as a sufficiently rational institution built into the socio-economic system that under existing conditions and constraints could not develop normally without the guiding role of the state. This system had a certain institutional balance, based on the principle of symmetric obligations: the peasants had to serve the nobles, who were supposed to serve the state. Further, Russian serfdom was never consistently distributed across Russia, even in its heyday. For a large part of the territory—the North, Urals, Siberia, and southern Cossack regions—serfdom was not established. Thus, while serfdom was an important facet of Russian political economy and shaped the logic of economic fields and institutions, materially it was confined only to the best agricultural regions of the country.

However, already at the end of the eighteenth century, serfdom was gradually but inexorably turning into an institutional obstacle to further economic development, at least in a Western manner. By that time the nobility was free from compulsory state service, at which point one part of the symmetry of obligations had vanished, which in turn created tremors in the institutional balance of economic relations. The serf system of the economy was deprived of a normative basis; its foundations were in elite interests and historical tradition. To this we must add another circumstance that grew increasingly significant with time and connected with the fact that serfdom became a serious barrier to further economic development—especially given growing economic dynamism in European countries (geopolitical competitors) and the delay in industrialization. This example is illustrative. If in 1800 10.3 million poods of pig iron were smelted in Russia, 12 million poods were smelted in England. By the beginning of the 1850s, English industrialists were smelting 140 million poods, while Russian industry was smelting only 15–16 million poods. The reason for such a sharp lag in the development of metallurgy was the use of outdated technology based on the use of charcoal, which in Europe was extremely expensive. In England, a more economical and productive method of smelting pig iron based on the use of coal (coke) was introduced. The rapid development of Western European metallurgical industry, based on more advanced and efficient technologies, meant that in the first half of the nineteenth century Russian exports of pig iron practically ceased—their volume exceeded 3 million poods of cast iron—and this became an important reason for stagnation in industrial development. In the following period, metallurgy restored its potential and in 1900, 177 million poods of cast iron were produced.

Unsurprisingly, the highest authorities in Russia from the end of the eighteenth century began to carry out preparations for abolishing serfdom, seeing it as a threat to the existing institutional system primarily because peasants rejected it and because it was a brake to the emergence of a fully-fledged industrial economy. However, serfdom had created not only its own logic but also its own interest groups. Rejecting serfdom was not popular among lower administrative orders, and especially among landlords—a vivid example that an institutional field can depend upon preceding development. It set up a confrontation between economic "routine" and "tradition" that protected existing interests and trajectories, versus institutional and field innovations aimed at shifting the economic trajectory. At the heart of such institutional conflict in the field of the economy was the existence of a group economically interested in preserving an obsolete economic institution that blocked necessary changes.

Another facet of this institutional conflict was the relationship between benefits and low transaction costs of the existing system and uncertain future results in the event of its demolition. Further and no less important, serfdom played an important role in relations of social control for a large country and a state that, while developing, was still too underdeveloped to control that vast territory and its population by itself. In fact, the state gained significantly from reduced transaction costs of managing the country through serfdom. Around 103,000 landlords and related peasant communities managed a population of 24 million people (only the male portion of the population) without receiving compensation from the state for such an arrangement, despite the fact that they were actually assigned the most important administrative functions, from maintaining law and order to conscription and tax collection. The serf system itself also had a certain inertia that contributed to its preservation. First, the serf economy at the beginning of the nineteenth century shifted to a more market-oriented orientation, the effectiveness of which was generally associated with advantages of a large-scale economy, and this enabled landlords and peasants to make better use of technical improvements. Further, low labor costs from serfs, when compared with hired labor, further supported the shift from a quitrent to compulsory corvée for agrarian profit (or the use of a mixed system).

Second, a serf economy based on corvée was economically more profitable than using quitrent or hired labor, because it could rely on traditions of the commune for managing peasants. The very organization of corvée labor with the use of livestock and other peasant inventories assumed the cooperation of equivalent economic units and therefore supported the communal tradition of leveling of peasant farms, that is, redistribution of land across different households. (The landlords, as a rule, did not have their own working cattle and agricultural technology.) This also allowed all peasants to fulfill all their obligations. This alignment was key to ensuring the survival of peasant communes in conditions of high dependence on weather and peasants' own experience and worldviews. To reduce the impact of unfavorable weather conditions on the development of peasant farms, the rule was to create common stocks of grain and other produce in communes as a reserve to be used in lean years.

Third, the economic potential for the development of serfdom also relied on effectiveness of forced labor with its inevitable "subsystem of fear of punishment." Undoubtedly, in the long run such a system would exhaust its capacity and collapse. Nevertheless, given the limited needs of the peasantry and their lack of sufficient incentives for the continuous improvement in production, forced measures brought results in the short run. Russian peasants, for objective reasons, reproduced "ethics of survival" rather than "ethics of success," as Max Weber suggested was the case in countries with Protestant culture. Even with the beginning of the twentieth century, A. V. Chaianov developed a theory of the peasant economy, according to which the latter increased its resistance to crises processes and was based on the fact that growth in peasant production was not based on profit, so long as consumption was balanced by an increase in burdens of labor (cf. Chaianov 1989: 114–143). That some institutionalists (e.g. Douglass North) have noted that the economic effectiveness of slavery in the United States could be higher than was commonly believed, the same was true for cases of serfdom. In both cases, the issue was not the effect of slavery or serfdom, but rather advantages of large-scale production organized to preserve traditional community relations in labor-based interactions.

Summing up, it can be argued that the reasons for reforms aimed at changing the old economic system were not always well connected with purely economic circumstances. Socio-political factors and other non-economic grounds were more significant, as well as internal and external threats. This only confirms the conclusion that the study of the economic system as a field of interaction from the standpoint of its autonomy and self-sufficiency is not justified. Economic processes exist inseparably with socio-political, legal, ideological, and other non-economic spheres. Therefore, although Russia's serf system had resources for use in the agricultural sector at the outset of reforms, this does not negate the fact that it was already a brake on further development, especially in view of the need to complete industrialization quickly and effectively (cf. Ryazanov 2011: 11).

That serfdom was an institution of social coercion, and that peasants under its shadow lacked many rights, played a special role exacerbating the crisis of serfdom—in particular, serfdom threatened to generate increasing social pressures from peasants themselves, and Russia did have a history of peasant revolts. This only increased the need for reform. This makes clear the warning Aleksandr II made at a meeting with leaders of the nobility in Moscow in 1856, after he had ascended to the throne: "It is much better that this [abolition of serfdom] happened from above, rather than from below" (Fedorov 1994: 85). In such a complex and acute situation, the state, in the person of emperor Aleksandr II, was only capable of carrying out agrarian reform radically changing the economic system in the country. It is important to emphasize that proposed reforms in 1861 were a compromise, aimed at taking into account the interests of both landowners and peasants. Like any other reform carried out from above, it could not but be a compromise and possibly half-hearted—otherwise this would be not reform, but a revolution from above. However, despite all its shortcomings, it was generally successful and helped to resolve acute social contradictions, while removing barriers to industrialization and expanding the field for the development of market relations in the countryside. At the same time, the complex problem of redistributing land ownership was realized without destructive shocks. Moreover, a serious reconstruction of economic life was not accompanied by a crisis, and so economic growth was preserved and even accelerated.

The advantage of the 1861 reform was that it did not blindly copy land reforms of European states but took into account the whole range of conditions, limitations, historical traditions, and peculiarities of the peasant class. Unlike agrarian reforms previously carried out elsewhere in Europe, which were accompanied by depriving peasants of land, Russian peasants retained land assigned to their communes. The innovative character of the reform was realized by its developers. Aleksandr II, speaking at a meeting of the State Council, noted: "We wish, by giving personal freedom to the peasants and recognizing land as property of landlords, not to make peasants a homeless people and therefore harmful to both the landlords and the state … We wanted to avoid what was happening abroad, where the transformation took place almost everywhere in a violent way" (Fedorov 1994: 195).

The most important historical lesson of the 1861 reform is that the success of any economic change is not based on simple reproduction of others' experiences but on careful study of existing conditions of governance in one's own country, taking into account historical traditions, behavioral norms, and so on (cf. Ryazanov 2011). Substantial preparation of reforms, participation in its development by several commissions with different participants, the gradual introduction of new forms of governance—all this contributed to the success of implementation, which could serve as an example for modern social transformations. Another historical lesson of agrarian reforms was the need to take into account the complex process of re-adaptation of economic agents' interactions in new emerging economic fields. The choice of an independent variant of agrarian reform was shaped by a number of considerations and circumstances. In 1816, experiments in agrarian reform were conducted in the Baltic provinces. Peasants were freed from serfdom but did not receive land, turning them into tenants of landlords—a variation of a normal European option. However, this experiment showed that peasants needed not only legal freedom but also land. Therefore, peasant demands were expressed through the formula "land and freedom." This combined economic interests of the peasantry as a class, and their own ideas about how economic activities should be organized extrapolated from their worldviews, experiences, and values. Reformers among the highest authorities were forced to take into account the mood of the peasantry in choosing reforms. Otherwise, the peasantry was ready for mass resistance, which could have led to social turmoil.

New forms of governance were by no means immediately more effective than older forms. Time and perseverance of reformers and society were required for providing necessary stability and avoiding possible disruptions and contradictions. Finally, a key lesson of 1861 was the authorities' ability to correctly time the course of reforms in the interests of successively removing acute contradictions and solving emerging challenges. Here we can find the weakest link in the reforms: there was too little consideration of the logic of Russia's economic development and the need to address remaining social contradictions. The first stage of reforms reduced the acuteness of the land issue, but did not remove it completely. The period between the first and later phases of reform should not exceed 25–30 years. Above all, the authorities ignored growing antagonism of the peasant and landlord classes and underestimated the peasants' principled rejection of the landlord class as unnecessary and useless for the development of agriculture—which only increased the rejection of the very institution of private ownership of land.

Reflecting on historical events in Russia in the subsequent period, we can assume that a more optimal way to solve the agrarian question that would not lead to devastating consequences could have been adopted. Yet no matter how painful it was for the authorities, it was necessary to choose a course for gradual elimination of the landlord class.[8] This was best done through the state's purchase of landed estates, while preserving the market-oriented sector of large-scale farms by creating state-owned agrarian enterprises or by transferring these lands to producers (primarily peasant communes but other possible buyers as well), while avoiding little-justified fragmentation. Historical experience has shown that the "black repartition" of the land, which happened after 1917, was another way to resolve the land issue that would also threaten the marketability of land and agriculture.

However, the authorities proved unable to develop the innovative potential that the 1861 reform might have facilitated. The state tried to do this after the revolutionary events of 1905, through a return to the Western European model of modernization of agriculture, by the Stolypin reforms that would break up peasant communes and introduce private ownership of land—which would encourage entrepreneurial peasant farmers, but also provide a fatal blow to the traditional peasant model of farming and even to the peasant class itself. In the end, the Stolypin reforms only strengthened overall peasant antagonism and resistance, which were further fueled by their traditional alienation from participating in the formulation of land reforms and reliance on administrative and bureaucratic methods. In the development of the reform of 1861, the authorities continued to hope for a shaky and deceptive balance, not caring about weakening the redemption and tax burden on the peasantry as the main productive class. Russia paid for this with the peasant revolution of 1905 and then by final collapse in 1917.

Conclusion

Economic life, regulated by market and non-market methods, was characterized by a complex socio-economic structure. The interaction of economic agents was determined not only by objective economic laws and economic interests. An equally significant role was played by subjective representations of economic participants, including their mentalities, behavioral traditions, and values. Moreover, even within the framework of one state and one ethnic community, these could differ, quite significantly, between different social groups. Traditional economic methods and analytical tools, for all their usefulness, cannot always provide a complete and accurate picture of economic life. In this respect, field theory not only complements the economic analysis of governance, it can also provide a synthesis of economic and social-behavioral approaches.

Even more important is the role of field theory in the study of the economies of countries in which market reforms are being carried out. In this case, its use allows us to rethink historical experiences, revealing such patterns that are outside the field of economic theory. This fully applies also to the history of economic development in Russia, in which during the nineteenth and twentieth centuries economic reforms were periodically carried out, with varying degrees of success. Success of reforms was more likely when reformers accurately took into account features of economic behavior and established traditions in Russian society. This lesson is worthy of the closest attention at the present stage of reforming the Russian economy.

Notes

1. In this sense, we are really talking about fields rather than institutions. Institutions created broad parameters and boundaries for strategies, but practices were realized in habitualized interactions between various actors—that is, how institutions were employed. Institutions provide schemas and incentives that can nudge actors in various directions, but the collective orientation to shared strategies and goals—which then induces others in these contexts to act accordingly (including acting in subversive ways)—is a field effect (Martin 2011).
2. This raises a serious theoretical question: if the state plays the central role in institutional design, does "field theory" revert to old-fashioned "political economy"? Fligstein's analyses involve pluralist societies (primarily the United States), in which state power is constrained by institutional fragmentation and constitutional limits. But what of other cases? I will not answer this question directly, as that involves a different project and context. However, I hope to imply possible answers by the end of this chapter.
3. Such isomorphic dissemination of the modern state form has accelerated at various historical junctures. The Treaty of Westphalia was, arguably, when the modern state was reified as a sacred political actor; the French

Revolution and Napoleonic Wars certainly spread the concept of "nation" and further reinforced the significance of the modern state; and decolonization following World War II spread the modern state form throughout the world.

4. In contrast, consider English history. England had better natural defenses that, while not perfect, did reduce possible threats of invasion by making them costlier. Once the English kinds lost holdings in France and gave up on claims to the French throne, English institution building could pick up its pace, as the monarchy and elites look inward to develop economic capacity.
5. According to S. Soloviev, from 1055 to 1462, Russia suffered 245 invasions. Moreover, 200 attacks on Russia took place between 1240 and 1462—that is, attacks occurred almost every year. Between 1380, the year of the Battle of Kulikovo, and 1917, Russia spent 334 years engaged in warfare, that is, almost two-thirds of the entire period.
6. In the Muscovite state in the seventeenth century, 65 years of the entire century suffered bad harvests, leading to famine and resulting to political consequences—in particular, the Time of Troubles (1605–1613). In total, for 830 years (1024–1854), only 120 bad harvests were recorded, including 10 general famines.
7. If many of the political innovations of Peter I were preserved until 1917, industrial manufacturing created by his initiative did not have such a prosperous destiny. In the list of 300 major factories still in operation in 1780, only 22 were survivors of Peter's time.
8. Note that one of the famous Slavophile ideologues, A. S. Khomiakov, in the end of the 1830s proposed a reform project providing for the state to purchase of landlords' land and then turn it over to peasant communes. In his turn, N. G. Chernyshevskii during the period of preparation of reforms came forward for the liquidation of landlord property and its use by peasants.

Works Cited

Chaianov, A.V. 1989. *Krestianskoe khoziaistvo: Izbrannye trudy*. Moscow: Ekonomika.

Fedorov, A.V. 1994. *Konets krepostnichestva v Rossii. Dokumenty, pisma, memuary, stati*. Moscow: Moscow State University.

Fligstein, Neil. 1990. *The Social Transformation of Corporate Control*. Cambridge: Harvard University Press.

———. 2001. *The Architecture of Markets*. Princeton: Princeton University Press.

———. 2013. *Arkhitektura rynkov: ekonomicheskaia sotsiologiia kapitalisticheskikh obshchestv v XXI veke*. Moscow: Higher School of Economics.

Martin, John Levi. 2011. *Explaining Social Action*. New York: Oxford University Press.

North, Douglass. 1990. *Institutions, Institutional Change, and Economic Performance*. New York: Cambridge University Press.

Polanyi, Karl. 1944. *The Great Transformation*. Boston: Beacon.

Rieber, Alfred. 1991. *Merchants and Entrepreneurs in Imperial Russia*. Chapel Hill: University of North Carolina Press.

Ryazanov, V.T. 1998. *Ekonomicheskoe razvitie Rossii. Reformy i rossiiskoe khoziaistvo v IX–XX vv*. Moscow: Nauka.

———. 2011. Reforma 1861 g. v Rossii: prichiny i istoricheskie uroki. *Vestnik SPbGU*, Seriia Ekonomika, 2: 3–17.

———. 2013. *Institutsionalnyi analiz i ekonomika Rossii*. Moscow: Ekonomika.

Tilly, Charles. 1990. *Coercion, Capital, and European States*. New York: Wiley-Blackwell.

Weber, Max. 1990. *Veber M. Izbrannye proizvedeniia*. Moscow: Progress.

Zysman, John. 1982. *Governments, Markets, and Growth*. Ithaca: Cornell University Press.

10

Neil Fligstein's Concept of Organizational Fields, Economic Processes and Dynamics, and Their Significance for Building Russian Markets

Svetlana Rumyantseva and Ainur Musaeva

In developed countries, states and representatives of capital and labor have created various institutional arrangements to facilitate economic growth and innovation. Neil Fligstein's theory of fields contains methodological conceptions to help describe conditions of institutional development necessary for innovation and growth. He convincingly shows how relations between economic actors are formed through crises or other economic challenges. This brings his approach close to theories of economic dynamics that account for cyclical crises to explain development.

Actors' comprehensions of their surrounding economic world, routines, practices, and social relations form fields that are a backbone for institutional relations, for which historical and cultural conditioning is characteristic. An investigation of fields of routines used by organizations

S. Rumyantseva (✉) • A. Musaeva
Faculty of Economics, Department of Economic Theory,
St. Petersburg State University, St. Petersburg, Russia

J. K. Hass (ed.), *Re-Examining the History of the Russian Economy*,
https://doi.org/10.1007/978-3-319-75414-7_10

brings field theory closer to an evolutionary approach that, when brought together with a theory of systems, might help us better understand how crises can beget new structures that help overcome initial crises. In current conditions after the crisis of 2008, this is particularly important because, as Fligstein notes, fields and organizations can create stable conditions for privileged classes. To systematize an investigation into economic change and practice, Fligstein proposes four types of rules at the heart of the creation and operation of markets: property rights, governance structures, rules of exchange, and conceptions of control. (As shorthand, we sometimes refer to these as "market rules.") The implementation of these rules in each country forms the basis for other institutions. In turn, the development of market rules depends on social stability, which is a condition for efficient operation of markets. This is especially important for Russia, where market rules have been shaped by the rugged quality of that country's institutional history.

Fields and Innovation

Exchange can never be stable only on its own accord, as a host of institutional scholars have shown. A core facet of field theory is that actors search for stability; however, system shocks, non-trivial changes in laws and policies, and institutional entrepreneurs external to fields can introduce uncertainty and change in field rules and boundaries (Fligstein 1990, 2001). These ideas are not so distant from those of systems theory (Saviotti 1986; Iberall and Wilkinson 1987; Osipov and Shurgalina 1994), in which an economy is viewed as a set of non-equilibrium processes. Disequilibrium is reproduced even though actors seek to create stability through systems of cognitive practices, including organizational hierarchies. One key problem of the sociological approach to sustainability of economic systems is to elucidate the capacity of managers to determine their positions when adapting to new economic conditions. As Fligstein points out, there are two opposing approaches to this question. In one approach, managers study and comprehend their environments and take actions to increase organizational effectiveness. In the other approach, the environment selects enterprises, and managers are not able to analyze problems and implement appropriate solutions sufficiently quickly. Therefore, an industry moves to a small set of practices, because the environment selects for those surviving enterprises.

This problem is directly related to issues of routines and organizational development in evolutionary economics. According to this approach, routines help strengthen internal structures of organizations, acting as a late stage of innovation in which routines function as organizational memory that sets down recurring ways of solving problems (Nelson and Winter 1985). At the same time, as shown by the institutional school of economic dynamics, such routines can lead to a "dinosaur effect," in which organizational growth and routinization of responses to the environment gradually deprive the firm of the capacity to react quickly to changes in economic contexts and to compete effectively with newcomers whose routines have not ossified and who can react more rapidly to environmental shocks or opportunities. The dinosaur effect creates a situation in which firms are no longer interested in improving their output even as consumer preferences change, and this can threaten long-run competitiveness and industry-level innovativeness.

In general, the theory of economic dynamics, as used in this chapter, is grounded in the ideas of Nikolai Kondratieff, who proposed long economic cycles lasting periods of 45–60 years (Kondratieff 1935), and which divided economic dynamics into streaming and cumulative processes that to varying degrees determine the evolution of economic systems (Kondratieff 1989). (One goal of this chapter is to begin a synthesis of Kondratieff's cycles with Fligstein's fields, as both deal with causes of crises and innovations in overcoming depressions.) Kondratieff showed that before the beginning of the upward wave of each large cycle, and sometimes at its very beginning, significant changes occur in conditions of economic life. Kondratieff cited a number of technological inventions introduced into economic practice in various branches of industry and that were connected with reorganizing production relations. In the first major cycle, Kondratieff also noted the United States' position on the world market before this cycle, and the expansion of American import and export operations before the second, as well as its entry into active trade with younger countries, that is, Australia, Argentina, Chile, and Canada. The beginning of large cycles coincides with the expansion of the orbit of global economic relations. Further, before the beginning of the major cycle that he studied, gold production increased, which had a powerful effect on economic dynamics under the gold standard. Thus, the beginning of large cycles is preceded by fundamental changes in the extraction of precious metals and monetary circulation and in major technological upheavals (Kondratieff 1925).

In contrast, and in continuing Kondratieff's approach, Mensch (1978: 98) suggested that contextual forces, not internal routines and structures, were more important when accounting for the rise and fall of innovation. Innovative activity slows not because of exhaustion of available technological knowledge but rather due to the economic environment of innovation activity, that is, the opportunity for entrepreneurs to obtain sufficiently high profit from using existing technology, as well as the concentration of production, rigid technological chains and management structures, and economies of scale. Perez (2009) showed that competitive pressures make deep and widespread changes possible due to technological revolutions, but new technical and economic paradigms are not easily assimilated, as they can stimulate intense resistance and require significant mechanisms to stimulate change. In the prosperity phase of economic cycles, institutional inertia usually grows because companies in advantageous positions vis-à-vis competition in the previous depression and recovery seize an increasing share of the market and form oligopolistic structures and routinize managerial practices that resist innovative practices. Perez (2009) suggests that complacency comes to large firms of the dominant techno-economic paradigm at the maturity stage, and society's institutional structure fully adapts to conditions of the victorious techno-economic revolution. Thus, non-trivial adaptation and innovation are an obstacle to further technological revolution.

This process is promoted by peculiarities of patent law—if a company has beaten competitors by introducing a significant innovation, a patent grants the monopoly to manufacture the dominant product for a number of years. Thus, Perez argues that in economic phases of late revival and prosperity, oligopolistic and monopolistic structures take root. Firms lose capacity to compete and instead reap the fruits of dominant market positions (i.e. rent-seeking) and increase the complexity of their organizational structures. Management systems develop conflicts of interests and tendencies to multiply incomes and preferences for employees, especially top managers. Institutional inertia grows: companies are not so much engaged in developing new competitive strategies, as in trying to consolidate market positions. To this we can add an increase in time for taking managerial decisions because of increasingly complicated organizational structures. Thus, the prosperity period in the economic cycle is associated with the emergence of monopoly and oligopolistic structures, and a decrease in overall economic competitiveness (Rumyantseva 2014: 152–153).

In sum, the theory of economic dynamics assumes that stable relations between market actors, achieved in the formation of the technical and economic paradigm, hinder the economic renewal based on innovation, and that only economic depressions that destroy established relations can open up the scope for new solutions. However, according to Fligstein, businesses require stable relations with suppliers, competitors, and labor to reap benefits of new technologies. Establishing such relations depends on establishing stable societal institutions, such as state structures and legislation. Moreover, according to field theory, such institutions as stable, non-rent-seeking states can facilitate systems of contacts, social security, money, and fundamental research for technological innovation. Societies blessed with developed fields of interaction have better chances for technological progress.

Thus, field theory, institutional theory, and theory of economic dynamics can intertwine in interesting ways. According to Douglass North, institutions play a key role in shaping whether and how economic innovations can occur (North 1990); as V. E. Dementiev claims, "The very nature of bifurcation creates a significant uncertainty as to how the selection of the realized variant of development takes place" (Dementiev 2016: 125). Institutional balance is a projection of existing forces on the institutional environment. This invites Fligstein's idea about correlations of forces in various economic fields hindering or contributing to innovations. In this respect, it is significant that institutionalists Acemoglu and Robinson (2010) translate the analysis of this correlation of forces into a distribution of political power, highlighting inclusive and extractive institutions. It is also important that institutions are more likely created to serve the interests of those in positions to influence the formation of new rules (North 2007: 33)—which is also at work in making fields. For example, Russian innovation processes, with large-scale developments in fields of scientific and technical creativity, are slow because economic elites have greater interests and incentives to move capital to safe harbors, rather than to invest it in innovation, because of political uncertainty. Small and medium businesses, which are often innovation oriented, face high tax burdens, narrow demand, and difficulty obtaining investment. Thus, property rights, extractive institutions, and conceptions of control in Russia have not yet aligned to support innovative development.

Fligstein notes that when creating markets, enterprises rely on both the state and collective actors from civil society. How various forms of capital are distributed then shapes possibilities for actors to create dominant market positions; alternatively, a balance of power forces firms to negotiate with each other and the state. The possibility of market power and domination of local markets, which can ensure market stability, also requires the existence of a local culture that contributes to field strategies, logics, and market rules. In particular, fields require rules and meanings roughly organized as concrete frames of interpretation that structure actors' interactions, boundaries, and menus of strategies. While a field framework has shown the significance of how firms, states, and markets are interconnected and create the conditions for effective functioning of enterprises and economic activity (in contrast to atomized neoclassical models), field theory still needs to address fundamental questions about where and how new markets emerge (Radaev 2007: 41). Further, the stress on stability and certainty can hinder analysis of change and innovation (unless we fixate on "institutional entrepreneurs" alone). Seemingly ambiguous positions about incentives for innovation raise the question of necessary institutional conditions for implementing innovations. In field theory, we argue, the development of relations between institutions, structures, and key actors (including the state) are crucial. However, the theory of economic dynamics suggests innovations that can overcome economic stagnation or depression require chaos and dissipative structures (as scholars of synergetics claim) that break established institutional structures and path dependence, thus allowing sprouts of a new economy to grow.

Note that field theory has tools to analyze the development of crises: Fligstein shows how such problems as monopsony, oligopoly, and obsolescence of goods are characteristic of turbulent markets—as we would say in economic dynamics, markets at their final stage life cycle of development—and these problems are most acute in early stages of market development, although they can also result from unexpected business downturns. At the initial stage of market development, pioneers can make enormous profits. When other actors note this opportunity and enter that market, prices and profits fall (Fligstein 2001). This assumes causes of crises are related to stages of market development and life cycle, as suggested in economic dynamics theory (e.g. that of Schumpeter). The recent economic recession in the United States and Russia has precisely this nature that mirrors dynamics of Kondratieff 's theory of economic dynamics and Fligstein's theory of fields. Technologies developed in the fifth (Kondratieff) wave (1999–2008) and in the field of communications and multimedia have reached the stage of oligopolistic market structure in the face of mature, saturated demand. In these conditions, the macroeconomic trend is potentially deflationary (Rumyantseva 2016),[1] which completes the innovation cycle and prepares conditions for a series of bankruptcies and, potentially, conditions for creative disorder, in which new forms of innovative entrepreneurship can be born.

If we split the innovation process into two stages—a preparatory phase (fundamental research) and implementation phase (when inventions become innovations)—we can overcome what seems to be a contradiction between field theory and theories of economic dynamics. Implementing innovations is possible only if there is a fundamental basis for research and application, for which there must be institutions and fields—that is, stable, rule-bound relations between economic subjects, including patent law—so seeds for innovation can grow. In the preparatory stage of the innovation process, slender and subordinated fields are necessary. On the other hand, economic development is (we claim) cyclical, moving through periods of periodic dissipation of organizational structures that generate institutional chaos in depressions of long economic cycles. At these times, when conditions of uncertainty begin to enter fields (i.e. their foundations and reproduction began to weaken), there is opportunity or even need to devise new practices developed while economic structures were stable and fundamental science could develop with state support. If development of scientific and technological knowledge in moments of field stability is insufficient, then in a period of uncertainty there will simply be nothing to implement. Note that we presume one effective driving force of innovation and growth is state support for fundamental research (a public good) private enterprise can apply to innovation and profit. At the same time, state policies might protect fledgling industries, just as they support field reproduction.

Fields in these conditions in the modern global economy can serve as national innovation systems designed to stimulate innovation by facilitating a network of institutions for innovative development. The concept of a national innovation system was introduced by one follower of Kondratieff 's theory of economic dynamics, Christopher Freeman (Freeman 1988; Freeman and Pertez 1988).[2] Learning and innovation correspond to modern changes in production, and competition requires a shift in focus, from distribution of resources under stable parameters, to creating new resources in a situation marked by constant change in technologies, preferences, and institutions (Lundvall 1996). In general, Nelson and Winter (1985), Freeman and Pertez (1988), Lundvall (1988), Kim (1997), and others reoriented this concept to thinking of the firm as a learning organization. At the same time, we can assume that in the formation of national innovation systems, one effect noted by Mensch—innovations automatically overcome depressions in the national economy—might be blurred by a new conception of control in the global economy. In the context of globalization and development of national innovation systems, individual enterprises no longer struggle for leadership; rather, the struggle for status is between countries and transnational corporations. Further, field theory and the theory of economic dynamics draw attention to a potential obstacle to constructing national innovative systems: field and institutional elites have to buy into the project. The development of Russia's national innovation system as a conception of control runs up against limitations in the distribution of property rights, when the lion's share of income is concentrated in the hands of 1% of the population that is more inclined to withdraw capital abroad than to invest in domestic innovations.

Fields and Innovation in the Global Economy

Constructing a new field order often means confronting legacies of previous fields and institutions, which should not matter in neoclassical economics. Field theory suggests that field rules and practices can persist even after institutional reforms provide new templates for criteria and strategies of legitimacy and success. It follows that depressions and similar economic or institutional shocks can weaken preexisting rules and structures and allow for Schumpeterian creative destruction. In fact, Fligstein (1990) makes such discontinuities central in his analysis of American corporate history: shifts in conceptions of control occur either after economic shocks (the Great Depression, 1970s stagflation) or acts by state elites and social movements to implement new legislation (e.g. the Sherman Antitrust Act). If a country's economic fields are embedded in economic and political history, we should ask whether they are embedded in other, broader fields, such as global fields. While current discourse presumes "globalization" has had a fundamental impact on national economies and has reshaped fields, we should open up this issue up to empirical examination. While we should not ignore global institutions and pacts (e.g. WTO) or unequal distribution of economic power across countries and elites (e.g. giving rise to capital flight), we should also note that national politics and fields can act as counterweights, pushing back against globalizing forces or acting as templates that alter how global forces are translated into local fields and practices. It might be that, at least for leading countries, markets are not as globalized as many believe. As Fligstein points out, national political and economic elites are vitally interested in controlling "globalization." To protect their national interests and economic stability, a country's political elites might create national innovation systems to promote innovation at home and thus to survive in international competition (Evans 1995). Hoping for depressions or shocks to trigger institutional innovations raises the risk of losing in international competition. In Russia, systemic shocks (1992, 1998) served less to induce economic adjustment, than to reintroduce the state into the economy and facilitate the return of patronage relations and to weaken the development of venture financing for innovative projects.

Regarding diversification and innovation, Fligstein (2001) argued that after the crisis of the 1930s, the world entered an era of crisis-free development. This was due in part to the dominance of the United States and American capitalism outside the socialist bloc, but also to a concept of control based on innovating with brands and products to avoid direct competition—a "marketing conception of control" (Fligstein 1990). Fligstein also noted that links between markets in large diversified economies can be weak, such that crises are localized (Fligstein 2001). However, the global crisis of 2008 showed that crises can be more than localized when financial relations are global. Diversification of a company's horizontal links can create conditions that reduce the risk of overproduction spilling from one industry to another in production chains. Thus, economic crises since 1987 have been financial rather than of overproduction. Yet the latter were the kinds of crises previous conceptions of control were designed to address, and so rules and relations within and between fields have been less helpful supporting market stability.

This suggests that an important facet of field rules and relations—one less studied but certainly important—is how interactions between fields within the national economy can facilitate breakthrough technologies that improve competitiveness in the international system of fields. In conditions of the modern networked economy, this is difficult, given that the network form of business has generated "open innovations." While this might provide some advantages, such as formal or informal pooling of resources and experience to aid innovation, it also reduces incentives for serious innovation, because open innovations make it more difficult to claim returns on investments—and intellectual property rights were a classic foundation for innovation.

The State as a System of Fields

States intervene in economies and elsewhere to create appropriate rules and ensure stability, although how and when states intervene, and optimal forms of intervention, requires more study. As Fligstein points out, the likelihood of intervention and success depends on the character of that crisis, the state's institutional form and history, and whether societal groups can resist or adapt to interventions (Fligstein 2001). Note that this general formula is more typical for Western pluralist democracies and economies; a cruder version of this formula was relevant to Russia in the 1990s, but today is far less applicable, given growing imbalances of power and scope of control between state and civil society. Further, interventionist policies (in contrast to regulatory approaches) risk creating conditions for rent-seeking, which can have devastating consequences for institution-building. Interventionist states that avoid rent-seeking and encourage economic innovation and growth tend to have bureaucracies in which professionalism is an important feature of organizational culture and bureaucrats themselves are well educated. Non-trivial economic innovation and growth might require not only operative markets but also states that can intervene in measured form to encourage positive market forces (cf. Evans 1995). Russia's wager on the creative capacity of a "wild" market in the 1990s faced not only preexisting cultural codes and practices that were anti-capitalist but also the lack of market rules and structures *and* a state that lost capacity to govern and did not have officials with appropriate knowledge and professionalism to intervene and build markets without risking capture by market actors (or capturing markets themselves). Stimulating technological development in Russia would be less likely to follow market mechanisms, and more likely to depend upon technocratic state support—if, and only if, patrons found innovations in their interests, which should not be taken for granted. Legacies of the Soviet era did not leave field rules and practices conducive to innovation, unless a technocratic elite could emerge and tame various fields, especially the field of power.

One fundamental institution, property rights, was formed with participation of the state in Great Britain and the United States, whereby joint-stock companies were legalized to ensure economic growth. If views of ownership rights, governance structures, exchange rules, and conceptions of control are unstable, it is entirely possible to expect systemic economic crises. When property rights, management structures, conceptions of control, and exchange rules are underdeveloped, as in Russia, competition has no limits and can approach a Hobbesian war of all against all, hindering innovation and social returns; innovation works with state patronage. Underdevelopment of basic institutions generates corruption, which fills in gaps in which there are no clear field rules. The state continues to pursue paternalistic policies with regard to development, while ignoring challenges of the emerging technological order and not encouraging incentives to innovative behavior. Uncertainty and risk can be even greater in conditions of a networked economy.

The state itself is a system of fields that can apply massive resources to create political spheres and shape economic fields, whether by direct influence (investment in particular sectors) or by shaping allowable strategies (regulatory laws). Stability, exchange, and innovation depend on the extent to which the state can find a common language with owners, managers, and organized labor. According to theories of synergetics, innovations appear in conditions of chaos, when new ideas do not meet strong resistance from established structures and pave the way for new trajectories. However, even in moments of chaos, there must some potentially orderly structures to facilitate development. Competition between structures ensures selection of the most stable development trajectories. Alternatively, control and management substitute for innovation, and governance structures of corruption arise and fields end up benefiting the most powerful. The splitting of Russian society into "above," that has few incentives for innovations, and "below," that has few opportunities, suggests political spheres as a system of fields in Russia have not emerged: there is a sphere of cosmopolitan wealth and a sphere of quiescent "poverty," and the latter has neither institutional nor financial resources for innovative behavior.

Russian society is among the more unequal for developed countries. For the first three months of 2017, the Gini coefficient showed a positive trend, increasing to 0.396. At the same time, incomes of the richest stratum of the population continue to grow, while incomes of the lowest strata continue to shrink. As a rule, the poorest strata have only wages as income, while the wealthy have multiple sources, including rental property, which further characterizes the Russian economy as "rentier." In these conditions, Russian economic fields are highly differentiated in terms of income, where the goal and meaning of benefits and attitudes to politics differ in principle. The historical background to this process can be found even in the Soviet era, when in Leonid Brezhnev's administration, remuneration of intellectual labor (scientists, doctors, teachers) by administrative methods was equal to the level of workers' wages, reducing the incentive for high-performance work. In the post-Soviet period, this trend continued, turning highly skilled specialists into outsiders in the distribution of economic benefits and giving high incomes to groups with the opportunity to rent-seek (Voeikov and Anisimova 2018). According to accepted international assessments, if the ratio of incomes of the most and least well-off exceeds 10:1, that country is at risk for social instability—for which Russia might be increasingly ripe.

Particularly dangerous in this regard is the tendency to marginalize the *intelligentsia*, which most sharply perceives injustices of income differentiation—not so much or always for themselves, as for other disadvantaged people. Thus, in Russia at least three fields of classes seem to be emerging: rent-oriented elites, who receive the lion's share of national income and have a cosmopolitan ideology; low-income intellectuals, mostly patriotic but with a tendency to make marginal decisions; and an apolitical class of blue-collar workers and service personnel, who are trying to preserve meager revenues and show little tendency to political participation. This is also evidenced by traditionally low turnout for elections in modern Russia. Perhaps the ascetic ideal of Russian Orthodoxy that has been rising of late will prevent a social storm if the material situation of the *intelligentsia* and workers worsens if a new crisis hits in 2020–2025—a crisis that might be on the horizon, if Kondratieff's long waves exist.

The Politics of Market Rules and Contemporary Russian Reality

Social structures of markets can be viewed as tools actors wield to protect themselves against incomplete information, thus increasing opportunities to maximize profits. In this process, the role of culture is important for field development. Local norms and practices can begin to prescribe how competition is conducted in relevant markets and to offer cognitive schemes for interpreting actions of other organizations. In the context of fields, Fligstein (1990) dubbed such norms and strategies "conceptions of control." As such, one important question is how such norms and practices, especially regarding relations of economic agents to the state and power, have shaped Russia's institutional fields—especially given that markets are sets of fields, and that the state not only interacts with fields but itself is also a set of fields. As Fligstein and others have noted, structures and relations of power underlie market organization. In this context, we argue that social and technological innovations both play important roles.

The role of culture must be taken into account in connection with acyclic discontinuities in the development of Russia's economy. While development in Anglo-Saxon countries after the Industrial Revolution went through cycles of crises and innovations, Russian crises were associated with a change in trajectories of state management of the economy. Instead of gradual (but punctuated) evolution of concepts of control in a dialectic of fields and the state, Russia experienced cultural breaks—but remnants of cultural codes of tsarist Russia, socialism, and so on are difficult to reconcile with efficiency imperatives. Fligstein argues that no market is automatically effective; only basic market rules can ensure stability for growth. Efficiency in the absence of market rules is unattainable, and any attempts to cultivate them accounting for persistent Russian cultural codes only lead to a depletion of resources and predatory competition because the rule of the boldest, rather than the optimization of outcomes, is the dominant economic logic.

At the same time, it is important to note that large-scale social change unfolds only as its demand arises. Fligstein notes that when new markets are formed, demand for new laws and rules grows rapidly (Fligstein 2001): when new sectors emerge or existing sectors undergo transformation, new rules arise in the context of the old rules. Perez agrees in her analysis of how new institutional structures emerge during a transition from old to new techno-economic paradigms. In a tension between breakthroughs and resistance of existing institutions and interests, new technological breakthroughs answer new technological needs, and new relations between labor and capital, labor and the state, and the state and capital correspond to practices shaped by new technologies. Yet the question remains about sequences of events. What comes first: demand for a new social organization as a result technological changes (e.g. the development of the internet), or is it necessary to create institutional conditions for new technologies to occupy new markets?

Given this overview, we turn to a key question: Why are there too few innovations in Russia? What are Russian market rules? This is probably a two-way process: there should be market rules and concrete institutional mechanisms to allow for innovative development, and new industries can then push for modifying those rules once they have grown and become stable. Let us consider what the Fligstein's market rules might look like in contemporary Russia. Markets are arenas of social action, in which relationships are structured and have a hierarchical pattern, and dominant actors create and enforce rules and economic meanings.

Property Rights

Perhaps the most important of all the market rules, property rights determine right of control and decision-making vis-à-vis scarce resources, how risks are distributed, and claims to returns. The centrality of property rights is a foundation to New Institutional Economics, going back to Coase (1960), Alchian (1965), Demsetz (1964, 1967), and others. Common to these works is that economic resources are not just as physical objects but objects related to rights and relations of use. Assigning different rights over economic resources to different actors risks conflicts of interests (e.g. externalities). Such conflicts, as shown by Coase and his followers, can be resolved through negotiations and redistribution of rights via the market. However, if transaction costs are too high, a potentially beneficial redistribution of property rights may not occur and move an economy away from optimal allocation of resources. One way to reduce transaction costs is to make ownership rights as clearly specified as possible and reliably protected.

In the post-Soviet economy, privatization in the first half of the 1990s was the most important process that influenced property rights. Research into stages, features, and economic and social consequences of privatization can help make sense of peculiarities of existing ownership structures in the Russian economy, as well as hierarchies and contradictions between elites and the bulk of citizens. Privatization included two fundamentally different stages: mass voucher privatization and monetary privatization. A distinctive feature of Russian privatization was that it was conducted under difficult economic and social conditions, in the absence of public consensus, and in the context of many groups with myriad interests. Legal uncertainty also exacerbated the situation, which was recognized at the official level. In 2004, the Accounts Chamber (*Schetnaia palata*) of the Russian Federation issued a report (Schetnaia palata 2004) that revealed numerous inconsistencies, lacunae, and outright legal violations committed during the accelerated denationalization of property, along with the absence of sufficient audits and independent control of the privatization processes. Using World Bank data, R. M. Nureev noted that by the end of 1996, 55% of assets of state enterprises had been purchased by employees (including managers) and 11% by the population through vouchers, while 34% remained with the state. More than half of privatized property eventually ended up concentrated among insiders, mostly managers (Nureev 2001: 301). This meant there was no real separation of property and management. That the population at large possessed only a relatively small share of property through vouchers was due to poor public awareness of the process and mechanisms of privatization, weak market culture, and the prevalence of unscrupulous intermediaries (e.g. voucher investment funds). The most popular forms of privatization were corporatization, sale at commercial auctions, and rent with the option of future purchase of the enterprise.

The name of the second stage of privatization reflects its essence: auction privatization. This was supposed to solve two core tasks: to replenish the state budget and to attract investment for development. The combination of heated politics and the social situation in the country forced the state to give priority to replenishing the budget, which meant the need to quickly sell off the most attractive and liquid state assets. This often led to faux auctions, shadow contract deals, and scandals. In the infamous loans-for-shares schema of 1995–1997, the largest financial empires provided loans to the state in exchange for shares of state-owned companies at several times less than fair market value. In several cases, the sum of loans received by the state from commercial banks was equivalent to temporarily idle state funds already deposited in those same banks (Schetnaia palata 2004: 58–62). In 1995–1997, under loans-for-shares, the state privatized Iukos, Sibneft, Surgutneftegaz, Lukoil, and Norilsk Nickel. The recipients solidified their positions as the new economic elite, and they further influenced the development of property rights and other Russian market structures in the second half of the 1990s and the early 2000s. Nureev (2001) noted a change in relative prices in two sectors most important in the Soviet-era economy: the fuel and energy complex and the military industrial complex. The end of the Cold War, the collapse of the socialist bloc led to a decrease in the relative value of output from the military-industrial complex. At the same time, a steady rise in prices and markets for fuels and commodities increased the relative value of output from these industries. Representatives of the fuel and energy sectors grew in the polity, while representatives of the military-industrial complex, including security officials, declined. Trends were reversed later on, associated with a decline in oil prices and expansion of Russia's foreign policy activity.

While privatization expanded diversity of ownership forms in Russia, it also left a number of issues unsolved, the most important being legitimacy of property rights. Results of privatization were perceived among many citizens as unfair. Private property, especially private ownership of large entities such as enterprises and corporations, did not gain widespread acceptance as a natural or normal form of economic relations. According to sociological surveys, there has been no significant positive change over time in popular perception of private ownership of a large scale of assets. According to the results of a large-scale survey published in 2006 by the Institute of Sociology of the Russian Academy of Sciences, almost 47% of Russians surveyed considered it necessary to renationalize most previously privatized enterprises.[3] Opinions were correlated with socioeconomic status: 55.1% of respondents in the bottom income quintile believed it necessary to renationalize previously privatized assets, and 36.2% of respondents in the upper income quintile agreed. Ten years later, Grigoriev and Kudrin (2016) noted that the situation had not changed much: legitimacy of private ownership of major enterprises remains weak. In 2002 and 2013, more than 90% of respondents believed that all or the most important major enterprises should be state-owned. In 2015, as in 2006, 55% of respondents considered an economic system based on state planning and distribution to be "more correct" than one based on private property and market relations. Interestingly, Russians demonstrate a different attitude to private ownership of land and housing or entrepreneurship. For these, Grigoriev and Kudrin give the following figures: in 2013, 35% of respondents included "inviolability of property, home" as among the most important personal rights, but when a similar question was asked with a bias to entrepreneurial activity, only 16% of respondents believed this to be a most important right.

Governance Structures and Control of Competition

Stable market fields imply an established relationship of competition and cooperation between enterprises. These relations are influenced by formal institutions and laws, such as antimonopoly law or tariff and trade policies, and by informal institutions and actors' shared beliefs about what measures are acceptable for participating in economic life. Alas, at present these features hinder free and effective competition in Russia. The most important of these trends are highlighted in the annual report by the Federal Antimonopoly Service on competition in 2016. The report emphasizes that problems of competitiveness are systemic and apparent in many markets and virtually all regions. The Antimonopoly Committee notes these main problems: preservation of state monopolies; cartelization; systemic problems of procurement legislation; mandatory bidding for rights to state and municipal property, including on a concession basis; and an imperfect tariff regulatory system (FAS 2016: 8).

Market participants themselves consider competition to be moderate. According to the results of surveys of 1100 business representatives in early 2017 (ACGRF 2017: 4), only about half of respondents characterize regular competition as strong or very strong, while 25% do not feel effects of competition or face weak competition. Two years earlier, the percentage of entrepreneurs and market participants who considered competition to be low or absent was 21%. Competition is weakest for those in distribution, especially of electricity, gas, and water. These are infrastructure industries that have a network structure of organization, in which a high degree of state participation persists. It is noteworthy that 40% of survey respondents consider anticompetitive actions by the authorities to be a prime reason for the drop in competition. This might indicate real state actions hinder or reduce competition, or that Russian businessmen hold stable views that competition, and hence effective markets, depend on whatever model of behavior the authorities support. The second most mentioned force acting on competition has been changes in the regulatory framework (37%), which are in the scope of the state. The withdrawal of Russian and foreign competitors from markets contributed to the drop in competition, as noted by 34% and 8% of respondents, respectively (ACGRF 2017: 11). A general perception of the state's role is illustrated by the fact that only 2% of respondents—a number that included no representatives of small and medium-sized businesses—pointed to the positive role of the state creating and maintaining a competitive environment. At the same time, 40% of respondents say that the state only hinders business, up from 33% in the previous year (ACGRF 2017: 14). Interestingly, 22% of those polled were interested in "proper" state intervention as an alternative to a "harmful" (predatory) state. Factors preventing entrepreneurs from developing their businesses, and that the state should address, include increasing available financing (including cheaper loans), reduction of taxes, stabilizing regulatory legislation, and combating corruption (ACGRF 2017: 12).

State-monopoly trends Russia are manifested in the state's increasing share of the economy. The volatility of property rights and ongoing processes of property redistribution have led to a situation in which, according to the Antimonopoly Service, the state's share of the economy is greater than 50% of GDP, compared to 25% before the 1998 economic crisis (FAS 2016). If we take into account the state's share in commercial enterprises, formally considered private property (e.g. shares of joint-stock companies), then the index of state participation in the economy is even greater. As of January 1, 2017, shares of 1356 joint-stock companies were held by the state; for 575 of these enterprises, the state was the majority shareholder, and for 81 joint-stock companies, the state held the "golden share" (i.e. a share that provides authority to outvote other shares or veto fundamental organizational changes). We should note that the number of joint-stock companies with state participation is declining, with a decrease of 13% in 2016. At the same time, the number of unitary enterprises is increasing, from 11,252 to 21,034 between 2013 and 2017.[4] Under the law, state or municipal enterprises can be unitary enterprises, a state-owned corporation that does not have property rights over assets used. Such enterprises have a negative impact on regional competition, since they function in a context of weakened market signals or incentives, for example, weaker budgetary constraints or "protected" demand for their products. While the main task of unitary enterprises is to implement state functions, in Russia they are often founded in competitive industries. For example, of the 1293 unitary enterprises with federal ownership on January 1, 2017, 201 are engaged in such highly competitive industries as construction, hotels and restaurants, wholesale and retail trade, and vehicle repair. The low efficiency of unitary enterprises has been the subject of attention by some of the authorities. A study conducted by the government's Expert Council discovered that labor productivity in unitary enterprises is 4.5 times lower than the average for enterprises with other organizational and legal forms. A number of unitary enterprises have not carried out any economic activity for years; the main activity for others is leasing property assigned to them. Privatization of unitary enterprises has not gone as fast as desired, and in preparation for privatization, it was found that more than half of federal unitary enterprises did not have properly registered ownership rights to real estate.

Anticompetitive agreements, especially cartels, also hinder efficiency in Russian markets. A specific feature of anticompetitive agreements is their ubiquity, as well as the authorities' deep involvement (ACGRF 2017: 15), and corruption as well as inefficiency is a problem. According to information from the Antimonopoly Committee, in 2016 more than 300 of nearly 700 cases initiated against anticompetitive agreements involved cartels. The most common type of cartel is bidding collusion, that is, during the process of placing bids for the purchase of goods and services for state needs—a type of cartel collusion that likely does not occur without participation or connivance of the authorities. The end result of such agreements is inefficient use of budgetary funds (e.g. in construction, where almost a quarter of all cartels are fixed, or in defense and security, which are the least transparent) and a significant increase in final prices for key goods and services (e.g. health and communal services make up another quarter of all cartel collusion). Further, public procurement is one of the biggest areas of anticompetitive practices. One can see here the connection between low competitiveness and unitary enterprises described above: until early 2017, unitary enterprises used a special procurement regime to avoid compliance with legal requirements for tenders for providing goods and services for state needs.

One not uncommon practice affecting competitiveness has been using trade restrictions against certain countries as an extension of foreign policy and measure of political pressure. Goods from Georgia, Ukraine, and Moldova have been banned at various times. Large-scale restrictions on imports (e.g. meat and dairy products, vegetables, fruits, fish) from the European Union, the United States, and several other countries were established in mid-2014 in response to sanctions against Russia over the annexation of Crimea. It is obvious that the artificial narrowing of suppliers had a negative impact on competition in food markets and a sharp increase in consumer inflation for food, averaging 28.7% for late 2013 to May 2015. Experts estimate that about 20% of this price growth was due to ruble devaluation; the rest was a consequence of inflation and restrictions on imports (ACGRF 2015a: 5). (Note that prices for other goods not hit by restrictions also rose.) Russian consumers also experienced a decline in quality of goods and an increase in counterfeit products, especially in dairy markets. Experts of the Antimonopoly Service and the Analytical Center for the Government of the Russian Federation drew attention to the fact that attempts to accelerate substitution of imported goods by domestic products to prevent a deficit and to control prices could conceal the reduction of competition and freeze it for many years. State policies for rapid import substitution include providing some economic actors (often affiliated with the state) with unreasonable preferences, which can induce quasi-monopolies requiring constant financial support.

Mergers and acquisitions (M&A) and bankruptcies—what we call "market concentration transactions"—also affect the structure and concentration of market actors, according to analyses by the Analytical Center into the Russian economy during the global economic crisis. During acute phases of the crisis (2008–2010 and 2014–2015), activity on the Russian M&A market declined, while the number of bankruptcies grew. If in 2007 more than 1100 market concentration transactions were conducted with participation of Russian companies, in 2009 there were only 541 such transactions. There was some recovery afterward, but the market has not returned to its pre-crisis level. By 2015, according to KPMG,[5] Russia's share in the global M&A market dropped to 1.3%, its lowest in 10 years. Authors from the Analytical Center attribute this behavior of Russian companies to several factors. First, during the period under review, the access of Russian companies to financial resources declined sharply—something noted in surveys cited earlier. The falling price of oil and the value of Russian fuel and energy assets could have increased their attractiveness to foreign investors, but political tensions and sanctions meant that, in terms of value, the share of M&A export deals fell to 4%, versus a post-crisis figure of 11% (ACGRF 2015a: 11). Finally, state regulations played a role: "de-offshorization" became one factor "to reduce activity of Russian investors (both legal entities and individuals) in relation to investments in foreign assets and cross-border mergers and acquisitions" (ACGRF 2015b: 6).

Rules of Exchange and Conditions for Transactions

Market stability requires that exchange occurs according to rules that are stable, predictable, understandable, and shared. These are formulated by the state when creating regulatory standards. An excessive number of regulatory norms, like their absence, can blur exchange rules. Russia's economy in the early 1990s was characterized by the lack of established exchange rules, but by the end of the 1990s excessive regulation had become a hindrance to economic growth. In their comparison of regulatory models in Russia and Poland, Frye and Shleifer (1987) concluded that the more liberal Polish model allowed the economy to stabilize and grow more quickly and successfully. In their study of process and consequences of regulating entrepreneurial activity in Russia in the early 2000s, Yakovlev and Zhuravskaia (2011) concluded that procedures for creating new businesses have not become more permissive. Some liberalization touched on new business registration, licensing rules, and inspections. In accordance with new laws adopted in 2001, the number of licensed activities was reduced from more than 200 to 120 (and in 2011 down to 50). Further, although the new regulations were country-wide, their impact varied by region. Regions with transparent governance saw growth in the number of small enterprises, while in regions with less transparency saw little change. The same formal institutions ultimately depended on local fields and local informal rules, in particular local governments.

For a modern market economy, stability of exchange rules facilitates compliance with contractual obligations—among the most important institutions, according to New Institutional Economics. Contracts act as channels for the exchange of property rights and can aid efficiency. Sustainability of contractual obligations is a guarantee that potentially profitable contracts will not be rejected because of risks and future uncertainty. If contract cannot cover potential risks—for example, when asset specificity is high and risks of "opportunism with guile" are real (Williamson 1985)—organizations can use property rights to augment control and reduce risks that contract cannot. However, contract and property rights are only as good as the willingness and capacity of states to recognize them. Here the judicial system—which gives economic actors the opportunity to act in their own interests and for recourse—becomes central to constructing institutions and fields. Judiciaries also can act as one venue for resolving disputes and lowering costs of action. Alas, we take the position that the quality of Russia's judicial system is questionable at best, in contrast to some findings that claim the opposite. For example, one of the more authoritative international rankings—the World Bank's "Doing Business" rating—ranked Russia in 2016 as 51 out of 189 countries. In that same year, Russia took fifth place (or eighth, according to adjusted data) in terms of "enforcement of contracts," which considers three parameters: time needed to resolve disputes, costs of court procedures as a percentage of the amount of the claim under scrutiny, and an index of quality of the judicial system (World Bank 2016: 229). However, a more careful consideration of these parameters allows us to draw additional conclusions. Russia's high position is explained by the low cost of using the judicial system and by moderate terms of dispute resolution. However, the quality of Russia's judicial system is inferior to those not only in highly developed countries (Great Britain, the United States, Germany) but also in such lesser-developed countries as Brazil, Turkey, and Georgia.

This seeming difference between ratings and aspects of reality is mirrored in another paradox of the judicial system, Russia's arbitration courts, specialized courts dealing with resolving disputes of entrepreneurial and other economic activities. The judicial system as a whole is consistently part of a set of institutions that Russians trust least. According to the Levada center, in 2016, only 22% of Russians believed Russia's courts "are quite trustworthy."[6] Yet the number of lawsuits regarding economic disputes brought to arbitration courts grows year by year. If in 2007, 1.1 million applications were submitted to arbitration courts,[7] 1.5 million were submitted in 2012, and 1.7 million in 2016.[8] For 2016, Russia's arbitration courts considered more than 1.5 million lawsuits,[9] of which more than 1 million concerned nonfulfillment or improper performance of contractual obligations, amounting to claims of 3.5 trillion rubles. Compliance with established exchange rules—not only with formal law and institutions but also with norms of what a "contract" means and its centrality to transactions—is the main subject of disputes and the main subject of appeals to the judicial system by Russian entrepreneurs.

Sociological research (Khakhulina 2011) on confidence in arbitration courts shows that more than half of enterprise owners or managers surveyed have experience using arbitration courts, and that there is variation in this experience. First, this indicator varies greatly depending on the size of the enterprise: from 27% among microenterprises, to 87% among large enterprises. Second, the level of trust in arbitration courts is higher for managers of enterprises with personal experience participating in litigation than among those who did not. Third, the level of trust is positively correlated with the size of the enterprise: the larger the enterprise, the higher the confidence in arbitration courts. It is possible that such variation arises because large enterprises can afford professional lawyers, and this helps them obtain favorable outcomes and generates positive perceptions of the system for dispute resolution. Among positive qualities of arbitration courts, respondents noted professionalism, openness, objectivity, and compliance with rules of proceedings. The courts received lowest assessment for independence, and entrepreneurs surveyed believe arbitration courts are under pressure from superiors, federal and local authorities, prosecutors and police, and even important individuals (Khakhulina 2011: 36). Judges are not perceived as playing an objective role in interpreting and enforcing legal norms. However, they do play a crucial role shaping market fields when their decisions empower or disempower particular sets of actors (e.g. larger enterprises), while they themselves are also products of professional and cultural fields—note the perception that lawyers and professionals do matter, even if they do not necessarily predominate in shaping ultimate court decisions. Additionally, these courts are embedded in fields of networks and property interests, which are obvious to those outside the system. As Kathryn Hendley notes, Russians are well aware of different categories of cases and probabilities of successful outcomes. They understand that despite formal laws and rules, there are informal norms that are even more important: Russians have a "culturally conditioned understanding of when corruption is possible, and for obvious reasons, in such circumstances, they avoid going to court" (Hendley 2012: 272). Regarding these perceptions of rules for the judicial field, Hendley (2012: 268–269) noted three motives that affect willingness of Russian entrepreneurs to resort to the judicial system: need, opportunity, and attitude to the law. The strongest is need, especially debt collection: a court order is often the only way to solve non-payments, and the relatively low costs of money and time for litigation make court procedures for resolving such disputes attractive. Further, the very fact of litigation between parties in a contractual relation does not prevent them from continuing to work together. Even doubts about effectiveness, integrity, and independence of the judicial system are not insurmountable obstacles to using courts. In sum, we see that "contract" and the like are embedded not only in institutions (judiciary) but also in fields of actors, statuses, and rules that shape what actors expect from institutions: for example, who can play the game, likelihood of success for different strategies, actors' statuses (e.g. lawyers), and so on. Measuring how market rules operate requires more than just turning to what is in the law or procedures and statistics; we have to tease out real perceptions and practice, which those in fields provide.

Conceptions of Control—Hierarchies and Status Ordering

Conceptions of control are basic field rules of normal strategy and structure, but these rules do not emerge of their own accord. Rather, they reflect a hierarchy and ordering of field elites and contingent interactions between groups of dominant actors and between elites and other actors (Fligstein 1990). Much as elite politics shape (but do not entirely determine) legislation and institutions, they also shape structures and rules of fields. And this issue—elite interests, positions, capacity, and fields—is of great interest to social scientists in Russia. In particular, one pressing issue is the degree to which the elite has attained social closure and constitutes a closed group, growing differences between interests of elites versus the broader public, and perhaps growing splits within the elite.

The origins and rise of today's Russian elites are inseparably linked with the distribution of property and property rights, and shifts in these have meant changes in the composition and relative power of different elite groups. One theory of the origin and hierarchy of Russian elites is the concept of "power and property," which is a modern reading of Marx's idea of the Asian mode of production. According to this concept, Russia historically has had systems of political and economic despotism, in which the state plays a predominant role in economic, political, ideological, and social spheres, without elites or norms such as a feudal contract acting as a counterweight. Path dependency (the persistence of autocratic structures and practices in fields) has hindered the importation of democratic and market institutions. The essence of "power and property" lies in the fact that power and access to material goods are determined by one's place in a status hierarchy and are guaranteed by personal loyalty to the ruler—a variation of Max Weber's "patrimonialism" and manifested in historical features of Russian society as state authoritarianism and communal collectivism (Latov 2004: 114). Such hierarchical social relations result from objective living and economic conditions, such as low productivity in land, high risks of farming, the need to reduce risks via centralized food funds run by the state (and their distribution in the event of famine), and guarding against external aggression across extended geographic borders.

The economic foundation for this system is restriction of property rights, state appropriation of basic means of production, and ultimately state monopoly over various modes of production (cf. Radaev and Shkaratov 1991)—features that do not seem entirely absent in Russia today. The Soviet command economy was characterized by a partial fusion of the Communist Party and Soviet state and their joint monopoly on political and economic power. For the post-Soviet period, some scholars distinguish between two stages (cf. Nureev 2001): an initial period in which this monopoly began to crumble and allowed the possibility to create Western-style property regimes (1992–2000), and a second period of rebuilding and reinstitutionalization of "power and property" (from 2001 to the present). In the first stage, a wide class of new private owners, made up of members of the old managerial *nomenklatura* and younger financial elites with support from reformist elites in the Kremlin, divided property among themselves and sometimes clashed over property and influence in higher circles of power. This led to oligarchic capitalism. In the second stage, state elites under Vladimir Putin set out to concentrate economic power around the state, especially in a new elite involving the security forces. Significant events in this second stage included the attack on Iukos, the state's purchase of Sibneft, nationalization of Bashneft, continuing low legitimacy of private property, construction of a "power vertical" around the presidency, and the growth of the public sector. Additionally, structures of "power and property" are characterized by parallel growth of corruption networks. Intensive development of such structures in this system is due to the centrality of patron-client relations between officials and businesses, as well as the need to demonstrate and maintain personal loyalty to superiors in the political hierarchy (cf. Pliskevich 2006). When disposal of resources depends on positions in this hierarchy, business representatives are forced to "buy" access to resources from officials and bureaucrats. Corruption networks also take shape within dominant groups, that is, between officials as well as between officials and business. Thus, not only is private money drawn into corruption schemes; state capital is as well, for example, from budgets. As a result, the dominant group seeks to protect the *status quo* from potential competitors or "supplicants," such as Russian citizens. In the context of declining budget revenue, this situation increases social tensions.

This resembles the model of "neo-patrimonialism" in political science, and there is more than a passing resemblance between that model and the political economy of fields and property in today's Russia. In a neo-patrimonial regime, with its mechanisms of political and economic governance, political and economic power is appropriated for private purposes of rent extraction (Gelman 2015: 37). The nucleus of such a system includes informal institutions and rules formed within the ruling elites, but hidden behind the façade of formal laws and institutions. Neo-patrimonialism is thus in contrast to "rational-legal" authority in Western democracies and manifests itself as a personalistic authoritarian regime in politics and crony capitalism in economic governance. Two important characteristics of neo-patrimonialism are an orientation to rent extraction and dominance of informal institutions. These traits shape cultural meanings shared by representatives of ruling elites, who seek to reproduce themselves and prevent challenges to the *status quo*. Rent-seeking in such a system inevitably leads to the creation and maintenance of "extractive" institutions that contribute little to growth (Acemoglu and Robinson 2006, 2010, 2012). Altering such a system grounded in patronage networks, such that field rules are personalized rather than institutionalized, is difficult. First, rent-seeking behavior and suboptimal equilibria they generate can be stable (Murphy et al. 1993). To move to a more efficient equilibrium, significant institutional changes are needed—but elites have few intrinsic incentives to do so. Only outsiders can force such changes not only in institutions but especially in fields constructed on a network, rather than institutional, foundation; prime candidates to impose such change include social movements and external elites (cf. Fligstein and McAdam 2012). At present, the regime has co-opted, cowed, or neutered civil society and challengers.

A second feature of neo-patrimonialism is the creation of informal rules and their de facto dominance over formal laws. Unlike formal law, informal rules are opaque, creating the risk of different or contradictory interpretations and the absence of universal sanctions for violations. The existence of informal rules is difficult to prove, as they are internal to relations of patronage and power and are invisible or incompletely apparent to field outsiders. Informal rules can be stable and their expansion difficult to stop, even if they are destructive. In modern Russian neo-patrimonialism, informal rules are expressed in exchange of patronage for personal loyalty. Informal institutions are often hidden behind a formal shell, including the infamous power vertical (cf. Gelman 2015: 40). However, this system of informal rules carries risks that become increasingly evident as rents shrink. In this case, resource struggles intensify, as insiders begin to perceive mixed signals, and in the case of impotent formal institutions, this can lead to paralysis of managerial activity, hindrance of communication between officials and society, and loss of management culture. When redistribution of spheres of influence and resources begins in the higher echelons of power, lower ranks become hostages, as the adoption of any decision from above is likely to lead to a breach of loyalty on one or the other side. Scholars at the Carnegie Center in Moscow call this a "deinstitutionalization trap"[10] and suggest Russia is falling into it—in turn aggravating institutional weaknesses, a forced transition to direct and personalized management when the institution of the presidency remains the only truly independent institution. The neopatrimonial system contributes to the formation of a worldview among officials that reduces economic activity not to creation but to distribution and redistribution of wealth. Socioeconomic modernization does not correspond to elite interests, and without such modernization it is impossible to innovate and improve society and economy.

Conclusion

Combining economic dynamics with field theory, we have argued that during periods of instability, existing links between economic actors weaken, boundaries of fields are modified and blurred, and prerequisites for emergence of new fields on the basis of innovations and new social practices arise. This is a process of weakening and strengthening of ties between field elements, which correlates with change of technical and economic paradigms and the long economic cycle. Stable, non-rent-seeking states in this case could create fields that, due to interactions between actors, promote innovative behavior. On the other hand, innovative activity, as economic dynamics shows, increases in periods of instability or downturn. The role of the state as a system of fields should consist in breaking up the innovation process at the stage of fundamental development and implementation. When the system is stable and fields are subordinated, conditions can support research and innovation by the state and firms, and in conditions of uncertainty, incentives are created for mass implementation of innovations.

Social and technological changes stimulate demand for new institutions. In this context, it is important how Fligstein's market rules in modern Russia have developed and transformed, and their relationship with the state and its elite. Property rights, constructed in Russia during the contentious process of privatization, created widespread cynicism and perceptions of injustice, hindering the legitimacy of property rights and opening the door to making property rights a tool of state power rather than economic gain and development. Existing governance structures create significant restrictions on competition, while even entrepreneurs await state managers to support business. Exchange rules are characterized by excessive regulation and low quality of the judicial system. Ultimately, current conceptions of control that govern field behavior have at their core neo-patrimonial power-and-property structures, which impede effective distribution and use of property, generate corruption, and strengthen incentives and opportunities to rent-seeking rather than innovation and overall social well-being.

Notes

1. The index of consumer prices in the United States dropped to zero in 2008 and has remained low ever since. Russian industrial production dropped in 2008 and has remained stagnant.
2. Golichenko (2014) has been developing a theory of national innovation systems in Russia.
3. http://www.isras.ru/INAB_2008_2_2_3.html.
4. This is according to data from the Federal Tax Service of the Russian Federation: https://www.nalog.ru/rn77/related_activities/registration_ip_yl/#t5.
5. https://assets.kpmg.com/content/dam/kpmg/pdf/2016/05/ru-ru-ma-survey-report-may-2016.pdf.
6. http://www.levada.ru/2016/10/13/institutsionalnoe-doverie-2/.
7. http://www.arbitr.ru/_upimg/9D21B607468CB87DE8953D0FF45866E3_02_itogiVAS08.pdf.
8. http://cdep.ru/userimages/sudebnaya_statistika/2016/AC1_2016_svod.xls.

9. http://www.arbitr.ru/_upimg/E71E1F5763D26D47E142A3F677BED00C_3.pdf.
10. http://carnegieendowment.org/files/CP_Kolesnikov_June2015_web_Rus.pdf стр 28.

Works Cited

Acemoglu, Daron, and James Robinson. 2006. De Facto Political Power and Institutional Persistence. *American Economic Review* 96: 325–330.

———. 2010. The Role of Institutions in Growth and Development. *Review of Economics and Institutions* 1 (2): 1–33.

———. 2012. *Why Nations Fail: The Origins of Power, Prosperity, and Poverty*. New York: Crown.

ACGRF [Analiticheskii tsentr pri pravitelstve Rossiiskoi Federatsii]. 2015a. Prodovolstvennoe embargo: importozameshchenie i izmenenie struktury vneshnei torgovli. Biulleten o razvitii konkurentsii. Moscow: http://ac.gov.ru/files/publication/a/6007.pdf.

———. 2015b. Kontsentratsiia na rossiiskikh rynkakh: tendentsii v period ekonomicheskogo spade. Biulleten o razvitii konkurentsii. Moscow: http://ac.gov.ru/files/publication/a/7470.pdf.

———. 2017. *Otsenka sostoianiia konkurentnoi sredy v Rossii*. Moscow. http://ac.gov.ru/files/publication/a/12824.pdf. Accessed November 6, 2017.

Alchian, Armen. 1965. Some Economics of Property Rights. *Il Politico* 30: 816–829.

Coase, Ronald. 1960. The Problem of Social Costs. *Journal of Law and Economics* 3: 1–44.

Dementiev, V.E. 2016. Dlinnye volny v ekonomike: institutsionalny aspect. In *Krizisy i prognozy v svete teorii dlinnikh voln*, ed. L.E. Grinin, A.V. Korotaev, and R.S. Grinberg, 123–139. Moscow: Uchitel.

Demsetz, Harold. 1964. The Exchange and Enforcement of Property Rights. *Journal of Law and Economics* 7: 11–26.

———. 1967. Toward a Theory of Property Rights. *The American Economic Review* 57: 347–359.

Evans, Peter. 1995. *Embedded Autonomy*. Princeton: Princeton University Press.

Federal Antimonopoly Service of the Russian Federation [FAS]. 2016. *Doklad o sostoianii konkurentsii v Rossiiskoi Federatsii za 2016 god*. Moscow. https://fas.gov.ru/documents/589904. Accessed November 23, 2017.

Fligstein, Neil. 1990. *The Transformation of Corporate Control*. Cambridge: Harvard University Press.

———. 2001. *The Architecture of Markets*. Princeton: Princeton University Press.

Fligstein, Neil, and Doug McAdam. 2012. *A Theory of Fields*. New York: Oxford University Press.

Freeman, Christopher. 1988. Japan: A New National System of Innovation? In *Technical Change and Economic Theory*, ed. G. Dosi, C. Freeman, R. Nelson, G. Silverberg, and L. Soete, 330–348. London and New York: Pinter.

Freeman, Christopher, and Carlota Pertez. 1988. Structural Crises of Adjustment, Business Cycles and Investment Behaviour. In *Technical Change and Economic Theory*, ed. G. Dosi, C. Freeman, R. Nelson, G. Silverberg, and L. Soete, 38–66. London and New York: Pinter.

Frye, T., and A. Shleifer. 1987. The Invisible Hand and the Grabbing Hand. *The American Economic Review* 87 (2): 354–358.

Gelman, V. Ia. 2015. *Modernizatsiia, instituty i "porochny krug" postsovietskogo neopatrimonializma*. St. Petersburg: European University Press.

Golichenko, O. 2014. Natsionalnaia innovatsionnaia sistema: ot kontseptsii k metodologii issledovaniia. *Voprosy ekonomiki* 7: 51–64.

Grigoriev, L., and A. Kudrin. 2016. Nereshenny vopros legitimnosti chastnoi sobstvennosti v Rossii. *Voprosy ekonomiki* 1: 36–62.

Hendley, Kathryn. 2012. Ob ispolzovanii sudebnoi sistemy v Rossii. In *Kak suii prinimaiut reshenii: empiricheskie issledovaniia prava*, ed. V.V. Volkov, 267–325. Moscow: Statut.

Iberall, A., and D. Wilkinson. 1987. Dynamic Foundations for Complex Social Systems. In *Exploring Long Cycles*, ed. George Modelski. Boulder: Lynne Rienner.

Khakhulina, L. 2011. Doverie biznes soobshchestva arbitrazhnym sudam i otsenka arbitrazhnoi sudebnoi sistemy. *Vestnik obshchestvennogo mneniia* 4: 32–46.

Kim, Linsu. 1997. *Imitation to Innovation. The Dynamics of Korea's Technological Learning*. Cambridge: Harvard Business Review Press.

Kondratieff, N.D. 1925. Bolshie tsikly Koniunktury. *Voprosy koniunktury* 2 (1): 28–79.

———. 1935. Long Waves in Economic Life. *Review of Economic Statistics* 17: 105–115.

———. 1989. *Problemy ekonomicheskoi dinamiki*. Moscow: Ekonomika.

Latov, Iu.V. 2004. Vlast-sobstvennost v srednevekovoi Rossii. *Ekonomicheskii vestnik Rostovskogo gosudarstvennogo universiteta* 2 (4): 111–133.

Lundvall, B.-A. 1988. Innovation as an Interactive Process: From User-Producer Interaction to the National System of Innovation. In *Technical Change and Economic Theory*, ed. G. Dosi, C. Freeman, R. Nelson, G. Silverberg, and L. Soete, 349–269. London and New York: Pinter.

Lundvall, B-A. 1996. *The Social Dimension of the Learning Economy*. Working Paper 96-1, Danish Research Unit for Industrial Dynamics, Aalborg University.

Mensch, Gerhard. 1978. *Stalemate in Technology: Innovations Overcome the Depression*. New York: Ballinger.

Murphy, Kevin M., Andrei Shleifer, and Robert W. Vishny. 1993. Why Is Rent-Seeking So Costly to Growth? *American Economic Review Papers and Proceedings* 83 (2): 409–414.

Nelson, Richard, and Sidney Winter. 1985. *An Evolutionary Theory of Economic Change*. Cambridge: Harvard University Press.

North, Douglass. 1990. *Institutions, Institutional Change, and Economic Performance*. New York: Cambridge University Press.

Nureev, R.M. 2001. *Ekonomicheskie subekty postsovietskoi Rossii (institutsionalnyi analiz)*. Moscow: Moskovskii obshchestvennyi nauchnoi fond. http://ecsocman.hse.ru/text/19204386/.

Osipov, Iu.M., and I.N. Shurgalina. 1994. *Perekhody i katastrofy: opyt sotsialno-ekonomicheskogo razvitiia*. Moscow: Moscow State University.

Perez, Carlota. 2009. *Technological Revolutions and Techno-Economic Paradigms*. Working Papers in Technology Governance and Economic Dynamics, The Other Canon Foundation. Tallinn: Norway and Tallinn University of Technology.

Pliskevich, N.M. 2006. 'Vlast-sobstvennost' v sovremennoi Rossii: proiskhozhdenie i perspektivy mutatsii. *Mir Rossii. Sotsiologiia. Etnologiia*. 3: 62–113.

Radaev, V. 2007. *Analiz rynkov v sovremennoi ekonomicheskoi sotsiologii*. Moscow: Makes Press.

Radaev, V.V., and O.I. Shkaratov. 1991. Vlast i sobstvennost. *Sotsiologicheskie issledovaniia* 1: 50–61.

Rumyantseva, S.Iu. 2014. Proportsii liberalizma i dirizhizma pri stimulirovanii innovatsionnogo razvitiia s uchetom karty ekonomicheskoi koniunktury. In *Neodirizhizm i modernizatsiia rossiiskoi ekonomiki*, ed. Iu.M. Osipov, S.Iu. Sinelnokov, and E.S. Zotova, 148–165. Moscow: RG-Press.

———. 2016. Teoriia ekonomicheskoi dinamiki N.D. Kondratieva i sovremennye dlinnovolnye protsessy. In *Krizisy i prognozy v svete teorii dlinnikh voln*, ed. L.E. Grinin, A.V. Korotaev, and R.S. Grinberg, 90–122. Moscow: Uchitel.

Saviotti, P.P. 1986. Systems Theory and Technological Change. *Futures* 18 (6): 773–786.

Schetnaia palata Rossiiskoi Federatsii. 2004. Analiz protsessov privatizatsii gosudarstvennoi sobstvennosti v Rossiiskoi Federatsii za period 1993–2003 gody. Moscow. http://nationalization.ru/biblioteka/Analiz-processov-privatizacii-gosudarstvennoj-sobstvennosti-v-rossijskoj-federacii-za-period-1993-2003-gody.pdf.

Voeikov, M.I., and G.V. Anisimova. 2018. *Politicheskaia ekonomiia neravenstva*. Moscow: Lenand.

Williamson, Oliver. 1985. *The Economic Institutions of Capitalism*. New York: Free Press.

World Bank. 2016. *Doing Business 2016: Measuring Regulatory Quality and Efficiency*. Washington, DC: http://www.doingbusiness.org/reports/global-reports/doing-business-2016.

Yakovlev, E., and E. Zhuravskaia. 2011. *Unequal Effects of Liberalisation on Diversification of Russia's Regions*. EBRD Working Paper No. 128 (April 2011). http://www.ebrd.com/downloads/research/economics/workingpapers/WP0128.pdf. Accessed November 6, 2017.

11

Economic Theory and a Constant Worry Across Time: Institutional Failures in the Development of Theories of Inflation

Aleksandr Protasov

This chapter attempts to apply two non-standard theoretical approaches to the analysis of inflation in Russia. The first is the "conflict" theory of inflation (sometimes called a "social accords" approach), which arose in the late 1970s thanks to the work of sociologist John Goldthorpe, Richard Erickson, and other scholars of social class (Hirsch and Goldthorpe 1978; Goldthorpe 1987). In this approach, inflation results from a constant conflict between different classes and social groups, connected with their struggle for redistributing income and increasing their share of its total value. One form of struggle in this redistributional conflict is inflation in the form of price increases. A second approach, developed in the tradition of modern economic sociology by Pierre Bourdieu (2005a, 2005b) and Neil Fligstein (2001), considers the market as social fields in which structuring of economic relations and formation of market structures results from competition that is not only market-based. This approach, in which elites struggle to

A. Protasov (✉)
Faculty of Economics, Department of Economic Theory,
St. Petersburg State University, St. Petersburg, Russia

J. K. Hass (ed.), *Re-Examining the History of the Russian Economy*,
https://doi.org/10.1007/978-3-319-75414-7_11

control markets, make profits, and ultimately dominate power relations in the economic field, substantially complements the conflict approach and deepens our understanding of inflationary processes.

Bringing together these two approaches, I explore a new aspect in the development of inflationary processes in the Russian economy from the beginning of market reforms in the 1990s to the present. In particular, we propose the hypothesis that the nature of inflation in Russia is rooted in redistributional conflicts, and its dynamics, having a cyclical nature, are determined by "conflict potential" in the market. This chapter begins with a consideration of various approaches in economics to inflation in general and in Russia in particular and then analyzes the cyclical dynamics of Russian inflation using conflict theory and field theory.

The Problem of Inflation in the Field of Russian Economics

According to Bourdieu, a field as a social construct is a relatively closed and autonomous space of social relations between various agents who occupy certain positions in that given space. In other words, the field is a structured space of positions of individual or group agents, differing from each other by the strength of power, influence on other parties, gain obtained, and capital accumulated in its various forms (economic, cultural, social, symbolic). From this point of view, a field of science is part of a social space in which social interactions between scientific groups, schools, trends, or their individual representatives are expressed in the struggle of scientific ideas claiming to occupy a monopoly position of scientific authority (Bourdieu 2005b: 474, 571).

Among existing competing concepts of inflation in the field of Russian economics, several approaches to the conceptualization of this subject can be singled out. The first approach (functional) is characteristic of modern Western economic thought, which often simplistically interprets the essence of inflation and emphasizes the analysis of its functional dependencies with other phenomena of social and economic life. In accordance with this approach, the essence of inflation manifests itself either in a rise in the general price level (Mankiw 2009) or in depreciation of money (or reducing the value of a monetary unit) (Timberlake 2000; Salerno 1999), or as a synthesis of the first and second, that is, as a process of depreciation of the monetary unit, accompanied by a steady increase in the general price level. The drawback of this approach is that it is not able to adequately describe inflation in countries with non-market economies. Indeed, if we proceed from this approach, it becomes incomprehensible how inflation can exist in a rigid price framework that does not change for a long time. It is known that in such economies, the level of tension is a means of expressing inflation, but then it turns out that the essence of inflation is in no way connected with the dynamics of the general price level.

The second approach (deterministic) is associated with the tradition of searching for root causes and contradictions of any phenomena and is more characteristic of Marxism and traditions of Russian economic thought. Within the framework of this approach, the essence of inflation is revealed through its connection with disproportions of social reproduction. Inflation is a process of violating the law of money circulation, manifested in the depreciation of money and uneven price growth, which are caused by disparities in social reproduction and the policies of the ruling classes (Matiukhin 1984: 7). This approach is closely related to the reproductive (or systemic) approach developed by L. N. Krasavina, according to which inflation is a multifactor process that combines its main causes, immediate factors, forms of manifestation, and social and economic consequences. Disproportions in the process of social reproduction and ineffective economic policy are identified as the main causes of inflation, and various internal and external macroeconomic conditions (monetary, credit, fiscal, institutional, etc.) play the role of direct factors (Krasavina and Pishchik 2009). Undoubtedly, such an approach has powerful heuristic potential, as it allows us to disclose the systemic nature of inflation and to seize its most profound essential features. However, determining inflation by combining causes, factors, forms of manifestation, and consequences does not disclose but rather obscures the essence of this phenomenon. To a certain extent, the methodological drawback of the reproductive approach is compensated by the concept of the "dynamic inflationary process" proposed by M. Iu. Malkina (2006) in relation to economic systems with emerging markets.

Inflation is one form of the movement of economic systems in the process of eliminating spontaneous imbalances and internal imbalances, and the inflationary process includes two components: the accumulation of inflation potential, allowing the redistribution of inflation over time, and open inflation, its apparent manifestation.

Finally, we should note the origin of the so-called synergetic trend in studies of modern inflation, which can be regarded as a further development of the reproductive approach in the system of other methodological coordinates. The idea of inflation is based on a nonlinear theory of large and complex self-organizing systems, where it is a mechanism for the formation and realization of dynamic market potential, which appears as a hierarchy of markets integrated into a strategic reproductive model of the economy (Evstigneeva and Evstigneev 2008). The advantage of this approach is that it reveals a new role of modern inflation as a regulator of market expansion, rather than prices of different markets, and also uses elements of the information economy. However, this approach is still in its early stages.

Let us note that all of these approaches reflect the internal logic of the development of economics, which, in its historical evolution, obeyed laws of the movement of knowledge from phenomenon to its essence, and reveals the multi-order nature of the content of inflation and multiple forms of its manifestation. For example, if we take the first approach, in which the essence of inflation was treated as depreciation of money, then from the perspective of the second approach, the depreciation of money no longer appears as an entity, but as a manifestation of inflation, and its essence is disproportions of social reproduction and policies of the ruling classes. This does not mean inflation has many entities—its essence is one, but there are many forms of its manifestation. In the course of its historical development, inflation manifests itself through other social phenomena and acquires many forms of existence, in each of which one of the sides of its essence is highlighted. In this connection, discussions about what inflation is turn out to be incorrect, in the sense that each of the approaches reveals one or several moments of its essence. This very phenomenon, representing these moments in its internal structural interrelations that form their substantial unity and integrity, reveals itself in a multitude of connections and relations with other phenomena.

Discussions of the Causes of Inflation in Russia: Monetarists Versus Non-monetarists

After the collapse of the USSR in late 1991, significant changes occurred in the field of Russian economics. Under the impact of radical economic changes promoted by the IMF, the Western concept of inflation prevailed in Russian economic discourse, with its emphasis on identifying functional dependencies between price increases in the economy and various macroeconomic factors that determine their dynamics. In conditions of unprecedented price increases in the last decade of the twentieth century (prices ranged from 2608% in 1992 to 120% in 2000),[1] the main issue in disputes over inflation in Russia was its nature (monetary or non-monetary). The main competing parties in this intellectual dispute were supporters of monetary interpretation of the causes of inflation (monetarists) and defenders of the theory of its non-monetary origin (non-monetarists). Probably one of the most consistent monetarists was A. Illarionov, who at that time headed the Institute of Economic Analysis and the monetarist camp. His intellectual adversary was R. Greenberg, who represented the Institute for International Economic and Political Studies. This opposition is most vividly reflected in their polemics and polemics of their supporters in the pages of journal magazine *Politekonom* (Illarionov 1996; Grinberg 1996).

The central position of monetarism is that the value of money is ultimately determined by supply and demand, as for the price of any other good. Over the long term, within the framework of the quantitative theory, all interrelationships are simplified, the demand for money is dependent on supply (which is exogenous), the stability of the demand function for money (i.e. stability of the velocity of money), and finally stability of the real GDP at full employment. As a result, the dynamics of the price level can be explained exclusively by shifts in the supply of money. In Russia, both "hard" and "moderate" monetarists adhered to the monetarist interpretation of price increases. The former consider inflation to be exclusively a monetary phenomenon, while the latter approach this issue more flexibly, recognizing that the fall in the rate of price growth below 4% per month compels authorities "to take seriously the cost factors of inflation."[2] Additionally, the interpretation of the monetary nature of Russian inflation among Russian moderate monetarists differs significantly from traditional views of their foreign counterparts. In particular, they view inflation as a "monetary phenomenon, that is, as a phenomenon that can be managed through monetary policy." However, the post-Keynesian view of inflation as a non-monetary process does not exclude the possibility that it can be influenced by monetary policy. Thus, representatives of the moderate wing objectively go beyond the orthodox version.

The empirical justification for monetary models is that inflation inertia and rates of money supply growth are factors of price dynamics. In Russia, inflationary expectations are the basis for the monetary model of inflation. The main problem is to determine the lag with which the money supply affects prices. Estimates of its duration by different economists are similar, although they do not entirely coincide. According to experts at the Institute of Economic Analysis, the lag is influenced by several factors: "level of development of the financial system, the state of the payment system, frequency of revenue payments, level of dollarization of the economy, prevailing duration of contracts" (Illarionov and Sachs 1995: 46). This purports to explain a time lag of four months in 1992 which then, as market infrastructure developed, gradually increased to five months in autumn 1993; six months in summer and autumn of 1994; and eight by spring 1995. M. Deliagin gives similar estimates for 1992–1994, but believes that in 1995 the lag was reduced to 4–5 months. Finally, another study showed that in 1992–1995, the lag varied within 4–6 months, and its steady increase or decrease was not observed.

Non-monetary concepts of price growth, however, were more widespread in Russian economics discourse. In its institutional version, inflation is explained by institutional imperfections, as revealed when comparing Russian and foreign experiences. At the first stage of market transformations, these were many: the unclear status of the Central Bank, the inefficiency of traditional monetary policy instruments, the disrupted budget process, non-market management of the public debt, special economic relations with other former Soviet countries, unsuccessful choices of privatization strategies, and uncertainty of property rights. Many of the listed institutional causes of inflation remain relevant even now.

A related non-monetary view of inflation in Russia is the structuralist approach, the birth of which dates back to the late 1950s, when American economist E. Schulze put forward the hypothesis that inflation is caused by a shift in demand from one sector of the economy to another, which is reflected in the increase in prices in the sector with growing demand and stickiness in the sector with decreasing demand.

Structuralism received further supported from the "Scandinavian model" of inflation developed in the 1970s, according to which exchange rate regimes play a key role in inflationary processes. This model proposes that in the case of a fixed exchange rate, the level of inflation depends on the relative value of goods produced in the national economy and intended for sale in domestic and foreign markets. For example, if the value of goods destined for domestic consumption decreases with respect to the value of same goods exported, inflation will accelerate. It follows that the exchange rate regime causes endogenization of the money supply, which is forced to adjust to the existing level of inflation, and that inflation itself is an exogenous factor in relation to the money supply and the exchange rate (cf. Malkina 2006). Without going further into details, we note two fundamental features of this approach. First, this is an analysis of causes of inflationary processes in terms of their relationship to structural changes in the economy (sectoral and sectoral shifts), which is absent in monetarist and Keynesian theories of inflation. A second important feature of structuralism is the search for causes of inflationary price increases, not in circulation, but in production, which is important in analyzing modern inflation in Russia.

Various forces shaped the structuring of the Russian economy. First was allocation of monopolized and competitive sectors; monopolists are a significant contributor to inflation in the final analysis. Rent-forming and "ordinary" branches of the economy are a second engine of inflation, from desires to redistribute the rent. Third, opposite orientations of export and domestic production could generate inflation by a relative increase in prices for primary resources, which a technologically backward economy cannot withstand. Obviously, there are no rigid boundaries between these different forces, and inflationary mechanisms in all cases are related the manner Schulze noted. Differences between the proponents of non-monetary inflation concepts are mainly in assessing whether current price changes are evidence of progressive or negative trends in the development of the economy.

In debates between monetarists and non-monetarists on the nature of Russian inflation, monetarists' arguments were correct, but they did not refute alternative theoretical constructions. Arguments by opponents of monetarism often revealed a lack of understanding of the details of the monetary explanation of inflation. The weak point of structural concepts is that inflation is directly derived from disproportions in sectoral and cost structures of the Russian economy. Some economists did note ambiguity in the relationship between structural imbalances and prices. Although this thesis was not widely discussed, it allowed monetarists to reject the structural interpretation on the grounds that it did not explain either significant fluctuations in the rate of price growth for short periods of time, or cross-country differences in inflation rates for similar economic structures. It seems these shortcomings are due to a simplification of reality, but it is possible to overcome them by synthesizing the two branches of Russian "non-monetarism," the institutional and the structural.[3]

Representatives of both positions appeared inflexible regarding their widely known "invention." Believing that structural and institutional features of the economy are inertial, proponents of the non-monetary approach advanced the concept of inflation-induced "natural," or "background," characteristics given by these data. It seemed that sustained maintenance of a rate of price growth at 20% per month during 1992–1993 supported this approach. However, as the inflation rate steadily decreased, estimates of the inflationary potential in the Russian economy began to shift. At one point, Illarionov noted with some irony: "For two years, state officials[4] claimed that 20% monthly inflation is a kind of 'natural' Russian level. Now they act as advocates for 10% per month, believing that stabilization is already enough" (Illarionov and Sachs 1995: 39). The monetarist approach, with adaptive inflationary expectations and a model predicting gradually decreasing "normal" inflation as actual growth slows, seems more flexible and convincing. Against the monetary approach, one argument was that in Russia, the growth in the amount of money in circulation over several years in the 1990s exceeded inflation; in the first half of the 2000s, against the background of growth in the money supply, there was a stable decrease in its pace.

Moreover, in the 1990s, the reduction in real money supply outpaced the decline in production, as a result of which the economy was saturated with money. Taken by itself, this fact does not refute the monetarist model—after all, it assumes that as a result of inflation, the demand for money decreases, as evidenced by the reduction in the level of monetization of the economy. Critics of this approach also try to take into account the velocity of money, while admitting this indicator is too simplistic. The increase in alternative costs of keeping money leads less to acceleration of payments and settlements, and more to the replacement of the national currency by foreign currencies, barter, and so on. Thus, this argument against monetarism is unconvincing.

However, it turned out that circumstances did not receive their proper attention. Until the late 1990s, the slowdown in inflation was accompanied by a decrease in the level of monetization of the economy, although the logic of the reverse process was observed (Table 11.1). It is not by chance that monetarists, when citing the monetization coefficient of GDP as an indicator of the demand for money, appealed to the inverse relationship between rates of inflation and rates of change of this coefficient, that is, not to the function of the relative demand for money, but to its derivative. From their point of view, inflationary dynamics are the decisive factor in the demand for money, and in connection with the slowdown in the growth of the latter, this demand should increase. Western scholars noted that the velocity of money in Russia initially did not respond to the acceleration of inflation, and there was no adequate response to its slowdown. Careful econometric analysis showed that in Russian conditions of the 1990s, the dependence of the velocity of money on the rate of inflation was the reverse of what monetarists postulated.

Monetarists faced a serious contradiction. On the one hand, the monetization ratio would have to increase due to a slowdown in the rate of inflation. On the other hand, data showed the opposite. Consequently, if we consider monetarist views to be true and assume that the demand for money increased, then we must recognize the existence for a number of years of excess demand over supply, that is, unmet demand for money.[5] But the deficit of money and even the most insignificant monetary inflation are incompatible. In the monetarist scheme, such a deficit must be met with a reduction in the price level. The lack of money, with a nominal

Table 11.1 Macroeconomic indicators of the development of the Russian economy, 1991–1998

Indicators	1991	1992	1993	1994	1995	1996	1997	1998[a]
GDP in current prices, billions of rubles	1300	19,900	162,300	61,1000	1,630,000	2,256,000	2,675,000	2684.5
Monetary mass aggregate, M2, billions of rubles	928	7120	34,650	97,826	216,700	297,400	380,000	452.5
Level of monetization, M2/GDP	0.714	0.358	0.214	0.160	0.133	0.132	0.142	0.169
Inflation by consumer price growth per year, %	750	2510	840	220	130	21.8	11	84.4
Economic growth by GDP each year, %	–5	–14	–9	–13	–4	–5	0.9	–4.6

[a]Data for 1998 take into account ruble devaluation

increase in the money supply, can be due only to excessive price increases. Hence the conclusion that Russian inflation, along with its monetary factors, is indeed non-monetary in nature. The proof of this is not just a decrease in the level of monetization, but the fact that it took place in an economy with declining inflation.

Ultimately, the demand for money for most of the 1990s was met by the use of various monetary surrogates. In the first approximation, they can be classified into three types: state loans in the form of treasury bills, foreign currency and credits in barter transactions and other types of private surrogate money (the value of which is not tied to "live" rubles), and private surrogate money denominated in rubles, primarily non-payments. If the first two types of money surrogates still fit into the monetarist scheme—state debt as a variant of exogenous money, foreign currency and barter as a means to avoid paying an inflation tax—then inter-enterprise arrears (mutual non-payment) in principle cannot be explained by the logic of monetary inflation.[6] In fact, the spread of non-payments did not exempt one from paying the inflation tax, which, according to monetarism, tried to extract issuers of exogenous money supply. The very existence of non-payments testifies to the existence of a non-monetary component of inflation.

Thus, the confrontation over the nature of inflation in Russia in the last decade of the twentieth century, between monetarists and non-monetarists in the field of Russian economics, was a zero-sum game. None of the parties could convincingly explain the nature of inflation in Russia's post-socialist economy. In itself, inflation at this time was much more complicated than either party could admit, because it was caused simultaneously by a whole set of factors, both monetary and non-monetary, and also by the peculiarities of Russia's transitional economy.

At the beginning of the twenty-first century, as the rate of price growth steadily decreased from 118.6% in 2001 to 5.4% in 2016, the focus in debates about inflation shifted to prioritizing macroeconomic policy goals. All actors in the field of Russian economics actively participating in such discussions can be divided into two opposing groups. In the first group were those who believed that the goal of economic policy should be a steady decline in inflation and maintaining it at the lowest level acceptable for Russia (6–8% in the mid-2000s and 3–4% in 2010). Low inflation is necessary for ensuring a stable rate of economic growth and social well-being. To achieve low inflation requires tight monetary policy, as inflation always and everywhere remains a monetary phenomenon. This view prevailed earlier and is currently prevalent in the financial and economic bloc of the Russian state, including the Central Bank and a number of governmental and non-governmental organizations in Russia, such as the Center for Strategic Research, headed by former Minister of Finance, Aleksei Kudrin. For the second group, Russian inflation is multifactorial in nature and determined by both monetary and non-monetary factors. Overall, the latter are predominant in inflationary processes, and so controlling inflation by limiting the money supply is inefficient and even harmful for economic development. This camp prefers to reorient economic policy, from curbing inflation to monetary stimulation of economic growth by using a soft monetary policy and an active industrial and neo-industrial policy. One representative of this position is S. Iu. Glaziev, presidential adviser on issues of regional economic integration, along with a pool of experts in an informal organization called the "Stolypin Club," the Institute for Economic Forecasting of the Russian Academy of Sciences, and other organizations, as well as some well-known Russian economists (B. Titov, A. Klepach, et al.).

As in the 1990s, today opposing sides in the polemic on ensuring stable growth proceed from methodological prerequisites of either a monetarist or non-monetarist theory of inflation. Each side provides sufficiently convincing arguments, including all kinds of econometric models. Further, inflation over the last 15 years has been steadily declining against the background of steady growth in the money supply. Thus, total increase in the money supply for the M2 aggregate exceeded the total increase in the consumer price index (CPI) for 2000–2015 by more than 20 times (Ershov 2017). The same is observed when comparing rates of inflation and growth rates of the monetization coefficient, which is an indicator of the demand for money (Fig. 11.1). An exception to this pattern is the acceleration of inflation in 2007 and 2008, as well as in 2014 and 2015. It is difficult to find an explanation for this behavior of Russian inflation in the framework of modern monetarist models. It is also impossible to get a satisfactory answer to why we observe cyclical acceleration of inflation in 1992, 1998, 2008, 2014, and 2015.

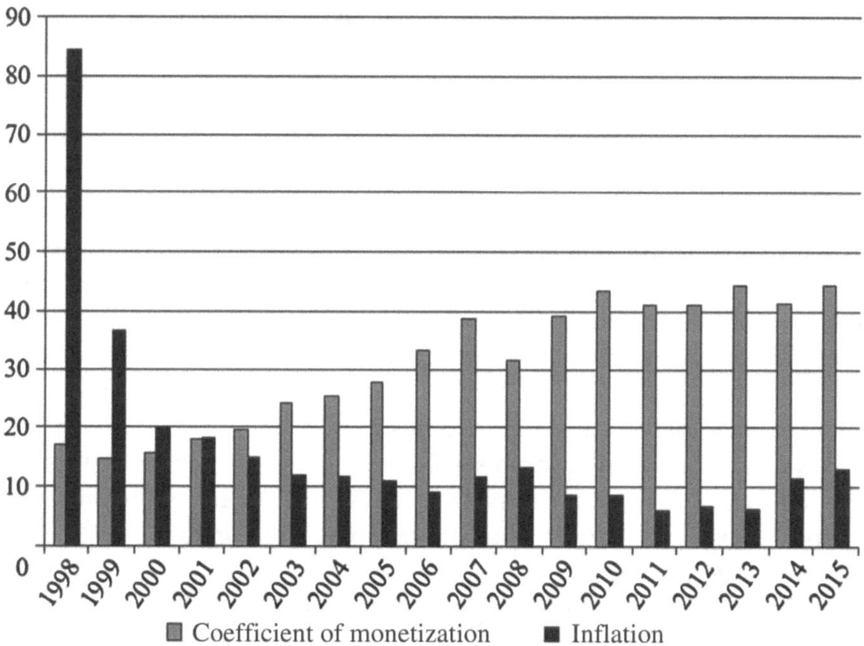

Fig. 11.1 Dynamics of the coefficient of monetization and inflation in Russia (%) (Source: Darovskii (2017: 58))

We can assume that this is because the technocratic approach prevails in economics, according to which inflation is outside the framework of social processes occurring in markets, and is described as a mathematical function of certain macroeconomic variables. Note that modern econometric analysis has become a tool for manipulating and fitting results of a particular model to predetermined goals. Given this circumstance, it is clear why researchers with different views on inflation obtain different results in their inflation models. If we consider inflation not only to be a function of certain macroeconomic variables, but a result of complex social interactions in market fields, then it becomes possible to overcome limitations of the economic approach to analyzing inflation and to broaden the notion that it is a social phenomenon.

Inflation, Distributional Conflicts, and Field Theory

One approach to inflation is the "conflict" theory of inflation. In this analysis, inflation arises when existing distribution and redistribution of national income between sectors, as well as between different classes and groups, cease to correspond to their interests. Any inflationary process means certain subjects of economic relations begin to claim a higher income than that created under given economic conditions. This circumstance leads to an inflationary price increase. Inflation is a result of a distributive conflict. This theory originated in sociological studies of changes in Western class structures. One early economic study in this vein was by American economists Rosenberg and Weisskopf (1981). In this approach, inflation is caused by excess requirements for income from various social classes over national income. Among the "conflicting" classes are, in general, workers seeking to increase wages and (broadly) capitalists aimed at maximizing profit (Gordon 1981). In this confrontation between labor and capital for share of national income, workers in powerful unions seemed to be "winners" asserting their rights to higher wages; in response, owners and top managers of the largest corporations easily include workers' rising salaries in prices for manufactured products, spurring a wage-inflation spiral and simultaneously increasing or at least maintaining a relative rate at profit. At the same time, "that commodity whose price usually rises last is labor" (Anikin 2009: 73), that is, relative prices for consumer goods rise first, and only then do wages rise.

The value of such studies is that, in their analysis of inflation as a result of distributive conflicts, these scholars included the state as one beneficiary of inflation, as well as its self-sustaining nature, conditioned by the institutional environment in which economic agents interact. In particular, if uncertainty about future development grows and risks of loss or decline in real incomes are increased, economic entities, including the state, cease to regard them as their private, local risks, and accordingly they refuse to take them on themselves, instead transferring them to others by assigning a greater mark-up on goods (including labor). "Rising prices in these conditions return to their initiators in the form increased costs, which deprive them of expected profits, and the process self-starts again" (Aglietta and Sapir 1995: 97).

In the post-Soviet space, the distributional nature of inflation began to be discussed in the first half of the 1990s. Analyzing inflation in Russia's transition economy, French economists M. Aglietta and J. Sapir characterized it as inertial inflation, which was subsequently called "conflict inflation." The logic of conflict inflation is that "all economic agents, or part of them, tend to change the distribution of incomes to their advantage," thereby exerting an upward pressure on prices (Aglietta and Sapir 1995: 98). Of the provisions of the conflict theory of inflation, one important consequence is connected with the fact that "inflation is both the starting point and the result of conflict interactions. Each agent has a moving 'target'; each one in turn is a 'hunter' and 'aims' at others." In other words, distributional conflicts are the primary cause of inflationary processes and simultaneously their consequence (Sheviakov 2009: 43–44).

We can see the distributional nature of inflation by comparing strikes and inflation, which shows a direct correlation between these indicators (Figs. 11.2 and 11.3).

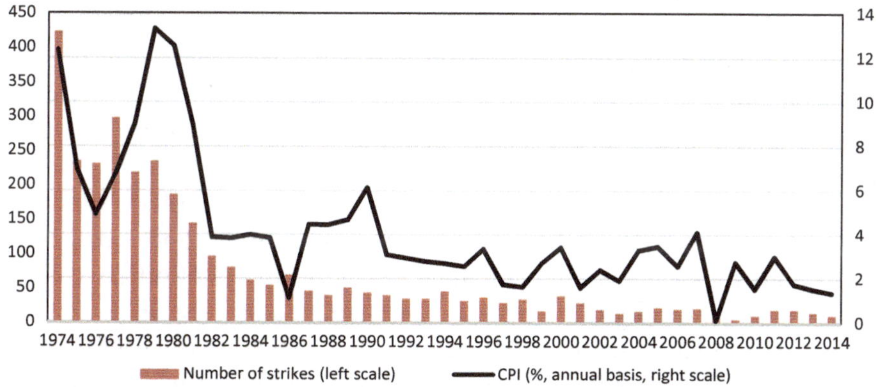

Fig. 11.2 Number of strikes and tempo of growth in the USA (1974–2014) (Source: https://www.statbureau.org/ru/united-states/inflation-tables (data for 4/15/2016); Bureau of Economic Analysis U.S. Department of Commerce (URL: http://www.bea.gov/national/ data for 4/16/2016))

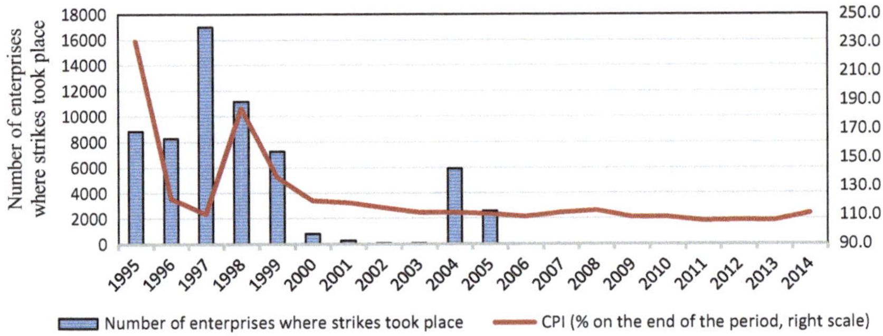

Fig. 11.3 Inflation and strikes in Russia (1995–2014)

It should be noted that distributive conflicts, which are built-in mechanisms of the inflationary process, can take various forms. In past periods, as a rule, they took the form of power conflicts: wars or revolutionary upheavals, accompanied by price-raising processes and depreciation of currencies. At present, such conflicts are less direct, realizing themselves in the price mechanism. For example, conflicts between salaried personnel and large corporations in relation to wages took the form of shifting increasing costs to labor and to the consumer through raising prices for finished products. This was possible due to monopolistic and oligopolistic market structures, where firms had the opportunity to manage prices. Another example is manipulation of interest rates in the framework of strict fiscal policy. It is interesting to note that a process that, according to monetarism, is designed to combat inflation (high interest rates) actually can lead to a rise in the cost of credit transferred to the consumer through prices, stimulating inflation.

The foregoing shows that inflation is rooted in social conflicts over the distribution and redistribution of the national product. Socioeconomic conditions for such a conflict are unequal opportunities of different economic agents, social groups, and countries defined by existing economic and institutional structures of access to the national product. However, even conflicts such as wars have their own rules and even structures; if competition and conflict over power drive inflation, then might there be structured rules that shape such conflict, and by extension facilitate inflationary processes (or not)? In this vein, Pierre Bourdieu introduced the notion of the "social field," further developed in neoinstitutionalist scholarship, especially by Neil Fligstein. While fields can seem static and deterministic, Bourdieu himself built possibilities for contention into fields. In Bourdieu's framework, economic fields are sets of interacting capital unequally distributed among agents. Economic fields are fields of struggle between these agents for preserving or changing their positions, and their success depends on their positions vis-à-vis rules of distributing and valuating capital in all its forms. Market power, shaped by the concentration of resources and technologies, allows actors to realize their interests in raising the rate of profit, for example, by raising prices (Bourdieu 2005b: 147–149). Overall, Bourdieu shows that economic fields are really not fields of perfect competition. In markets there are monopolistic and oligopolistic associations of large enterprises that influence, if not set, prices in market and, at their discretion, set wages at a level that provides an acceptable level of profit.

In Fligstein's (2001) related framework, social fields are arenas of interaction between market participants, where the main structuring factor is not only the concentration of capital with large companies but actions of companies relative to each other. Economic fields are formed not only as a result of the mutual positioning of enterprises but also as a result of constant social interaction between market agents that constantly reproduce their status in the market hierarchy. From sellers' perspectives, markets are always potentially unstable, because they function in conditions of uncertainty and risk associated with the possibility of losing their market position. If we use Fligstein's approach to analyze inflation, the social nature of monopolized markets that systematically can generate inflationary impulses in the economy becomes understandable. Further, if we consider competition as an expression of market conflict, then markets breed conflict potential that, under appropriate conditions, leads to inflation (Strebkov and Bondarev 2014). So, for sellers, conflict unfolds for control over markets, which leads to a price war and a drop in profits for all participants. The strongest enterprises remain on the market and market monopolization is reproduced, leading in turn to distributive conflict and inflation.

This said, we should pay attention to specifics of organizational fields in which inflationary impulses are born. In economic theory, the goal of market agents is to maximize their corresponding objective function: profit for enterprises, utility for consumers. In mainstream economic theory, competition should lead to Pareto-efficient allocation of resources and maximum satisfaction of consumer needs. At the same time, in real economic practice, market agents do not always seek to compete with each other, because competitive struggle threatens market stability and survival of market participants, for example, because of the threat of falling prices as a result of price wars. Field theory can overcome such economic determinism and more realistically describe agents' behavior by widening the range of market players' goals—in particular, by including risk aversion and maintaining market stability. This stability is attained in two ways: by establishing a market hierarchy between dominant and subordinate players and imposing a market strategy for all; and by developing such a concept of control (formal and informal rules of market behavior) that avoid unfettered competition and risky conflict (such as price wars) and keep less influential players in a subordinate position (cf. Fligstein 2001). In this sense, social structures of markets are a system of power through which dominant enterprises use strategies to ensure stability and preserve advantages.

Inflationary impulses in the economic field arise when leading players begin to violate previously established conceptions of control to obtain income greater than that actually produced in this market, or rules of control are ineffective because players get the opportunity to redistribute unearned income in their favor. Such actors may be monopolists, associations, or the state. For example, economic fields in Russia have a high concentration of managers inherited from the Soviet era. In the USSR, practically all significant enterprises were part of the rigid command economy ministries and departments. In today's Russia, a significant portion of production assets is integrated into large businesses that occupy dominant positions in markets and are simultaneously under direct or indirect control of the state. Studies evaluating the role of big businesses in Russia often note their negative effect on the economy, often because of monopoly positions and state participation in the corporations. Particularly problematic have been questionable schemes for transferring assets previously owned by the state to private individuals, as well as precedents of large-scale state support for large companies, including those controlled by private capital, and aimed not at modernizing production but to pay off corporate debts and to acquire assets, including those abroad (Nefedkin 2016).

One negative result of such monopolistic activity is inflationary price increases. As a rule, large Russian corporations are concentrated in basic branches of the economy: electricity, oil and gas, and housing and communal services. The opacity of activities of such giants leads to artificial overstatement of their production and operating costs, which is one manifestation of distributional conflicts and violation of field rules, and one reason for cost inflation. Confirmation of this thesis comes from the calculation of the Herfindahl-Hirschman index (HHI) based on the data about revenues of the 200 largest Russian corporations for 1994–2014. This index is used to assess the concentration of market power and degree of market monopolization and is calculated as the sum of squares of the shares of sales of each enterprise in that industry or market (Fig. 11.4). Comparison of values of this index with rates of inflation reveals a close correlation of these two indicators (coefficient of linear correlation r = 0.799). This suggests a field approach to inflation complements and expands the conflict approach.

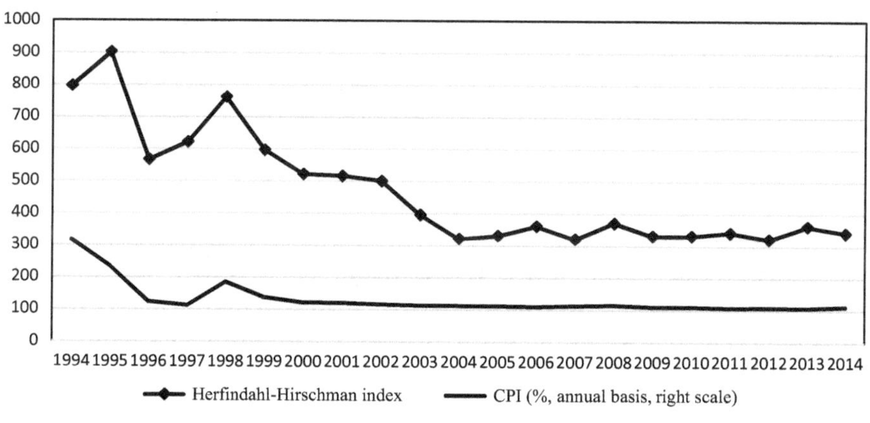

Fig. 11.4 Herfindahl-Hirschman index and inflation in the Russian economy

Cyclical Dynamics of Inflation and Distributional Conflicts in the Russian Economy

The application of conflict theory and field theory makes it possible to uncover cyclical dynamics of inflation in Russia that developed in the 1990s.[7] Since the initiation of market reforms, four cycles of inflationary acceleration have been revealed. The first cycle of inflation coincided with the transformation crisis, which hit the Russian economy with the collapse of the Soviet economy and implementation of radical market reforms. During this period, inflation took the form of stagflation. The second cycle of accelerating inflation was observed in 1998 and coincided with the currency and financial crisis in the Russian economy. The Russian economy experienced a third cycle during the global financial crisis in 2007–2008. The fourth cycle was observed in 2014–2015. All these periods of inflationary acceleration were accompanied by clashes of collective economic interests, which were manifested in large-scale distributional conflicts. In particular, the deepening of structural imbalances, catastrophic decline in production in the 1990s, rapid growth of the transaction sector, capital flight, voucher privatization, payment arrears, the devaluation of the ruble, and similar negative phenomena were manifestations of distributional conflicts in the Russian economy.

We can call the first inflation cycle of 1992–1996, connected with the transformation of Russia's economic structure, *the transformational inflation cycle* (TIC), the key feature of which was the combination of a dying mechanism of "Soviet" inflation (Protasov 2015: 75) and elements of an emerging market-type inflationary cycle (Fig. 11.5). The logic of this cycle was that in the initial phase of market building (such as it was), basic forms of accumulation of inflationary potential that were inherent in the Soviet economy persisted. First was the existence of a certain share of forced savings among the total amount of savings of the population, which were of an inflationary nature and constituted a certain part of surplus reserves of commercial banks that arose in conditions of high credit risks and political instability during this period. Second was the deepening of structural imbalances in sectoral and technological systems,

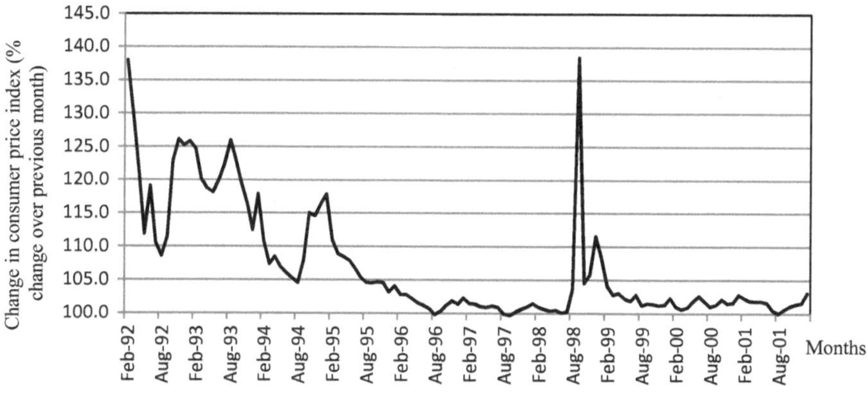

Fig. 11.5 Inflationary cycles in the Russian economy, 1990s

as well as in the system of relative prices that were the legacy of the Soviet economy. Rapid institutional and structural changes, and consistent redistributional conflicts caused by transitional processes, created new forms of accumulation of inflationary potential: obsolescence of fixed productive assets, payment arrears, money surrogates, accumulation of domestic and external debt, and so on. The accumulation of inflationary potential associated with increased disparities in the transitional system, after a certain period of time, reached a limit and turned into an accelerated rise in prices.

In the period of market transformation in the 1990s, large-scale distribution conflicts testified to the extreme imbalance in the Russian economy, expressed in high inflation and a decline in production, as well as in the formation of basic macroeconomic proportions during confrontations of competing actors. The distributional conflict was influenced by five groups of factors: structural imbalances, production decline, growth of the transaction sector, capital flight, and other institutional factors (Kamenetskii 1998: 57–58). Deep imbalances inherited from the Soviet command economy caused significant distortions in sectoral and technological structures of the Russian economy. The destruction of the Soviet institutional template in 1991 and corresponding curtailment of centralized allocation of resources led that hidden competition carried out in the "bureaucratic market" (i.e. the shadow economy) to come out into the open. As a result, Russian enterprises began to compete openly for access to limited resources, using price mechanisms. Such competition against the backdrop of serious price imbalances created incentives for self-sustaining price increases.

The *decline in production* in many sectors of the economy, which triggered the distributional conflict in the transitional economy, had a serious impact on the inflation cycle, in particular creating stagflation. Note that decline in production continued almost until 1999. Until the mid-1990s, decline in output was due to supply-side factors: the lack of normally functioning market institutions, collapse of the single socialist market, and disruption of economic ties between enterprises. As is known, supply shocks lead to significant price fluctuations. Hence, it becomes obvious why in the first phase of the inflationary cycle there was such large amplitude dispersion in the price dynamics. An important factor in the distributional conflict was the *rapid growth of the transaction sector*. For example, if 5.9 million people were employed in trade in 1990, in 1995 the number of officially employed people exceeded 6.5 million. In the banking sector during the reform years, there was record growth in the number of commercial banks: from several tens in 1990 to 2600 in 1997. The decline in production in the final demand sectors in 1992–1998 was 63% in engineering, 87% in light industry, 50% in the food industry, and 80% in textiles (Lipina 2004: 20–21). The accelerated development of the transactional sector was natural under conditions of fundamental economic reformation. However, its development aggravated the crisis of material production and, therefore, strengthened the latter's struggle for survival in new market conditions and, in the final analysis, contributed to the acceleration of inflation. *Capital flight* was also a factor of distributive conflicts in the Russian economy. The outflow of capital was a natural result of Russia's transformational crisis: capital flowed either to sources of better returns or lower risk. *Institutional factors* affecting distributional conflicts included "shock therapy," which accelerated the liberalization of prices and trade, voucher privatization, reduction in savings, and the formation of corporate networks.

If we reproduce the chronology of how the transformational crisis unfolded in the 1990s and compare key events in the institutional environment with phases of the inflation cycle, then we see many parallels

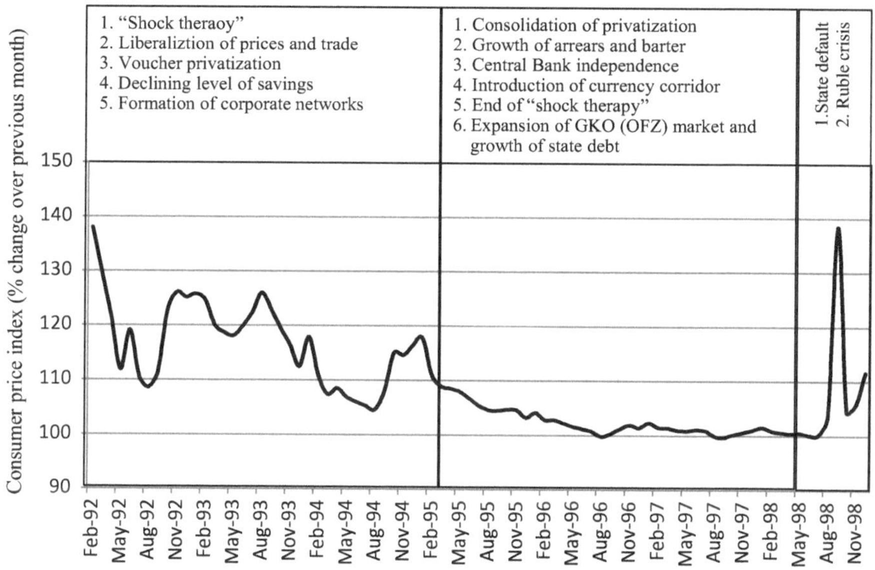

Fig. 11.6 Inflation and institutional changes in Russia, 1992–1999

(Fig. 11.6). From August 1992 to August 1993, inflationary price increases were influenced by mass voucher privatization and the process of forming corporate networks in the institutional structure of the economy. The battle for property, and the emergence of various networked alliances within business and between business and the state, proved to be almost the most important in the rise of conflicts associated with a large-scale redistribution of property and the flow of financial resources into the informal sector of the economy. Shock therapy, primarily involving the liberalization of prices and trade combined with extrication out of the planned centralized economy, to a new model of macroeconomic regulation instead of control, also contributed to distributional conflicts that led to a price shock of early 1992—the state no longer could contain conflict within the parameters of centralized, hierarchical control of resources and decision-making.

The same factors played a role in changing the nature of distributional conflicts from 1995 to the first half of 1997. The consolidation of voucher privatization, increased competition in emerging internal markets, changes in the status of the Central Bank, and other institutional innovations meant that distributional conflicts acquired uglier forms than manifested in so-called collateral auctions during voucher privatization. Its effect was temporarily transferred from pricing to the financial sphere, where market agents, including the state, realized their interests through payment arrears, barter, and money surrogates. Given tight monetary policy aimed at combating inflation, the level of economic saturation fell sharply—the monetization ratio fell from 0.358 in 1992 to 0.142 in 1997. In accordance with the law "On the Central Bank," the Bank of Russia ceased to directly finance the federal budget. The budget deficit was covered by loans in the financial market via short-term high-yield liabilities (GKO). The yield on these securities was high due to the fact that the GKO market faced a competitive currency market, where, under conditions of high inflation, market agents could store wealth in less risky investments (e.g. the US dollar) or make money through speculation. The highly profitable GKO market diverted a significant portion of financial resources from the economy. At the same time, the national debt was growing, accumulating inflation potential. Compression of the money supply led to the expansion of alternative settlement methods (again, payment arrears), which were used as tools in distributional conflicts. Thus, the decline in inflation in this period was a fiction: it simply changed its form and was less open or apparent, and distributional conflicts took on an even greater scope, as practically the entire population was drawn into them.

The inflationary crisis in 1998, related to imbalances in public finance, the GKO pyramid, and currency restrictions occurred against the backdrop of sharply aggravated distributional conflicts, when accounts receivable increased several times, with forced cancellation of the currency corridor.[8] Exporters, anticipating a rise in value of the dollar, deliberately delayed transferring foreign exchange earnings from abroad, causing an acute shortage of foreign currency and contributing to a rapid increase in its rate and accelerated inflation in August–September 1998.

Comparisons of cyclical dynamics of inflation processes with factors that contributed to the spread of distributional conflicts in the transition period confirm the thesis that one important foundation to inflation is mass distribution and redistribution of national income and social wealth. Yet attempts to manage the latter exclusively by monetary methods are doomed to fail. The formation of an inflationary cycle of the market type

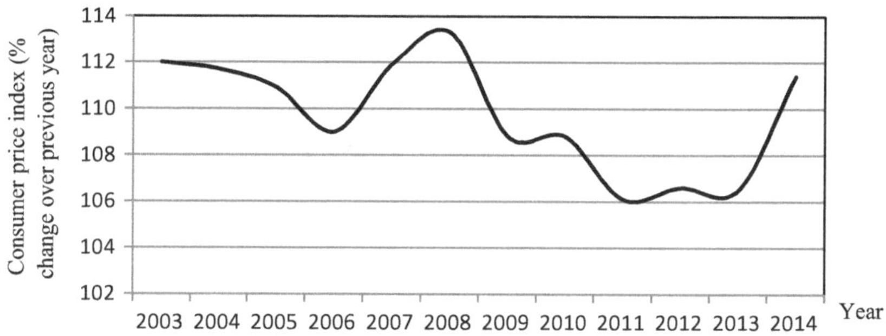

Fig. 11.7 Inflation in Russia, 2003–2014

in the early twenty-first century (Fig. 11.7) indicates the launch of a monetary and credit mechanism, typical for countries with developed markets. In particular, the dynamics of market interest rates, structure of the money supply, and corresponding monetary regulation by the Central Bank began to play the main role in changing the phases of acceleration and slowing of inflation. Without dwelling on the specifics of this mechanism (cf. Protasov 2007), we note that the observed acceleration of inflation in the global financial and economic crisis of 2007–2009 and in the context of aggravated political confrontation with the West is also associated with distributional conflicts. In particular, the crisis of 2007–2009 was associated with the so-called devaluation conflict that arose between the Central Bank and the government, the essence of which was reduced to conflicting goals within the state. In particular, if the government aimed to stimulate economic growth and increase competitiveness of Russian goods by devaluing the ruble (thus sacrificing price stability), the Central Bank opposed devaluation, which threatened to increase liabilities of the banking sector in foreign currency (Malkina 2006: 119–120, 137–140). The conflict was resolved in the government's favor, and devaluation led to a rapid price increase in 2014 and early 2015, as the Central Bank tried to reorient monetary policy from supporting a currency corridor to inflation targeting and a sharp fall in oil prices (another indicator of an acute distributional conflict). Overall, with the sharp drop in oil prices, decline in corporate profits, political confrontation between Russia and the West and resulting sanctions regime, and capital flight, corporate and other elites again resorted to compensation via higher prices.

Conclusion

This brief survey of inflationary dynamics in Russia's post-socialist economy using a field framework and conflict theory of inflation suggests that inflation is one manifestation of contentious economics and that distributional conflicts contribute to a cyclical dynamics of inflation. This suggests several important consequences. First, combating inflation using monetary management tools (e.g. raising interest rates) might not always be the most effective way to combat inflation. In particular, this overview suggests that, in Russia's case at least, raising interest rates leads to a rise in the cost of credit resources and an increase in the alternative production costs, which, sooner or later, leads to an acceleration of inflation through the price mechanism. Second, in modern conditions, inflation has evolved from a universal spontaneous way of eliminating macroeconomic imbalances into an instrument of conscious struggle for a share in national income or wealth. Third, distributional conflicts are not unique phenomena; they are inherent in distributional relations in economies in general. However, they can acquire the character of inflationary distribution and redistribution of national income between the spheres of production, branches of the economy, and different social groups when all other forms of its distribution and redistribution are exhausted.

Finally, it should be noted that field theory significantly expands and complements our understanding of inflationary processes in general and in Russia in particular. If standard macroeconomic theory considers inflation a function of certain economic variables, ignoring social factors and microeconomic foundations, field theory allows us to describe inflation as a product of social practices carried out in frameworks of market interactions of players. From a methodological perspective, field theory has value for developing scientific knowledge of inflationary processes, in that it allows organically combining macro- and microeconomic approaches to analyze inflation and outlines directions for further interdisciplinary studies of this phenomenon.

Notes

1. Data from Rossat (http://www.gks.ru/).
2. At the beginning of 1993, after a major defeat over policies aimed at financial stabilization, Yegor Gaidar (as if to justify himself) expressed more categorically: "Cost inflation is a powerful influence. All behavior of economic entities is fully cost-oriented."
3. It is characteristic that monetarists' criticism of the institutional approach is completely inconclusive—it is based on the correlation of shares of state and private property in various economies. Obviously, it is wrong to reduce a whole variety of institutional characteristics to one.
4. This could include economists who did not take a liberal approach to economic reforms.
5. Andrei Illarionov, as a consistent supporter of monetarist views on Russian inflation, eventually acknowledged the lack of money in circulation in reference to payment arrears. This camp has offered a far from indisputable explanation for monetary deficit: the concentration of resources on the market for state debt. Other sources often mention the criminal nature of financial flows in the Russian economy. This probably has little significance in the end: in M. Deliagin's apt remark, "an indication of the reason for the shortage of money does not eliminate that shortage."
6. At one time, non-payments appeared to be due to distrust of economic entities in stabilization efforts of the government, that is, their expectations of monetary emission or simple mutual offset of debt claims (one-time acceleration of the velocity of money). This might have been true only in relation to the initial stage of the payment crisis in Russia in 1992. Subsequently, the state officially refused to carry out a global clearing of accounts, and inflation gradually decreased—but non-payments persisted into the early 2000s.
7. In this analysis, I examine correlations between inflation and institutional context in *time*. While this does not show causation, I argue that the correlation is strong enough to warrant some validity for the framework presented here. This is in a similar vein to Fligstein's (1990) analysis of fields using *timing* of corporate mergers and acquisitions: usual market and other explanations cannot explain either the form or timing of mergers and acquisitions in a way that a field analysis can.
8. Instituted in 1995, the currency corridor was a maximum exchange rate for the ruble. The goal was to stabilize the ruble, and two effects were reducing the capacity of banks and similar actors to profit from currency speculation and to reveal which banks were in worse health.

Works Cited

Aglietta, M., and J. Sapir. 1995. Inflatsiia i defitsit v perekhodnyi period. In *Denezhnye i finansovye problem perekhodnogo perioda v Rossii: rossiisko-frantsuzskii dialog*, ed. V.I. Ivanter and J. Sapir, 44–101. Moscow: Nauka.

Anikin, A.V. 2009. *Istoriia finansovykh potriasenii. Rossiiskii krizis v svete mirovogo opyta*. Moscow: Olimp-Biznes.

Bourdieu, Pierre. 2005a. Principles of Economic Anthropology. In *The Handbook of Economic Sociology*, ed. Neil Smelser and Richard Swedberg, 2nd ed., 75–89. Princeton: Princeton University Press.

———. 2005b. *Sotsialnoe prostranstsvo: polia i praktika*. St. Petersburg: Aletheia.

Darovskii, I.Ia. 2017. Ot stagflatsii k ustoichibomu rostu: imperativy rossiiskoi makroekonomicheskoi politiki. *Ekonomicheskaia politika* 12 (3): 38–79.

Ershov, M. 2017. Ob usloviiakh ekonomicheskogo rosta: o valiutnom kurse, valiutnoi stabilnosti i dlinnikh dengakh. *Obshchestvo i ekonomika* 2: 5–13.

Evstigneeva, L., and R. Evstigneev. 2008. *Inflatsiia v kontekste evoliutsii deneg. Nauchnyi doklad*. Moscow: Institut ekonomiki RAN.

Fligstein, Neil. 1990. *The Transformation of Corporate Control*. Cambridge: Harvard University Press.

———. 2001. *Architecture of Markets: An Economic Sociology of Twenty-First-Century Capitalist Societies*. Princeton: Princeton University Press.

Goldthorpe, J.H. 1987. *Social Mobility and Class Structure in Modern Britain*. Oxford: Clarendon Press.

Gordon, D.M. 1981. Capital-Labor Conflict and the Productivity Slowdown. *American Economic Review* 71 (2): 30–35.

Grinberg, R. 1996. Inflatsiia v perekhodnykh ekonomikakh. K diskussii o prirode i sposobakh regulirovaniia. *Politekonom* 2: 17–19, 23–24.

Hirsch, Fred and John Goldthorpe (eds). 1978. *The Political Economy of Inflation*. Oxford: Martin Robertson.

Illarionov, A. 1996. Rost i padenie rossiiskoi inflatsii. *Politekonom* 2: 14–18, 20–22.

Illarionov, A., and J. Sachs. 1995. *Finansovaia stabilizatsiia v Rossii*. Moscow: Progress-Akademiia.

Kamenetskii, I.A. 1998. *Inflatsiia v perekhodnoi ekonomike Rossii (1992–1996 gg.)*. St. Petersburg: Nestor.

Krasavina, L.N., and V.Ia. Pishchik. 2009. *Regulirovanie inflatsii: mirovoi opyt i rossiiskaia* рocmy *praktika*. Moscow: Finansy i statistika.

Lipina, S.A. 2004. *Sotsio-ekonomika Rossii perekhodnogo perioda (1991–2003)*. Moscow: URSS.

Malkina, M.Iu. 2006. *Inflatsiia i upravlenie infliatsionnymi protsessami v rossiiskoi i zarubezhnoi ekonomike*. Nizhny Novgorod: NGU.

Mankiw, N. Gregory. 2009. *Macroeconomics*. New York: Worth.

Matiukhin, G.G. 1984. *Rost dorogovizny i inflatsiia v mire kapitala*. Moscow: Mezhdunarodnye otnosheniia.

Nefedkin, V. 2016. Ekonomicheskaia vlast krupnykh korporatsii i razvitie regionov. *Voprosy ekonomiki* 3: 99–114.

Protasov, A.Iu. 2007. Tsiklicheskaia dinamika inflatsii. *Vestnik SPbGU, Seriia 5: ekonomika* 2: 39–48.

———. 2015. Tsiklicheskaia dinamika inflatsii v ekonomike SSSR: istoricheskie trendy i raspreditelnye konflikty. *Problemy sovremennoi ekonomiki* 2 (54): 73–79.

Rosenberg, S., and T.E. Weisskopf. 1981. A Conflict Theory Approach to Inflation in the Postwar U.S. Economy. *American Economic Review* 71 (2): 42–47.

Salerno, J.T. 1999. Money and Gold in 1920s and 1930s: An Austrian View. *The Freeman: Ideas on Liberty* (October): 31–40.

Sheviakov, A.Iu. 2009. *Neravenstvo, ekonomicheskii rost i demografiia: neissledovannye vzaimosviazi*. Moscow: M-Studio.

Strebkov, A.I., and V.G. Bondarev. 2014. Sotsialno-politicheskii mekhanizm snizheniia konfliktnogo potentsiala rynka. *Vestnik SPbGU, Seriia 17* 1: 55–64.

Timberlake, R. 2000. Austrian Inflation, Austrian Money, and Federal Reserve Policy. *The Freeman: Ideas on Liberty* (September): 38–43.

12

Fields of Russian Finance: State Versus Market

Aleksandr Bartenev

The intentional allusion of the title of this chapter to that of a chapter from John Bogle's (2012) *The Clash of the Cultures: Investment vs. Speculation* is not accidental. Conclusions about the financial market and the state, taken from the position of a broader system analysis, are in the spirit of the ideas of Bogle's book about the homogeneity of the positioning of players in the financial markets. Given the essential unity of the hierarchy of social fields of state, quasi-market capitalism, the disposition of state regulation, and the market economy seem illusory, and Russia's market economy acquires features of a state monopoly dominating the nation's economic life.

Fields and Markets

Methodological holism, characteristic of most social sciences, makes it possible in principle to expand the analytical and conceptual apparatus of the theory of financial markets. The holistic approach of social system

A. Bartenev (✉)
Faculty of Economics, Department of Economic Theory,
St. Petersburg State University, St. Petersburg, Russia

J. K. Hass (ed.), *Re-Examining the History of the Russian Economy*,
https://doi.org/10.1007/978-3-319-75414-7_12

theory implies that the transformation of social institutions is intertwined with changes in dispositions. Any person is inextricably linked with social relations, and dispositions are formed in the process of interpreting experiences gained via social exchanges. In their turn, values and norms established experientially in practice reflect actors' goals and behaviors. Financial markets as historically established forms of organizing public exchanges are characterized by the ability to influence people's behavior through formal and informal rules that emerged in the course of functioning. Conversely, dynamics of changes in players' dispositions brings changes in the structure and functional qualities of market institutions. The concept of social fields, proposed in economic sociology, allows us to conceptualize integral sets of interrelated systems. Taking into account personal interests and a vision of surrounding reality by participants in exchanges as a source of self-determination facilitates modeling a more realistic picture of reality in its social dimension.

The basis of the field, constituting its homogeneity and internal properties, is a diffuse understanding of what takes place in this context and what is at stake, and it is akin to a genetic link between social communities and mental structures of their participants (Bourdieu and Wacquant 1992). The objective reality of the social field is made up of the subjective realities of its participants. At the same time, the heterogeneity of social space is associated with formal and conventional (informal) constraints that separate and/or combine participants' behavioral practices in social exchanges into social institutions, which in turn are organized as structured hierarchies in which agents and groups can occupy a dominant or dependent position. Convergence and conventionalization of norms (the transition from formal to informal) in the development of institutions is associated with the formation of systems of fields, the participants of which are situated in positions by accepting legitimacy of the positions of other participants and relatively shared goal setting. Fields differ in degree of proximity, measurable by mutual influence they exert on each other. Adjacent fields often have subordinate or two-way interactions, while remote fields are disconnected and their participants have different social practices. Another important difference between fields is their mutual spatial positioning: horizontal fields do not have hierarchy and do not affect adjacent fields, while vertical fields can be lined up in complex multilayer structures, connected by different relationships that can create pressure from within, causing stress within the field.

In networks of social fields, actors follow a combination of formal and informal rules, the proportion of which depends on their position in the hierarchy. Dominant players at the top of the field hierarchy, who are also its main beneficiaries, use administrative and financial resources to establish the rules of the game for subordinate players, and use pressure of legislative procedures, secured by the police and armed forces, to make fields operational. These active social groups, expanding their influence within the hierarchy of existing social fields, draw into the system other adjacent fields that previously developed autonomously. The impact on adjacent fields can be physical or psychological: in the first case it is geographical or political expansion of influence, and in the second case it affects mental and behavioral attitudes of actors by translating the value system and incentives that allow integration into the hierarchy of social fields on a cognitive level. The introduction of prescribed behavioral programs into the conscience of the recipients and the legitimization of role practices allows the consolidation of structural elements (public institutions) at each level of the system of interactions.

Fields and Russian Financial Institutions

The 30-year history of Russia's market economy is, in theory, grounded in new liberal values of freedom of economic activity. This provides grounds to accept as reasonable the convergence of socio-economic systems—that is, convergence of goal setting and behavioral practices of new generations of Russians socialized under new norms in which money is perceived both as a synonym for personal success and as a social and psychological phenomenon. The transformation of dispositions associated with a transition to commodity-money relations in interactions was noted by the German sociologist Georg Simmel: money was a potential leveler of other social hierarchies, as it became a new medium of interactions (Wolff 1950). This isomorphism of worldviews homogenizes the sociocultural environment, which facilitates links between participants in social exchanges that are reducible to monetary relations and constitute a hierarchy of fields in the global economy and local states. The possibility of exchanging material capital for social capital fixes positions of political elites in this "power-property" system. Financial security is a fundamental value in the contemporary Russian economy, as it helps reduce uncertainty. The state guarantees the national currency and preservation of savings and social guarantees, contributing to rationalization of expectations in our vulnerable material world.

However, Russians' acceptance of liberal values, coupled with Russia's state structure, leads to paradoxical consequences. Having gained experience of national currency devaluations in the course of "shock therapy" in the 1990s, many Russians prefer foreign (mostly dollar denominated) liquidity to avoid risk, which exacerbates the money sovereignty of Russian national currency and indirectly degrades the investment process by immobilizing money outflow and savings. Dollar liquidity preference, which is individually rational, can lead to the devastating consequences for the national economy. Russians who figure on the Forbes list of richest individuals happen to be there after the privatization of state property. They are affluent individuals, while the country to which they owe their wealth has not received sufficient investment for 30 years, and during the same period capital flight has been a constant problem.[1] According to the Global Financial Integrity survey for 1980–2012, the net outflow of capital from Russia amounted to $1233.4 billion.[2] This fits the spirit of age: "forget all but yourself," with privately owned dollars extracted from Russia taken to quieter harbors. Currency exchange operations are becoming the most popular service in Russian financial markets. After the ruble devaluation in 2014, the annual volume of currency trading on the MICEX market increased by more than 100 billion rubles, to 228,546 billion rubles and then rising steadily to 347,671 rubles in 2017. In the same period, the volume of trading in shares slowed down from 10,283 to 9,185 billion rubles. It seems like currency speculations are fueled by MICX, the major Russian financial market.[3] According to calculations from *The Economist*, the Russian ruble is undervalued by almost 70%.[4] The drop in the ruble exchange rate below parity reduces real incomes and makes average Russians increasingly dependent on state provision, while also reducing investment activity by the private sector. The reasons for devaluation included a drop in export revenues by which state-owned hydrocarbon exporters received income under international contracts before the collapse of oil prices. The state budget compensated for the quantitative losses of petrodollars due to currency dumping.

Russia's monetary sovereignty and the managed pricing regime on the domestic financial market make it possible to formulate a budget even under unfavorable external conditions. It is curious to follow the built-in mechanism of social rent, which helps the state budget compensate for decreases in tax revenues from exporters. Fiscal and foreign exchange policies work together to achieve flexible tailoring of the state's financial plan. Tax revenues amount to approximately nine-tenths of budget revenues. According to the statistics of the Ministry of Finance, for January 1, 2015, 14,296.8 billion rubles were accumulated in the federal budget against 8905.48 billion rubles of consolidated budgets of subjects of the Russian Federation. The main source of tax revenues for the federal budget, 5463.4 billion rubles, was from foreign economic activity, more than 80% of which were export payments.[5] A second source of revenue was indirect taxes on imported goods and for the sale of goods (works, services), which added up 4524 billion rubles. The third most important source of tax revenues, the mineral extraction tax, was 2884.6 billion rubles. Given the existing structure of exports, the generator of revenue has been state-owned resource giants, making up more than two-thirds of export revenue. The distribution of funds from the federal budget works through budget transfers, appropriations, salaries to state employees, pensions, and benefits distributed to the periphery. Incomes received in the form of wages and social benefits are returned to the budget of federation subjects in the form of personal income tax, which is the main source of regional tax revenues. Another part of payments is indirect taxes. Thus, the budget receives the bulk of tax revenues from external sources (export, customs duties), and the main taxpayers are state-owned enterprises that export hydrocarbons. Federal expenditures become local revenues, through the payment of personal income tax by citizens living in the regions. Through the payment of indirect taxes, levied on expenses in the regions, budgetary allocations are partly returned to the federal budget. Ruble devaluation thus brings additional and projected revenue through the expansion of ruble proceeds of exporting state corporations and the growth of tax deductions to the budget.

A pyramidal structure of social redistribution has arisen, in which the political regime is supported by social subsidies in favor of an increasingly expanding field of state-financed sections of the population: employees of state ministries and departments, employees of state corporations, and pensioners. Market competition has turned into a struggle for access to administratively distributed financial resources. The "middle class" is small—only 4.1% of the population—and its share of national wealth corresponds to its share of the population. These indicators have been decreasing since 2000 because of growing centralization brought to life by the building of a "power vertical" that is in reality growing monopolization of political and economic life. A built-in mechanism of social rent is working: with devaluation of the national currency and/or shifting the tax burden from exporters to end users, Russia's economic system has turned to internal sources; real incomes of citizens dropped sharply, while elite wealth has increased, in part because their foreign currency-denominated savings are at low risk. Further, the surplus from the sale of natural resources is returning to safe-haven countries that issue foreign currencies.

If we use Fernand Braudel's world-economy model, the hierarchy of state fields can be viewed as a network corporation dominated by a central state apparatus that orients external and internal financial flows around itself. The parent companies of the largest oil-producing and financial firms are located in the federal center, and a high degree of centralization of financial and administrative spheres leads to a merger of political and economic elites. The power vertical structures the entire hierarchy of state fields at the expense of social rents, which accumulate at the top of the state hierarchy and provide a golden parachute, and is then distributed downward to sustain the regime. Economic consequences of this financial architecture are manifold: the domination of the center over the periphery, a non-egalitarian distribution of wealth, and unequal exchange between the extractive, productive sector of the economy and the financial and administrative services sectors, which in turn hurt long-term development through underinvestment. The concentration of income in Russia is traditionally high and similar to that of the United States.[6] The average monthly salary in the financial sector is 73,209 rubles, for the mining sector 65,199 rubles, and for manufacturing and agriculture 35,018 and 21,616 rubles, respectively.[7] The rate of deprecation is increasing from year to year for extracting and manufacturing industries.[8] The isomorphism of goal setting between the political and business elites, for example, fetishizing material well-being, is expressed in copying institutional rules typical for top management of transnational corporations that have similar positions in network structures of countries in the global economy. National sovereignty cannot then be fully associated with national identity, as copying borrowed patterns of behavior places certain limits on goals and activities of politicians and businessmen. Returning to the interdependence of the whole (society) and the individual (its part), isomorphism of goal setting of actors striving to maximize personal welfare at each level of state fields leads to the realization of individual or group goals in the political arena, which is indifferent or even contrary to the goals of the nation as a collective. This logic has been leading to the elimination of demarcation between market and state, and political decisions will be correlated with goals of the social community to the extent that they meet the goals of the leaders of big business and the state.

The obvious convergence of socio-economic field systems on an international scale, as a process called "globalization," is explained by convergence of state institutions and transnational corporations that have become major players in private capital markets. The adaptation of political elites to new global order by imitating rules of the game led to the formation of a pseudo-market economy with institutional monopoly. The similarity in mechanisms governing the state system and finance is due to the uniformity of approaches in the sphere of collective consciousness—a form of normative isomorphism (DiMaggio and Powell 1983). Intensification of information flows and a significant expansion of communication channels bred a routine social order based on cultivation of public consciousness through print and broadcast media, news agencies, and analytical reports (with greater legitimacy from their technical language, a form of cultural capital). Ironically, much of the way for this convergence of fields via globalization and new cognitive boundaries was made possible by democratic mechanisms during the 1990s. As Bernays (1928) noted, democracy requires a taming of citizen subjects, which can open the door for manipulation of habits and tastes and can put up a façade covering less liberal uses of power. Institutions created by the state exercise power over citizens in various forms: police and armed forces (hard power) and cultivation of public opinion (soft power). Although coercive powers are not implemented as widely as in the past, in a democratic society, people acquire new behavioral patterns through state-run social networks. Such formation of mass consciousness violates basic premises of the demarcation between state and market. Institutions for using information and manipulating public opinion are built into state fields, to create and to reproduce necessary events and images for developing stereotypes, establishing cognitive boundaries, and transmitting particular goals and practices as "normal" and thus aligning interests.

The concentration of such means for propagating information and discourses can affect popular senses of power and powerlessness and expectations of what kinds of strategies are possible, for example, by shaping dispositions and available knowledge. In Russia, according to reports on television and radio broadcasting for 2015, free television and radio channels are predominantly state-owned or quasi-state-owned.[9] In the United States, the institutional owners of TV channels are a group of large mutual funds. Vanguard, Fidelity, State Street Corporation, and BlackRock Institutional Trust Company own large holdings of the largest American banks and media.[10] In both cases, the concentration of media in the hands of business and political elites can provide the potential to narrow the focus and angle of information provided, the scope of discourses, and available ideas that then can shape perceptions, expectations, and sense of legitimacy that can support existing field structures. A media that does not question arrangements of state and financial institutions or of elites, for example, models common perceptions by erecting cognitive boundaries and practices.

This returns us to issues of financial institutions and fields: those institutions (organizations and formal rules and laws) are both guidelines and tools for practices, but *how* they are employed, altered, or resisted depends upon how actors perceive not only those rules but also how other actors in particular positions understand and employ those rules. For financial fields, what constitutes *legitimate value* is a central facet of strategies of reproduction, competition, and change. According to Aristotle, the value of any object is measured by its passive ability to meet the people's needs. This might have made sense at a time when physical objects such as food and clothing made up much exchange; need and status were two key components to value. The growth of institutions and fields in modernity, when embodiments of the symbolic have become as important as the material, reveals the limits to such logics of valuation—in particular, that they are not driven by material need and scarcity alone. Assets that are not physically incarnate are more difficult to evaluate, and so a potential buyer needs to be notified of the value of the object and risks of revaluation (i.e. sudden drops in the value of shares or currency, which are symbolic rather than material and so embody collective meanings). This creates challenges convincing actors to "buy" and "own" these assets, and here signals and discourses gain paramount importance. (Land will produce certain crops regardless of our perceptions of them—but currencies, shares, and the like do not "produce" of their own accord.) A wide range of listeners and viewers receive financial information regularly: via various media (e.g. special online publications or reports via television or newspapers), analysts' reviews and investment recommendations are used to assign to assets particular attributes that satisfy not existing human needs (cf. Menger 1992). The effectiveness of these estimates and expectations can be measured indirectly, for example, the study of the *Wall Street Journal* Dart Board Contest leads to interesting conclusions regarding investment prospects of professional financial consultations versus a "random walk" (to use jargon from macroeconomics): an investment portfolio arranged by a monkey throwing darts at a list of issuers exceeds the average dividend yield of professional portfolio analysts by 1.2% versus 3.1% for the Dow Jones Industrial Average.[11] Is that this "modeled reality" is only a work of misconception created by professional financial advisors, or is it a conscious manipulation? Information flows have unidirectional movement from the top to the bottom, that is, the goal setting and the performance of global players can be known to the average person to the extent that the informer needs it. No less ambiguous are assessments of management quality and calculation and forecasting of managerial risks, the implementation of which might contribute to crises. Political campaign budgets rise every year, contributing to the shaping of public discourses about available and normal policies, including those of financial practices and institutions.

The spread of free market ideas at the international level, the exploitation of unprecedented conveniences of information exchange, transborder capital flows, and the expansion of financial intermediaries led to fruitful cooperation between consumers and producers of financial and management services for the benefit of the latter. New opportunities for information technologies facilitated the standardization of practices and policies, shaping perceptions of opportunities and constraints. Political and financial markets, therefore, are largely mental projection imposed by leading participants, fabricated by standardizing codes of social interaction. Consider the fate of Russian democratic freedoms, or lack thereof. The process of post-Soviet restructuring was hindered by the establishment of a puppet political regime because of the monopolization of the power of one political party and the prerogative of elites, although the adoption of stereotypes of conspicuous consumption speaks of the importation of rules of the game from consumerist economies in the West. It seems Russian statesmen were guided to maximize personal well-being without constraints of decorum that earlier might have balanced elite strivings with the need to "pay back" something to their societies.

The abovementioned isomorphism of key social exchanges, namely, replacement of commodity-money relationship for personal relations, inextricably led to rent-seeking behavior by people in authority at various levels in the social hierarchy. Therefore, the excess of executive power and managers is the new normality, not the exception. Indeed, even hypothetically perfect forms of law form the structures of social fields. To be egalitarian they would have to be dispositive, while the very existence of hierarchy has led to fundamental inequality of participants in different strata. Rights and liabilities would be asymmetrically distributed among individuals in this imaginary state. In the real world, unequal redistribution of wealth can only be fueled by external inflows of liquidity. If for the most Western countries this is due to expansion of national or multinational debt (as in case of European Central Bank), for Russia this is mostly through hydrocarbon revenues, as Russia has no international money sovereignty (international contracts values in foreign currency) and is now cut off from foreign capital markets. Local currency depreciation cannot provide sustainable growth and degrades the economy via underinvestment, low labor efficiency, extensive state control, dependence on the export of natural resources, and unilateral economic specialization that is detrimental in the long run.

The topology of capital markets already has a corporate core at the peak of the financial Olympus, where the amount of capital under their management is a multiple of the volume of assets of the largest Russian quasi-state banks. If political and financial markets in the United States are controlled by the largest transnational players, in Russia the largest player in the financial market is the state, represented by pseudo-public banks under its control, which top the list of financial institutions, as confirmed by data from a report by the Russian Academy of Sciences project, "Monitoring Global Competitiveness of the Financial Market of the Russian Federation and Analysis of Measures to Raise It." The dynamics of basic indicators of the Russian financial market in the global economy are disappointing. For 2015, these are percentages of all positions presented in the report. Russia's share decreased 3–5 times in comparison with 2006 and 23 times in volume of global Initial Public Offering/Secondary Public Offering (IPO/SPO).[12] Unfortunately, this reflects the fact that the current trajectory of policies and institutional practices is leading to an increase in uncertainty, an escalation of tension, and rising threats of the destabilization of the system of social fields, with potentially sudden and irreversible changes in their structures and interrelations.

Notes

1. http://investorschool.ru/ottok-kapitala-iz-rossii-ctatistika-po-godam.
2. *Financial Flows and Tax Havens: Combining to Limit the Lives of Billions of People* (Centre for Applied Research, Norwegian School of Economics, 2015): 8 (http://www.gfintegrity.org/wp-content/uploads/2016/12/Financial_Flows-final.pdf), accessed October 24, 2017.
3. https://www.moex.com/ru/ir/interactive-analysis.aspx (accessed March 18, 2018).
4. "The Economist": dollar dolzhen stoit 23 rublia." *Vesti.ru*, January 11, 2016 (http://www.vesti.ru/doc.html?id=2706834), retrieved October 24, 2017.
5. Ministry of Finance of the Russian Federation: http://info.minfin.ru/fbdohod.php.

6. OECD data on inequality, https://data.oecd.org/inequality/income-inequality.htm (accessed October 24, 2017).
7. *Glavnaia kniga,* http://glavkniga.ru/situations/k502289 (accessed October 24, 2017).
8. http://www.gks.ru/free_doc/doc_2017_rusfig-rus17.pdf.
9. *Economy Times*, http://economytimes.ru/sites/default/files/25-06-2016-doklad-ranhigs.pdf (accessed October 24, 2017).
10. As of 2015, 64% of Google shares belong to this group: JPMorgan Chase & Co. 6.56%, Goldman Sachs Group 5.72%, Bank of America Corporation 6.2%, Berkshire Hathaway 8.68%, and Wells Fargo 5.99%. http://www.investopedia.com/articles/markets/011516/top-5-google-shareholders-goog.asp#ixzz4eUtO1wpT (accessed October 24, 2017).
11. http://www.investorhome.com/darts.htm.
12. http://economytimes.ru/sites/default/files/25-06-2016-doklad-ranhigs.pdf.

Works Cited

Bernays, Edward. 1928. *Propaganda*. New York: Horace Liveright.

Bogle, John C. 2012. *The Clash of the Cultures: Investment vs. Speculation*. Hoboken, NJ: John Wiley & Sons.

Bourdieu, Pierre, and Loic J.D. Wacquant. 1992. *An Invitation to Reflexive Sociology*. Chicago: University of Chicago Press.

DiMaggio, Paul, and Walter Powell. 1983. The Iron Cage Revisited: Institutional Isomorphism and Collective Rationality in Organizational Fields. *American Sociological Review* 48: 147–160.

Menger, K. 1992. Osnovaniia politicheskoi eonomii. In *Avstriiskaia shkola v politicheskoi ekonomii: K. Menger, E. Bem-Baverk, F. Vizer*, ed. V.S. Avtonomov, 31–242. Moscow: Ekonomika.

Wolff, Kurt H. 1950. *The Sociology of Georg Simmel*. Glencoe, IL: Free Press.

13

Fields in Flux: Post-socialist Reorganization of Property and Power

Jeffrey K. Hass

Fields and Economic Restructuring

When socialism fell apart in East Europe and the then soon-to-be-former USSR after 1989, pundits and academics focused primarily on actors: elites' backgrounds, interests, alliances, and possible strategies. As Boris Yeltsin's administration attempted to build a democratic polity and capitalist economy, scholars returned to institutions in a simplistic fashion. While "shock therapy" infamously garnered attention in the halls of power and elsewhere, critics emerged as well, such as economists espousing a slower "gradualist" approach and political scientists warning that "institution-building" would be central to the entire endeavor—democracies and capitalism could not simply be legislated. Commentators over

J. K. Hass (✉)
Department of Sociology & Anthropology, University of Richmond, Richmond, VA, USA

Faculty of Economics, Department of Economic Theory, St. Petersburg State University, St. Petersburg, Russia

J. K. Hass (ed.), *Re-Examining the History of the Russian Economy*, https://doi.org/10.1007/978-3-319-75414-7_13

the decade oscillated between actors and institutions: some presumed that elite coalitions and interests mattered, while others paid attention to legal codes that could be used to create or contest the new economic order (e.g. Åslund 1995; Frydman and Rapaczynski 1994; Blasi et al. 1997; Gaddy and Ickes 2002). Common to early analyses was a presumption, implicit or explicit, of rational choice: that actors involved sought to maximize utility, and that they had fairly clear interests. All that matters to analysis are an actor's interests and rules of the game. Strategies that lead to "institutions" are driven by power and profit.

Except such approaches to economics and institutions failed at worst, and were incomplete at best, especially for Russia. That interests, profit, and power mattered was no surprise, and analysts ignore these at their own peril. Yet fixating on such a troika at face value has its own risks, in particular, oversimplifying the nature of social relations and meanings, which an economy ultimately is. Just what actors' interests might be (and their sources), their conceptions of acceptable relations of power, and what they imagine a "normal" or "acceptable" economy to look like—issues as much of legitimacy and worldviews, as of vague "utility functions" (problematic) and instrumental rationality (less problematic but potentially oversimplifying)—have been missing in most analyses (but see Herrera 2005). Even as it became clear that post-socialism was not so straightforward a process as "interests + institutions," this was still a dominating (if implicit) logic of many analyses. Scholars turned to high taxes, legal barriers, rent-seeking elites, and weak contract law to explain why making markets was so difficult (e.g. Hellman and Schankerman 2000; Hendley et al. 1999, 2000). However, such explanations beg the question of what "contract" or "property" is in the first place. Whether institutions "work" or not risks theoretical circularity that too often is written off as a result of corrupt or incompetent officials rather than a serious question to be pursued rigorously. And "interests" are too often either taken-for-granted or presumed to be something we can extrapolate from choices, despite the fact that those supposed interests are far less clear and observable than actors' own practices and claims—and consistencies and contexts of the latter point more readily to dispositions and knowledge, than choices point with any accuracy to "revealed preferences."

The problem is an instrumental approach to *institutions*, which in so much social science are assumed to exist just like chairs, lathes, or guns. In a stable, developed economy, this might be forgivable: "institutions" as scripts and schemas are sufficiently assumed as real by enough actors that the analyst (like actors) can treat them as real for grounding the analysis. However, such cases of radical change as Russia's in the 1990s do not allow this—in such historical moments, these shared assumptions are under assault, such that we cannot invoke "institutions" that themselves are objects of contentious reconstruction. An instrumental approach that treats institutions as invariant and eternal, and thus that shape costs and benefits (that actors calculate) in a constant manner, oversimplifies a complex reality (Woodruff 2000) and ultimately informs bad policy (such as shock therapy). To posit that "institutions" are solutions to problems of transaction or agency costs (Williamson 1985; Chandler 1977) avoids asking what these social compacts are. While political scientists add elite struggle and bring power in to the equation (e.g. Barnes 2006), discussions end up oriented to "institutional design" and reduced to agents' interests, without explaining their origins (Friedland and Alford 1991). Even more problematic is that institutional approaches implicitly or explicitly posit that actors orient to rules, rather than to *how those rules matter in practice*. It is as if actors are accountants or tax lawyers who read the law and act accordingly. Yet most actors are not accountants and do not follow their advice all the time. Rather, "institutions" matter as actors perceive, interpret, and apply rules. Economists and political scientists either miss this important facet of institutions in everyday life, or they decouple theory from empirical reality, creating one framework and then telling a different story.

In this chapter, I employ and explore a field analysis to make sense of such contentious institution-building. My basic motif is that pre-crisis (late Soviet) institutions situated different actors in different contexts of rules, relations, and habits, bequeathing particular knowledge and dispositions (*habitus*), social relations both concrete and "ideal" (abstract types of relations perceived as legitimate), and capital. In setting to remake some semblance of economic order that would restore some certainty and benefit, actors set out to reconstruct not only institutions but also (and especially) arrangements of relations, other actors, and rules (including institutions). That is, once institutions weakened between 1989 and

1992, fields as sets of actors, relations, and collective practices persisted, and actors used these field templates to remake institutions. Further, not all actors had similar experiences and capital, and so reconstructing institutions exhibited both variation and contention.

To develop this theme, I use the case of privatization and restructuring in the first decade of Russian post-socialism, when rules and structures were in flux (Hass 2011a, b). One important insight from a field framework is that it can make sense of *collective configurations* of economic strategies and structures. Usual political economy and institutional economics take for granted that "institutions" work as rules actors "rationally" interpret and employ for individual interests. Yet were this the extent of the story of economic strategy and structure, we would likely see more variation, especially in single cases. As Dobbin (1994) noted, as one pulls back from specific details to a broader picture, one sees unusually stable patterns over time, for example, public policies. Even changes in strategies and structures retain core logics that are not reducible to interests or institutions alone (Fligstein 1990). This is an important issue for analyzing Soviet and Russian economic history. The stereotype of post-socialism is a "wild east" and on the brink of chaos and policies as hostages of competing elites. Such stereotypes have some truth, but the reality of post-socialism is not so limited. Within the politics and contention of rather uncertain Russian post-socialism, there were more stable patterns of field logics embedded in *relations* and elite *habitus*. These would constrain these elites' perceptions and responses (but would not rigidly determine them) and become the tools for strategic action and claims-making for legitimation. Russia's post-socialism was a case of fields in flux.

Legacies and Economic Orders: From *Habitus* and Fields to Groups and Clans

Recall from Chap. 1 the three components to Bourdieu's framework: *habitus* (knowledge and dispositions), capital (categories of resources), and fields (structured and contextualized relations of goals, rules, and practices). All three were crucial in the politics of economic reform, in particular, remaking enterprise structures and property, because, it turns out, different sets of actors emerged from different field positions in the late Soviet era. This combination of shared *habitus*, institutional locations, and networks created three proto-classes, which set about at different times to reorganize relations between enterprises, devise strategies to access property and finance, and to make claims to authority. These three sets of actors were Red Directors (managers of Soviet-era enterprises), financial entrepreneurs and "oligarchs" (many of whom began in the Komsomol and institutes), and state elites (located in the state bureaucracy, including security forces). Their positions and formative experiences shaped their economic knowledge and dispositions (*habitus*), which contributed to interpretive frames through which they perceived and understood their economic surroundings and templates for devising strategies for survival and gain. Further, these sets of actors were linked in particular networks (e.g. production relations between enterprises), which reinforced a sense of shared interests, identities, and economic dispositions.

These actors and the post-socialist contexts in which they operated did not emerge in a vacuum. Their experiences in Soviet-era institutional configurations inculcated particular dispositions and knowledge that allowed them to survive and succeed, and these were templates to which they turned for remaking post-Soviet fields and institutions. The Soviet era bequeathed not only knowledge but also particular social relations, from production relations to personal networks, that constrained and enabled action after the collapse of communism. These three groups embodied different interests, dispositions, and conceptions of normal economic relations and practices: Soviet fields embedded in *habitus* and acting as a template for reconstructing post-Soviet authority, fields, and practices. While *habitus*, capital, and practices coalesced into multiple meanings and facets of field structures and logics, I focus on three dimensions related to conceptions of the normal economic order and important for making sense of post-Soviet Russia's economy. The first is perceived core risks that threaten the existence of norms of legitimate economic relations and practices. Positing or implying risk simultaneously constructs a general sense of what might be normal and legitimate. The second facet is source of authority: the crucial form of capital or position that grants an actor significant rights to make legitimate judgments and decisions. The third facet is rights and limits of property governance. Property governance involves normal, natural claims to resources. As such, assumptions of fundamental rights regarding the use of property, as well as limits, underpin models of how power or authority should be distributed in a normal economy. Note that this does not rule out opportunism, such as claiming property rights in order to rent-seek or guard one's privilege. Property and governance rights are claims for legitimacy. Unless it is a purely naked grab for resources—such naked grabs are rare generally, even in post- Soviet Russia—opportunism ultimately is cloaked in these claims of normality, which in turn constrain the behavior of those who make these assertions by making them accountable to their claims. (Note that the same might be true for such opportunistic behavior as "theft" and "corruption"—these also have rules actors follow.)

The three groups that eventually came to organize and battle for remaking the logics and rules of Russia's post-socialist economy were these: (1) Soviet-era enterprise managers, the "Red Directors"; (2) rising financial elites, especially self-proclaimed and eventually infamous "oligarchs"; and (3) security officials and elites (*siloviki*) and state managerial officials running state-owned enterprises outside the industrial sector. These groups set out to institutionalize these templates for practice and relations in two forms of relations: informal *political clans*, with elite representatives in President Boris Yeltsin's inner circle,[1] and financial-industrial groups (*finansovo-promyshlennye gruppy*, FPG), enterprises interrelated through structured property ownership. While clans had been developing over time, in 1995 Thomas Graham, working in the American embassy, published his analysis of Kremlin clans in *Nezavisimaia gazeta* understandings of how the politics of economic policy and privatization in Russia worked under Yeltsin (Graham 1995). These clans and the interest they represented correlated with concrete positions, capital, and economic worldviews. The finance clan was led by Anatolii Chubais and his allies from St. Petersburg, and it included heads of major banks (Oneksimbank, Inkombank, Menatep) and reformers in the Kremlin (e.g. Sergei Beliaev, Alfred Kokh). Financial elites ("oligarchs") were situated in this clan. Presidential aides and former cabinet members Aleksandr Korzhakov and Oleg Soskovets represented military-industrial firms (VPK, *voenno-promyshlennyi kompleks*) that de facto represented broader industrial interests, creating an industrial clan. A natural resources clan that included Gazprom, Sibneft, and Rosneft emerged, and Prime Minister Chernomyrdin, former Gazprom head, championed them in the Kremlin.

At this time there was no true "state-centered clan" that would correlate with our third set of field logics and actors (i.e. state-centered, *dirigiste*): this would have to await Putin's project of bringing the security elites back into the Kremlin and rebuilding state power. The military-industrial clan did not openly encourage concentrating industry around the state; rather, they preferred protecting older Russian industries and enterprises. The natural resources clan would eventually form the core of the state-centered clan, given that Gazprom and Rosneft were state-owned and would later become important vehicles for Putin's state-rebuilding project.[2]

If clans were informal (if still very real) vertical political patron-client relations from the Kremlin on down, members of these groups were also setting out to formalize relations between organizations, mostly through different structures of property ownership, to cement particular forms and logics of economic relations, authority, and practice. Soviet-era Red Directors, linked to the Korzhakov and Soskovets clans, tried to create Defensive FPGs: groups of enterprises linked in production chains and aiming to defend their capacity to produce and employ workers. The new financial elites, especially the oligarchs represented by Chubais and the finance clan, founded banks and created Financial FPGs, diverse holdings to optimize profit and reduce risk. The less well organized but still embryonic state managerial officials, who were linked indirectly through the state-owned energy enterprises (e.g. Gazprom), waited in the wings, and when their chances came and following orders from on high (i.e. Vladimir Putin and the Kremlin *siloviki*), they set out to reorganize firms into state-centered FPGs that defended particular sectors deemed important to national security.

These groups did not always coexist peacefully. Competition emerged both over property and political power and over the normal post-Soviet economic order. Each group laid claim not only to wealth and power but also to status bequeathed by being the face of the legitimate economy. As these groups jockeyed for autonomy and influence, they were shaped by their earlier positions and experiences, and in turn tried to reshape emerging post-Soviet institutions and property arrangements. While these actors were not perfect carbon copies of their Soviet-era positions, they did carry those dispositions and experiences forward as resources for navigating post-socialist politics and constructing claims and strategic positions—and in turn influencing the logics of economic fields and field politics (Table 13.1).

Table 13.1 Groups of actors and logics of economic organization

Actors	Clan	Type of FPG	Economic logics: 1. Perceived core risks 2. Nature of authority 3. Property governance
Red Directors	Aleksandr Korzhakov, Oleg Soskovets	Defensive FPGs: 1. Oriented to banks owned by members 2. Sense of voluntary association Energomash	1. Collapse of supply, and thus of production 2. Expertise in production, care for labor 3. Property as rights to residuals, but production and related expertise paramount
Financial elite (oligarchs)	Anatolii Chubais	Financial FPGs: 1. Diverse holdings 2. Centered on bank or financial institution Rosprom (Menatep) Interros (Oneksimbank)	1. Uncertainty and financial loss 2. Acquiring property and financial capital 3. Property as decision-making authority and control
State officials (especially *siloviki*)	Viktor Chernomyrdin	State-owned FPGs: 1. Holdings (conglomerates) in strategic industries 2. Centered on financial or trading firm or industrial firm Gazprom, Rostekh	1. Threats to state power, national security 2. State office and national security 3. Property secondary to state and national interests

Red Directors and Defensive FPGs Versus New Capitalists and Financial FPGs

In the late 1980s, vertical relations of control central to the logic of the Soviet command economy began to dissolve under the twin assaults of stagnation and Mikhail Gorbachev's *perestroika* (and with unraveling of political institutions taking place in parallel, both through *glasnost* and *demokratizatsiia*, and rising ethnic movements and conflicts). In this context, Soviet industrial directors began to use new freedoms to restructure their enterprises to increase their autonomy and material gain, for example, creating cooperatives or transforming shop floors and other subdivisions into semi-autonomous "rented firms" (*arendnoe predpriiatie*), through which these directors could circumvent state limits on pricing and increase their profits through speculative resale of cheap state inputs (cf. Hass 2011a). In other words, in the last years of the USSR and its command economy, Red Directors were drawing on what they knew—production and enterprise relations—to create their own new worlds in which they would gain greater authority and wealth (for themselves and possibly for their enterprises). At the same time, a younger generation of Soviets, with less experience in production and more in navigating bureaucratic structures (especially those of the Party) for financial gain, saw the opportunity in the late 1980s to create their own independent financial institutions, which they would then use for different strategies of gain. To them, a world of finance beckoned.

By January 1992, industrial and financial elites were emerging to stake claim to the post-Soviet world. They were armed with different logics and practices of legitimate economic structure and practice, different justifications for a post-Soviet economy, and different capital and resources. While they did not start out competing with each other, by the middle of the new post-Soviet decade, they would come to blows—sometimes directly, sometimes indirectly and in the abstract—over Russian economic fields and outlines of the new economy.

Rise of the Red Directors and Defensive FPGs

Socialized and on the job in Soviet industrial enterprises in the Soviet era, Red Directors as a rule were knowledgeable primarily about the process of production. Supply of necessary inputs, coordination of labor and provision for the labor force, and ensuring output were key to Red Directors' survival and success in the Soviet era. Those individuals who best played the game of using networks and maneuvering the Plan were most likely to rise to the ranks of general directors. Long experience in the politics of production and the Plan shaped their *habitus* and logics and knowledge of practice. For Red Directors, the core risk to normal organization was the disruption of supply and labor—resources that were key to production and to political legitimacy, especially because "success" in the Soviet era meant fulfilling or overfulfilling output norms and providing for the enterprise's workers. The source of authority for Red Directors was their organizational knowledge and position. Atop the enterprise structure, they could see the entire production process and apply knowledge they had gained through experience to guarantee proper running of the enterprise, so that it could fulfill its productive function. While Red Directors came to accept that property rights allowed claims to residuals (i.e. profit), it did not allow total interference within the enterprise if this would hinder production and employee provision. Red Directors often articulated a "moral economy" of the enterprise as a sacred collective that could not be hurt by the mere pursuit of profit, because such naked pursuit was little different from rent-seeking and theft (Hass 2011b; see also Boeva et al. 1992; Dolgopiatova and Evseeva 1994; Gurkov and Avraamova 1995).

In the Soviet system, fields centered on the state bureaucracy, from Moscow-centered ministries to local-level *glavki* that coordinated distribution, production, and enterprise functioning. This provided one template for post-Soviet industry. Formally, ministries and *glavki* lost their influence over enterprises as the Plan was replaced by *goszakazy*—state purchase orders that enterprises had to fulfill first (but then could follow with private outside orders)—and late 1980s reforms of enterprise relations gave directors more authority to restructure their firms. This only increased as privatization became inevitable. However, this did not mean that ministries and *glavki* disappeared entirely. Their shadows remained in those personal networks created by constant Soviet-era enterprise-ministry/*glavki* interactions and in directors' own templates for economic relations. Older directors used to state-centered planning saw risk in collapsing exchange and distribution, finances, and unsettled rules of the new economic order. To reduce these perceived (and often real) risks and to reinforce an economic order they understood and from which they could benefit, directors turned the earlier ministry/*glavki* templates into voluntary associations. Exchange partners needed to group together to support exchange and supply of goods linked in production cycles, and voluntary associations initially seemed to be a useful way to support these production relations on which these enterprises depended. In other words, Soviet relations re-emerged, only without direct compulsion from Moscow and with profit as one (albeit not the only) goal of economic activity.

Red Directors learned quickly that privatization was for real and that property ownership would be important to authority and economic structure. In part to gain some popular support for what might be (and indeed became) a contested and not entirely legitimate venture, the Yeltsin administration gave workers and managers privileged positions in the first wave of privatization: in particular, either a majority of shares at full value or a minority of shares at discounted value. Managers could then try to convince worker-owners to transfer their shares to directors, either through purchase or transferring them to a third outside firm (cf. Blasi et al. 1997; Barnes 2006; Hass 2011a). Those directors who had joined their partners through networks and production chains into voluntary associations could now solidify those relations of mutual aid by transferring shares in member firms to a central holding company. This holding company would run member firms as majority shareholder, and shares in the holding company were split among FPG members, whose directors sat on the FPG board of directors. This would hold off outsiders who might want to buy member firms and sell them as scrap for profit or restructure them—either way threatening the employment of workers and managers. Unsurprisingly, the composition of some Defensive FPGs (among the first in Russia) reflected their Soviet past (Prokop 1995; Gorbatova 1995, Starodubrovskaia 1995). Defensive FPG Fin Prom was a coalition of a particular set of local state officials, enterprise managers, and the State Privatization Committee; member organizations included mining companies, real estate firms, a bank, and trading company. FPG Konsensus was formed from the Soviet Ministry of Light Industry. Petersburg's Energomashstroitelnaia Korporatsiia (EMK, or Energomash) was originally a voluntary association made up of mostly Petersburg (Leningrad) enterprises in heavy industry, much of whose output often went to each other.

The fundamental logic and structure to these new Defensive FPGs, then, involved interlocking property relations to defend members against outside control or accountability, and against the uncertainty of supply and finance in a context that was both unstable and increasingly involving some basic market principles, such as the need to sell goods with demand, make profit, and pay for expenses. Solidifying these relations through property ownership was not a perfect solution, but it did provide something of a safety blanket for the time being. In 1995, the manager of a chemical FPG in St. Petersburg angrily denounced contemporary Russian and Financial FPGs as bloodsuckers draining away Russia's wealth, while showing little concern for production, employment, maintaining Russia's stock of human capital (especially in the sciences), or general well-being that was not associated with immediate profit. The head of another Petersburg FPG, involved in producing furniture, shared this view of competing market actors in 1997. I proposed that perhaps their FPG should seek additional investment from banks or other financial actors; while this would mean granting some authority to those investors, it would also mean an injection of capital to help stabilize or even expand production and sales (and thus profit). This manager replied immediately that he did not trust bankers or other financial elites, and accused them of wanting to buy up other firms and squeeze any immediate value out of them. Both these managers confirmed the general Defensive FPG logic: fear or distrust of more market-orient actors who did not value production and employment, and conversely a focus on maintaining production and employment.

This manager-centered, production-centered logic eventually ran into potential problems. First, this logic of *habitus* and field increasingly contradicted "transition culture" (Kennedy 2002) that privileged "post-Soviet" over "Soviet" and finance (money, profit) over production and employment. Second, profits from sales of some products or attracting capital to the FPG's pocket banks did not always generate sufficient capital to develop or even survive for long. Third, once Defensive FPGs had been created, the director of the central holding company could turn the tables and take control of the member companies. Just this happened to Energomash (EMK) by Aleksandr Stepanov (cf. Hass 2011a). EMK's holding company guided the sale of member companies' output and the capital generated by such trade. Stepanov was able to keep sales profits with the holding company and thus deny them to member firms. As a result, Stepanov could force these member firms into bankruptcy and buy them out through an offshore company that he controlled. Stepanov's strategies were initially successful, but they also violated the core Defensive FPG principle of guarding member firms' autonomy. The managers and workers of one of Stepanov's later targets, Leningradskii Metallicheskii Zavod (LMZ), fought back, and LMZ and other allies (including EMK member Elektrosila) defeated Stepanov.

In the end, Defensive FPGs and Red Directors also lost status when their Moscow clan lost out to the rising financial elite and their clan. After 1995, the Kremlin increasingly used its authority to privilege this new elite and their empires against Red Directors. The contingent reduction in their position vis-à-vis competitors inside the Kremlin, their inability to generate sufficient capital to guard themselves against economic instability in the 1990s, and their problematic legitimacy in the era of transition culture ultimately were too much, and the Defensive FPG and its logic lost out to their rising competitors.

A Competing Field and Empire: Financial Elites and Financial FPGs

Financial entrepreneurs, younger than Red Directors, had their formative experiences in non-industrial bureaucratic organizations, such as the Komsomol and research institutes. Here they learned the art of navigating the rules of financial resources (e.g. dues, research funding) and maneuvering around legal obstacles involved. As reforms evolved in the late 1980s, people in these organizations saw the opportunity to make use of available money and opportunities to engage in domestic and even some foreign exchange, from which they could garner some profit. They were also more optimistic about the possibility of reforms providing further economic opportunities; in contrast, managerial elites (some of them bosses of these younger future entrepreneurs) were more cynical about any useful results from reforms, having seen various reform movements (e.g. the Kosygin reforms of the 1960s or reliance on technological advancements in the 1970s) (Hoffman 2002). Their experience was in using organizational resources for further gain, for example, from speculative trading during Gorbachev's economic liberalization or dealing with Party dues (which would translate later into setting up embryonic financial institutions), and this conditioned their dispositions and knowledge. Not production, but investment of capital for future returns was their logic. Along these lines, they conceived a core risk as losing control over capital and return on that investment. Authority, in turn, was grounded in knowledge of transactions (financial or otherwise). Finally, control of resources meant decision-making authority, and so the property rights to them meant authority to make decisions as well as control residuals. Their relatively monetized experiences meant that enterprises, like other goods, were commodities to be bought and sold. Economic capital was sacred and the ultimate end.

Because they did not run existing Soviet enterprises, and because their formative experiences were with financial machinations and transitions that had no real formal templates, these budding financial entrepreneurs had to innovate, for example, opening up private cooperatives that relied on relations to other Party cadres for access to finances. They also were not yet so steeped in Soviet economic life not to adapt to changing conditions, and they managed to learn a new Western market vocabulary that helped them engage Western investors. Rather than seeing such cooperatives as a means to siphon speculative capital to a mother factory—the logic of Red Directors—these younger cadres saw financial cooperatives as a source of profit in and of themselves. And so such experiences of Mikhail Khodorkovskii, Boris Berezovskii, and other future oligarchs led them eventually to build empires centered around banks and diverse holdings in whatever ventures might produce profit—and so the Financial FPG was born. These new commercial banks and similar financial institutions made capital by speculating on the ruble and Russian treasury bills, and they used profits to purchase shares in newly privatized firms, and as their property empires began to take shape, these financial elites reorganized their holdings into more centralized financial structures. Oneksimbank and finance company Mikrodin united as Interros, and with capital and a more streamlined structure, they acquired shares in such privatized enterprises as Norilsk Nickel and (temporarily) in car factory ZiL. Menatep, Alfa-bank, and Inkombank invested in oil, confectioneries, and metals.

Because they conceived of risk primarily in terms of control of and return on investment, their property empires differed from Defensive FPGs. First, this elite made sure to have a presence on boards of directors and managers in newly acquired properties. Menatep President A. Zurabov noted, "If we control management, the situation, capital streams, account transactions, and the like, then of course the probability of a return of such loans in order are higher than loans by a clientele on the side…" (Pappe et al. 1997: 52). Also, Financial FPGs were organized around diversification to reduce risk. As one manager at Financial FPG Sokol noted, "we operate on the principle of the submarine, where there are several compartments—this helps keep it afloat during difficult times when one compartment is suffering" (*Ekonomika i zhizn* #33 1994: 37). Menatep invested in a variety of firms, even if oil became central to this empire. And so on. Eventually, some financial elites invoked the South Korean *chaebol* and Japanese *keiretsu* as examples of new economic organization—and everyone knew that Korea and Japan were far more successful economically than the USSR (*Kommersant*, April 16, 1996: 18–19).

Two key moments for field development were the second stage of privatization in 1995–1996 and the 1996 presidential election. Contentious politics between these clans came to a head in 1995 and 1996 over loans-for-shares auctions and the looming presidential election. Chubais and oligarchs traded property in rigged auctions for financial support for Yeltsin's campaign, and they reaped rewards after Yeltsin's victory. Scandals about rigged voting after the first round of presidential balloting in July 1996 ultimately cost Korzhakov his position and status in the Kremlin and hurt his industrial clients. A 1996 cabinet reshuffle strengthened the finance clan, as Oleg Soskovets lost his position and Chubais' ally Boris Nemtsov was brought into the Kremlin inner circle. While Viktor Chernomyrdin remained prime minister and could defend Gazprom's interests, Chubais and Nemtsov could act as a counterweight.

The key to oligarch domination was the loans-for-shares auctions. Anatolii Chubais had staked Russia's future on market capitalism, but reforms had not taken deep enough roots yet, and he needed some way to end once and for all a possible return to Soviet socialism. Hence, he manipulated privatization policy to favor the new financial elite, in the hopes that they would eventually lead Russia into the new land of the market economy: "the fact that [oligarchs] would be the forces supporting their own private property, that they would defend their private property, and that in the political process they would be, by definition, against communism and pro-reform—that was 100 percent sure" (Hoffman 2002: 312). In the loans-for-shares program that Chubais and oligarch

Vladimir Potanin orchestrated (*Kommersant*, November 28, 1995: 47–50; *Ekspert*, September 12, 1995: 43–48), the state would exchange its shares in lucrative enterprises (often in resource extraction) in return for large loans from private banks and financial institutions (conveniently owned by these oligarchs). Should the state not repay the loans after one year, the shares would remain with the investors—which eventually happened, to no one's surprise. Firms privatized in loans-for-shares included metallurgical conglomerate Norilsk Nickel, oil firms Iukos and Surgutneftegaz, and Perm Motors, among others. In the end, Menatep won oil giant Iukos, Oneksimbank/Interros won Norilsk Nickel and oil conglomerate Sidanko. Menatep daughter Laguna, which won Iukos, sold the shares to Menatep daughter Monblan. Following their victories, these elites tried to consolidate governance over their gains, and sometimes this led to contentious economics. Managers at Norilsk Nickel challenged the new owners in court, although the latter was victorious. Mikhail Khodorkovskii also faced temporary challenges to his attempts to consolidate control over Iukos.

In 1998 the oligarchs seemed ascendant, until the 1998 ruble crisis delivered a blow to their new hegemony. Not only did oligarchs and Financial FPGs lose capital in the crisis; they also lost the halo of authority. They seemed not only to be parasites sucking the wealth of the Russian people; they were also vulnerable. Yeltsin himself seemed significantly tainted by the crisis, following already contentious economic reforms. By 1999 he handed power over to a shadowy figure who had risen quickly in his administration, a former KGB officer named Vladimir Putin. Boris Berezovskii and other oligarchs believed that Putin would guard their gains and those of Yeltsin's family. Putin had different ideas and set out to realize them.

The State Strikes Back: State Cadres and State-Centered FPGs Topple the Oligarchs

One group that had lost out in the 1990s returned to power after 2000: the security apparatus and state officials who continued to administer state-owned enterprises, especially those outside heavy industry. Their logic was state security and national interests,[3] and it was not inconceivable to claim that privatization and reforms generally had compromised the capacity of the state to defend the nation. The project of this new elite was an "adjustment" to privatization and property: renationalization in some cases and reassembly of enterprise into new state-run empires, or bringing oligarchs into line behind new state policies. This was not a return of the Soviet command economy; but it was a new post-Soviet *dirigisme* inspired in part by that model. In retrospect, it also seems as if Putin and his inner circle did have a master plan of sorts, for the strategies for taming oligarchs and their fields, and the Russian economy generally, did seem to unfold with a fairly coherent logic. Whether there was a master plan is impossible to verify or disprove, and certainly contingencies did matter, such as oligarchs' own responses and the temporarily high price of oil, that contributed to the state elite's status. However, that there was some coherence suggests a relatively structured and shared worldview at the top inside the Kremlin and seemingly by local security authorities as well. As for the other two groups and their fields, institutional embeddedness in the Soviet era inculcated particular shared *habitus* and relations that became a template for another facet and stage in reorganizing the post-Soviet Russian world.

In a sense, this third group of actors and their logics of economic organization were a residual in the first decade of post-socialism. Red Directors also ran state-owned enterprises initially, but they lobbied for privatization on their own terms to gain autonomy from the state and other competing groups and elites. The boundaries and logic of this new group became clearer after 2000, but murmurs of this logic were apparent earlier. For example, the contentious plan to privatize Rybinskie Motors elicited pushback from Rybinskie's own managers, who claimed that the American firm that wanted to purchase Rybinskie would threaten Russia's capacity to produce engines for Russian-built airplanes (the IL-62, IL-76, and TU-154). In 1996 Yeltsin's communist competitor for the presidency, Gennady Zyuganov, was riding a wave of resentment against market reforms, and selling property to foreigners had become part of that critical discourse (Hass 2011a). Rybinskie, therefore, was part of a set of industries that were of national interest and should not be privatized, at least for the moment. In a moment of political expediency,

Boris Yeltsin decreed that a set of enterprises were of "strategic significance" and were too important to be privatized, and he included Rybinskie on this list in October 1996.[4]

While Rybinskie managers and Yeltsin might have used this national security argument as a fig leaf for good old-fashioned interests, the latent logic of state-centered elites and organization centered on state authority and security; production and profit were secondary to the prerogative of the state and the security and prestige of the nation for which state officials spoke. Correspondingly, the main risk was loss of state influence (not necessarily ownership) over economic activity. Further, if total state economic control of the Soviet era had seemed to be disastrous, strategic state control was not. Majority share ownership, rather than total control, had benefits: minority shareholders could buy shares and invest capital in return for a cut in profits, even if this meant no real say in governance. (In fact, Gazprom and similar firms invited foreign investors to go for the ride when hydrocarbon prices were high in the mid-2000s.)

At the apex of this new group was *siloviki*, literally a "power elite" (Kryshtanovskaya 2005). Who comprised this power elite has been subject to some debate. Perhaps it has included as many as 6000 people trained in the KGB-FSB and later holding positions in the Ministry of Internal Affairs and FSB.[5] While there has not been a perfect alignment of interests and identities between various elites and agencies,[6] there was a striking degree of coherence and corporate identity (Kryshtanovskaya and White 2005). Early in his reign, Putin instituted a "power vertical" of regional officials answerable to him to act as a counterweight to regional governors and elites, and Putin et al. made Edinaia Rossiia (United Russia) the regime's vehicle in the Duma, facilitating the centralization of real power around the executive. Further, this elite quickly identified property as a potential basis for competitors to gain power. Rebuilding state power and consolidating elite positions and privilege, and a state-oriented *habitus* and logic, combined to breed a new *dirigisme* (Shevtsova 2007: Chaps. 9–11).

This project began early in Putin's first term, when it became clear behind closed doors that the oligarchs still considered themselves Russia's ruling class. In their toolbox were two important instruments: *kompromat* (legally compromising materials, not difficult to obtain, given Russia's vague and often contradictory laws) and state-owned enterprises that could purchase existing firms, especially those of *siloviki* competitors. In particular, tax audits were a powerful weapon that the regime had little trouble employing. Regarding state-owned giants, the first step was reining these in, as managers of some such firms had come to enjoy independence under Yeltsin. An early favorite was Gazprom, and the regime set out to strengthen control over the natural gas monopoly. CEO Rem Viakhirev had turned Gazprom into his personal empire, and he had altered the charter to make it difficult to oust the CEO, but Putin et al. were able to stack the board of directors, such that by July 2000, five of the 11 board members were Putin loyalists, and in 2001 Aleksei Miller became the new CEO.

Their first target was Vladimir Gusinskii, whose media holdings were too critical of the new regime and its actions in Chechnya. Gusinskii's empire Media-MOST owed Gazprom $473 million, for which Gazprom held 40 percent of Media-MOST shares as collateral. Out of nowhere, Gazprom announced it would transfer those shares to Gazprom-Media, and to drive home that this was a serious policy, the police raided Media-MOST in May 2000. Gusinskii was charged with embezzling $10 million, and Gazprom-Media sued Media-MOST to recover debts. In November 2000, Gusinskii transferred Media-MOST to Gazprom-Media and fled Russia. The next target was Boris Berezovskii, who owned television station ORT. The Kremlin reopened investigations of fraud at Aeroflot, which Berezovskii happened to own, and the latter was forced to turn over ORT. (Much of the rest of his empire he sold at cut-rate prices to his former protégé, Roman Abramovich.) Rather than sell off new gains, the Kremlin kept them under Gazprom ownership, giving the state increased media power for shaping public discourse.

These early attacks on Gusinskii and Berezovskii worried the oligarchs, who called for a truce with Putin, which ended abruptly in 2003 after the pro-Putin Council for National Strategy claimed that the oligarchs were organizing some form of opposition to the Kremlin.[7] The next attack was against oligarch and oil magnate Mikhail Khodorkovskii, who had openly hinted at turning to politics. Khodorkovskii was soon arrested for tax fraud, and after a lengthy investigation and trial, he ended up in jail. In retrospect, the next step was unsurprising: in autumn 2004, Putin suggested the state might take Iukos in lieu of tax debts. In December 2004 state-owned oil firm Rosneft announced it would take Iukos' most important daughter firm, Iuganskneftegaz. *Siloviki* increased its holdings of hydrocarbon firms and profits, as well as tools for symbolic power in new media holdings and the image of an elite able to crush its opponents.

This was not the end of the new project of remaking state power in the economy. With the oligarchs cowed, the Kremlin set about creating its own version of the French national champions with state-owned FPGs, the new state corporations (*federalnoe gosudarstvennoe unitarnoe predpriiatie*, FGUP). For example, one state-led consortium, OAK (Obedinennaia Aviastroitelnaia Korporatsiia, United Aircraft Corporation), was created to secure state control over various facets of airplane construction as an industry of strategic national interest. By bringing together under one corporate roof the likes of aircraft producers MiG, IlIushin, Sukhoi, Tupolev, and Irkut, the state could guard domestic airplane production and not become dependent on the United States and European Union (i.e. Boeing and Airbus), while also setting up the possibility of challenging them in the global market in the future (e.g. with the Sukhoi Superjet). Perhaps the most impressive national champion was Rostekh (Rossiiskaia Tekhnologiia), which began its life as a daughter firm of the state weapons exporter, Rosoboroneksport. Founded in April 2000, Rosoboroneksport expanded its holdings from helicopter production (via Oboronprom), to auto giant AvtoVAZ, to various other corporations involved in metallurgy, machine-building, and production of goods (trucks, jets) of "strategic value" to the military and military-industrial complex. The last included titanium giant VSMPO-AVISMA, Petersburg shipyards Severnaia Verf and Baltiiskii Zavod, and various jet engine manufacturers. Eventually, the Kremlin reorganized this new empire around daughter firm Rostekhnologiia, led by Putin's ally Sergei Chemezov, who believed that the market was not too kind to Russian heavy industry (as was the case in the 1990s). While this imperial project faced some opposition inside the state, by 2008 new president Dmitrii Medvedev officially created Rostekhnologiia and transferred 426 enterprises to its control (*Rossiiskaia gazeta*, July 16, 2008).

While these state-centered FPGs have, like Defensive FPGs, helped defend production from market forces, that defense has not been oriented to maintain supply relations or managerial authority, as was the case for Defensive FPGs; and profit remains important, but not as central as for Financial FPGs. Rather, this new *dirigisme* has reoriented the economy to state strategies and profit. The state-centered field grounded in a logic of security did not have to come to this. It is possible that Putin's ruling elite are simply using new centralized power for their own gain. Alternatively, or perhaps simultaneously, power corrupted: the perceived need by those in the Kremlin to concentrate power in their hands so as to counter the damage of the 1990s habits of a "Wild East" economy made temptations to rent-seek even greater. It is also possible that, in the absence of sufficient professionalism in the state and too weak a sense of esprit de corps, Putin et al. had little choice but to allow some degree of rent-seeking and corruption to keep those further down the bureaucratic order in line.

Whatever the case may be, this new *dirigisme* carries potentially dangerous political consequences for the regime, the economy, and perhaps the political order. In the mid-2000s, this seemed a winning strategy: record profits from oil and natural gas were channeled from Gazprom and Rosneft into Rostekh and similar FPGs, creating the image that Russia's economy was on the move. Once the price of hydrocarbons dropped, and then when Western sanctions took effect after the events in Crimea and eastern Ukraine, this *siloviki* version of *dirigisme* began to run into trouble. If earlier it was possible to blame Yeltsin and the oligarchs for the pain and injustices of the 1990s, it is increasingly difficult for the state to avoid some blame for growing economic woes. Further, this regathering of the commanding heights might hinder improvement of the Russian economy. While states do not inevitably do a worse job than the private sector in cultivating entrepreneurial innovation and growth, as was the case in some of the East Asian dragons a few decades earlier (cf. Evans 1995), there is no guarantee states will get it right, either. State professionalism necessary for effective governance does not appear sufficiently developed for this new *dirigisme* to move beyond rent-seeking and control alone. Further, Putin's *dirigisme* also seemed aimed at insulating the commanding heights from market accountability and politicized the economy, risking the rise of neo-patrimonialism. It has, somewhere in its core, a defensiveness and aversion to accountability. This might be, in part, because the rumors of corruption and massive wealth stashed away abroad have some truth; it also is likely a legacy of the politics of security agencies, in which one investigates others but guards one's own practices as part of a game of security and advantage. That defensiveness has advantages for war, as it is a logic of conquest; but for pragmatic adjustment, such a collective logic runs into problems, for the shared goal at the heart of organized striving is not so much economic performance, as economic power. Reforms that challenge power are always at risk of being still-born; does Putin have the wherewithal, or even the stamina and wellspring of innovative ideas, to carry through the changes necessary to help Russia catch up once again with its competitors (to which we can now add China and, before long, India)?

This drama is far from over. Red Directors seemed in charge initially once they emasculated shock therapy; oligarchs seemed in charge after privatization. To assume *siloviki* and their *dirigisme* are eternal would be foolish as well. For if history teaches us one thing, it is that culture is ever-changing—even in economies.

Conclusion

Analyses that rely on efficiency and market dynamics to make sense of Russia's post-socialist experience ultimately succeed only by forcing square pegs of some data into round holes of economic theory and then by cherry-picking said data. Economies are another form of politics and culture. This does not mean profit and politics are unimportant. Red Directors, oligarchs, and elite state cadres all needed inputs, whether money (e.g. investment, loans, income streams) or in kind (e.g. barter for electricity or the like). They merely presumed the best strategy was to obtain such income streams from different sources: loans from pocket banks and profit from traditional clients (Red Directors), profit from financial speculation or from firms in lucrative sectors bought on the cheap (oligarchs), or petro-dollars flowing from the state budget (state elites).

Importantly, this story has involved not just elites and alliances but also sets of actors with sets of shared interests, perceptions, and practices who had come from particular contexts and who set out to create similarly structured contexts in the new post-socialist world. This *organized* and *collective* facet of this story suggests something more than individual interests and *habitus* were at work—that Soviet-era fields conditioned different sets of actors in clearly structured ways and that those actors set out not create not only new structures that would serve their interests but also sets of rules and norms of legitimate economic strategies and structures. In short, this was a story of fields weakened and reborn, even if their newer versions were not carbon copies of the old. Further, an important part of Russia's story is that field reproduction and generation are not devoid of politics and contention, often missing in neoinstitutionalist field accounts. Fields might be negotiated as various actors encounter new laws or economic landscapes; but struggle can emerge as well, especially when the new context does not have either a sufficiently powerful state that can mediate field-building and the economic landscape is uncertain enough that threat and opportunity are both present—creating the perception, and perhaps reality, of real risks that must be combated.

The power politics at the center of usual accounts in political economy and political science—the stuff of traditional political economy—are not absent here. In fact, I have invoked such a time-honored narrative. However, in such accounts, elite interests drive alliances and conflict, and the strongest institutionalize their gains in property and policy. Institutions provide tools and boundaries for contentious politics and act as foundations for further institutional evolution. None of this is in question—but leaving the narrative here would leave us with an incomplete picture. First, the *structure* such institutional crystallization takes is not as contingent as political economy might often imply, and nor is it dependent on institutions and individual actors alone (even if they are in alliances). Interests and relations at the heart of alliances and the reproduction of institutions do not come from black boxes. How one conceptualizes or understands those interests, the ways to fulfill them, and the obstacles that might be out there in the economic night are not random. Rather, they are conditioned by structured experiences—experiences embedded in concrete institutions and institutional fields, replete with their logics and structural arrangements. In fact, the Russian case has an important element of fields: "organized striving" (Martin 2011: 252–254), whether by Red Directors, financial elites, or *siloviki*. Post-Soviet remaking of strategies and structures was not only politics of power and property; it was also a clash of particular *economic aesthetics*.[8] And this brings us to another important lesson of Russian post-socialism for field theory: the past is not unimportant (as mainstream macroeconomics and microeconomics cannot begin to imagine), yet nor does it repeat itself mechanistically and deterministically or without contention (an impression some work invoking "path dependency" might make). The content of actors' *habitus* and structural arrangements (especially networks) bequeathed by fields might be recombined in the politics of economic restructuring and reform. It was not inevitable that the Red Directors would lose in the end, nor that the state would return with a vengeance. Yet it was improbable that post-socialist Russia's economy would come to resemble that of the United States or Taiwan, as these legacies of *habitus* and fields were too strong, and their carriers remained alive and networked, able to attempt some collective politics. Even when institutions were in flux and their design still up for grabs, the past survived to inform the present; fields as an echo of earlier institutional configurations persisted in *habitus*, network relations, and capital, and continued to reverberate.[9] Their altered sound meant that the present became the past, but imperfectly. Fields, capital, and *habitus* were the mechanisms of path dependency *and* of innovation, of negotiation *and* contention.

Notes

1. The existence of clans suggests shared interests and structured positions and relations in the broader economy *and* in competition with other groups. That is, the clan battles of the 1990s were not simply about gain but also about positioning vis-à-vis others: implicitly constructing a "Them" to define "Us."
2. Additionally, there was a Moscow clan defending the interests of business and other elites in the capital, and this was headed by Moscow Mayor Iurii Luzhkov. An "agrarian clan" existed but was not as strong as other clans and by the mid-1990s was mostly impotent, as the Agrarian Party could not gain traction in electoral politics and the Duma. One could also include a "Family clan" consisting of Boris Yeltsin and his gatekeepers, daughter Tatiana Diachenko and presidential administrator Aleksandr Voloshin—but as they were at the center of Kremlin power, I am no longer sure they should have been considered a "clan" in the same sense as the others.

3. To repeat, this does not negate the possibility that these actors were also playing political games for personal gain, as Dawisha (2013) hints in her work on the early years of Putin and his Petersburg allies.
4. "Vysshii arbitrazhnyi sud RF otkloniaet isk AO 'Rybinskie motory' k pravitel'stvu," *Segodnia*, January 19, 1996; "Vosvrat: 'Rybinskim motoram' ne udalos' vykupit' 5% svoikh aktsii," *Kommersant*, March 23, 1996; "Podopleka: 'Rybinskie motory' ostalis' na shee u gosudarstva," *Segodnia*, October 25, 1996.
5. The top *siloviki* include Vladimir Putin, Sergei Ivanov, Igor Sechin, Viktor Ivanov, Nikolai Patrushev, Sergei Naryshkin, Vladimir Yakunin, Sergei Chemezov, and Viktor Cherkesov. Some fell out of favor in recent years. "Chekists in the Corridors of Power," *Novaia gazeta*, July 2003; translated and reprinted in *Johnson's Russia List*, #7255, July 18, 2003. See also "Twelve who have Putin's Ear," *RFERL* Oct 15, 2007; Kryshtanovskaya 2005: 256–279.
6. "'Nel'zia dopustit', chtoby voivy prevratilis' v torgovtsev,'" *Kommersant Daily*, October 9, 2007.
7. Lilia Shevtsova, "Implications of the Yukos Scandal for Russian Domestic Politics," September 16, 2003 (https://carnegieendowment.org/2003/09/16/implications-of-yukos-scandal-for-russian-domestic-politics-event-643).
8. This is something entirely absent from mainstream economic and most of the rest of the social sciences: the idea that underlying relations and practices is something more than interests or ideologies, namely, a logic of wholistic perception and evaluation, with non-utilitarian judgments and more than instrumental rationality at work as actors deliberate and respond. As Martin (2011) notes, a field framework, having a wholistic logic to it, can encompass this aesthetics and make humans human again in social science, rather than being the equivalent of a walking, talking Excel spreadsheet.
9. In this sense, post-socialist recombination of property in Russia shows remarkable continuity from the past, which Stark and Bruszt (1998) downplayed in their assessment of privatization and recombination in East Europe.

Works Cited

Åslund, Anders. 1995. *How Russia Became a Market Economy*. Washington, DC: Brookings Institution.

Barnes, Andrew. 2006. *Owning Russia: The Struggle over Factories, Farms, and Power*. Ithaca: Cornell University Press.

Blasi, Joseph, Maya Kroumova, and Douglas Kruse. 1997. *Kremlin Capitalism. Privatizing the Russian Economy*. Ithaca: Cornell University Press.
Boeva, I.N., Tatiana G. Dolgopiatova, and Viacheslav Shironin. 1992. *Gosudarstvennye predpriiatiia v 1991–1992 gg: ekonomicheskie problemy i povedenie*. Moscow: Moscow Institute of Economic Policy.
Chandler, Alfred. 1977. *The Visible Hand*. Cambridge: Harvard University Press.
Dawisha, Karen. 2013. *Putin's Kleptocracy: Who Owns Russia?* New York: Simon & Schuster.
Dobbin, Frank. 1994. *Forging Industrial Policies*. New York: Cambridge University Press.
Dolgopiatova, T., and I. Evseeva. 1994. Ekonomicheskoe povedenie promyshlennykh predpriiatii v perekhodnoi ekonomike. *Voprosy ekonomiki* 8: 40–50.
Evans, Peter. 1995. *Embedded Autonomy*. Princeton: Princeton University Press.
Fligstein, Neil. 1990. *The Transformation of Corporate Control*. Cambridge: Harvard University Press.
Friedland, Roger, and Robert R. Alford. 1991. Bringing Society Back In: Symbols, Practices, and Institutional Contradictions. In *The New Institutionalism in Organizational Analysis*, ed. Walter Powell and Paul DiMaggio, 232–263. Chicago: University of Chicago Press.
Frydman, Roman, and Andrzej Rapaczynski. 1994. *Privatization in Eastern Europe: Is the State Withering Away?* London: Central European University Press.
Gaddy, Clifford, and Barry Ickes. 2002. *Russia's Virtual Economy*. Washington, DC: Brookings.
Gorbatova, Larisa. 1995. Formation of Connections between Finance and Industry in Russia: Basic Stages and Forms. *Communist Economies and Economic Transformation* 7 (1): 21–34.
Graham, Thomas. 1995. Novyi russkii rezhim. *Nezavisimaia gazeta* (November 23): 5.
Gurkov, I., and E. Avraamova. 1995. Strategii vyzhivaniia promyshlennykh predpriiatii v novykh usloviiakh. *Voprosy ekonomiki* 6: 22–30.
Hass, Jeffrey K. 2011a. *Power, Culture, and Economic Change in Russia 1988–2008: To the Undiscovered Country of Post-socialism*. New York and Abingdon, UK: Routledge.
———. 2011b. *Rethinking the Post-soviet Experience. Markets, Moral Economies, and Cultural Contradictions of Post-socialist Russia*. New York and Basingstoke, UK: Palgrave Macmillan.

Hellman, Joel, and Mark Schankerman. 2000. Intervention, Corruption and Capture. The Nexus between Enterprises and the State. *The Economics of Transition* 8 (3): 545–576.

Hendley, Kathryn, Peter Murrell, and Randi Ryterman. 1999. A Regional Analysis of Transactional Strategies of Russian Enterprises. *McGill Law Journal* 44: 433–472.

———. 2000. Law, Relationships and Private Enforcement: Transactional Strategies of Russian Enterprises. *Europe-Asia Studies* 52 (4): 627–656.

Herrera, Yoshiko. 2005. *Imagined Economies: The Sources of Russian Regionalism*. New York: Cambridge University Press.

Hoffman, David. 2002. *The Oligarchs: Power and Wealth in the New Russia*. New York: PubllicAffairs.

Kennedy, Michael. 2002. *Cultural Formations of Post-communism. Emancipation, Transition, Nation, and War*. Minneapolis: University of Minnesota Press.

Kryshtanovskaya, Olga. 2005. *Anatomiia rossiiskoi elity*. Moscow: Zakharov.

Kryshtanovskaya, Olga, and Stephen White. 2005. Inside the Putin Court: A Research Note. *Europe-Asia Studies* 57: 1065–1075.

Martin, John Levi. 2011. *The Explanation of Social Action*. New York: Oxford University Press.

Pappe, Ia. Sh, V.E. Dement'iev, B.I. Makarenko, P. Venediktov, and A. Iu. Zudin. 1997. *Finansovo-promyshlennye gruppy i konglomeraty v ekonomike i politike sovremennoi Rossii*. Moscow: Tsentr Politicheskikh Tekhnologii.

Prokop, Jane. 1995. Industrial Conglomerates, Risk Spreading and the Transition in Russia. *Communist Economies and Economic Transformation* 7 (1): 35–50.

Shevtsova, Lilia. 2007. *Russia—Lost in Transition*. Washington, DC: Carnegie Endowment for International Peace.

Stark, David, and László Bruszt. 1998. *Postsocialiust Pathways*. New York: Cambridge University Press.

Starodubrovskaia, Irina. 1995. Financial-Industrial Groups: Illusions and Reality. *Communist Economies and Economic Transformation* 7 (1): 5–19.

Williamson, Oliver. 1985. *The Economic Institutions of Capitalism*. New York: Free Press.

Woodruff, David. 2000. Rules for Followers: Institutional Theory and the New Politics of Economic Backwardness in Russia. *Politics and Society* 28 (4): 437–482.

14

Structure in Bourdieu's Fields and Realities of Contemporary Russia

Aleksandr Shevelev

Fields and Russian Post-socialism

The situation in modern Russia at first glance does not allow excessive optimism or demoralizing pessimism. However, the structure of the economy does not provide hope for a clear way out of protracted stagnation.[1] International comparisons of state economic activity and regulation show unacceptably low quality.[2] Increasingly acute social contradictions are visible, and ambiguous processes taking place in Russia's spiritual sphere might cause alarm. The introduction of a general vision and rituals of institutionalization carried out by the state are not effective enough when belief in the inviolability of the state order is undermined, representatives of state authority disregard norms, and a gap grows between principles and deeds. Disorder of systems of norms and destructive processes in education confound attempts to create a shared national vision or national symbolic capital, without which it is impossible to consolidate goals for development.

A. Shevelev (✉)
Faculty of Economics, Department of Economic Theory,
St. Petersburg State University, St. Petersburg, Russia

J. K. Hass (ed.), *Re-Examining the History of the Russian Economy*,
https://doi.org/10.1007/978-3-319-75414-7_14

According to a survey conducted in 2017 by the All-Russia Public Opinion Research Center (VTsIOM), almost 60% of Russian youth aged 25–34 consider themselves cynics, almost 50% noted that their main quality is greed, and 52% that the main quality of Russian youth is aggression.[3] The growing scale and scope of the state apparatus, and redundancy of bureaucratic work in relation to productive strata of society, leads to underdevelopment and scarcity of symbolic capital and inflation and devaluation of social capital, and consequently the fragility of social ties and structures they mediate. Power in its many manifestations has become imitative, and in a situation when the crisis of controlling social processes might grow and low-level politicization seems inevitable, physical coercion increasingly becomes the most important form of capital.

Meanwhile, Russian society is increasingly divided by cultural and political preferences; it is even problematic to talk of "society" given colossal regional differences. Electoral mobilization is achieved through manipulation that provides imaginary consensus and fictitious choice. According to Rogov (2016: 92–93), authoritarian institutions use control of the media to distort or conceal opposing arguments, to manipulate the news agenda, and to increase distrust and activate insecurities, thus reducing the capacity to express varieties of positions and, as a result, distorting preference formation.[4] Propaganda is aimed in equal measure at mobilizing supporters and demobilizing opponents. Analyses by economists, sociologists, and political scientists reveal a significant "scatter" of perspectives and meaningful interpretations of social and economic reality that do little to clarify prospects for developing Russia's social, economic, and political institutions. To diagnose social processes, a more substantiated and convincing logic for understanding social reality is needed that can reflect its complexity and inconsistency.

If we follow the logic of Pierre Bourdieu, social space is formed and reproduced via objective structures of ordered relations determined by distribution of capital and mechanisms that ensure their reproduction and simultaneously by actors' dispositions in relation to reproduction. The latter generate systems of reproduction strategies: not simple implementation of established rules, but strategies of reproduction that determine the content and sequence of actors' actions. Social space is always relevant, and its configuration is formed by structures of social positions and relations, which manifest themselves as the interaction of forces.

Capital provides power to dispose of mechanisms of reproduction, products of activity, and material (and symbolic) gain. Capital itself is a force structure of the domination of some agents over others.

Different areas of social space form "power" fields (economy, polity, culture) subject to field-specific regularities and structures. Actors' interactions (individually or as groups) are determined by positions that differ in power or influence (Shmatko 2005). It is important to note that an analysis of social organization of fields should take into account not only social relations and built-in power hierarchies but also cognitive models and practical meanings actors share. This is in addition to regulatory norms and rules. It is impossible to understand how social institutions are created and used if one abstracts from only one of these elements or relies merely on an appeal to "interests." These elements mutually penetrate each other. Practices by economic actors are carried out through ensembles of social relations, are regulated by means of norms, and are imbued with common meanings via field location. Further, a key concept in Bourdieu's theory of fields is *habitus*, a system of stable dispositions or socially constructed cognitive abilities and motivating structures. *Habitus* is a product of history and produces practices, individual and collective, on which reproduction strategies are based. It provides schemes of perception, thinking, and action that contribute to consistent practices over time more effectively than do formal rules and norms (Bourdieu 2001). Thus, the economic and social world has a polymorphic structure formed by cognition as well as social relations and the distribution of capital (economic, cultural, social, and symbolic).[5] Economic capital is only one form of capital and is a special case of exchange, which has received the most attention in economics and other social sciences. However, these other forms of capital can be transformed into each other and serve as a manifestation of power and not merely as resources for exchange (Bourdieu 2014).

Features of Russian Fields and the Field of Power

Taking into account the provisions and conceptual apparatus of Bourdieu's field framework, one can conclude that the structuring of economic and social space in today's Russia is shaped to a decisive degree by the absolute predominance of the political field. Market fields resulting from post-Soviet transformations did not turn into autonomous spheres of competitive interaction of economic agents pursuing private interests. Rather, markets are subordinate to a hierarchical bureaucratic field of politics. Political and administrative capital as resources of domination, coercion, and regulation are successfully converted into economic capital. We can think of an "administrative market" where status is converted into money and back again, and through status one "buys" a position in the power hierarchy (Kordonskii 2012). The distribution of resources is to no small extent shaped not by economic efficiency, but by arbitrary priorities of the bureaucracy and state elites, and established in the course of informal approvals in a closed order. Legislative processes, interpretation and enforcement of laws, exclusive benefits, favorable state contracts, and episodic pressure on private businesses are associated with struggles between elite clans for resources and rents. Power, in short, has become the most profitable business. As A. Zaostrovtsev (2008: 30) noted, "The abundance of oil and gas that was reflected in the structure of the economy, exports, and revenue for the state budget, was able to create powerful incentives for behavior that does not need formal restrictions or rules of the game, but is aimed at capturing rents through the seizure of the state, which makes it possible to change rules along the way." At the same time, all else equal, the replacement of property rights by state arbitrariness reduces resource rents on the basis of which the entire system of political and social relations is built.

The term "corruption" is generally accepted to mean illegal redistribution of income, but in Russia this does not quite accurately or exhaustively reflect the essence of this phenomenon. Rather, it is a form of status rents (perhaps "class rents," cf. Kordonskii 2008) expressing the structure of positions and dispositions of social agents, formed not by market forces, but by the choices of dominant players of the political field. These choices, realized in informal agreements, shape the status hierarchy, priorities in the distribution of income and allocation of symbolic capital by imputing significance and exclusive roles to certain actors. Political administrators at the highest level control much of the volatile balance of social forces, clearly indicating their positions and preferences as to what is important for "national interest." Such institutional arrangements and agreements push banal corruption to the background, replacing it with a completely legal schema and transferring privilege to social groups or classes, such as top managers of the largest state companies and representatives of law enforcement agencies, thus limiting and regulating access to valuable economic resources. Relative and absolute reduction of budget expenditures for the social sphere (which are not considered investments in human capital), taking place alongside countless declarations of increasing average life expectancy and the need to develop the digital economy, demonstrates the real priorities of the state elite.

In the system of power-market relations of Russian capitalism, the rent-seeking state dominates, and business is built on the principle of patron-client relations. Revenues from the export of raw materials and other savings are not sufficiently invested in production and environmentally friendly technologies, or in infrastructure (with a high multiplier effect) and developing human capital. Not producers of value-added or innovators, but agents of the transaction sector—resource traders, administrative intermediaries, and financial players—run fields. In 2016, according to income level, financial industry employees were the leaders, and their average salary exceeded 78,300 rubles, while the average salary in Russia was 36,700 rubles according to Rosstat. Entrepreneurs lack incentives to invest in technical re-equipment, due to this political economy and because cheap labor costs do not correspond to qualifications. If those in the field of power have set the rules for their benefit, and if economic elites have learned to play this game for personal or class advantage, the rest of the country has suffered. By 2019, income for approximately 20.5 million people will be below subsistence level, which is 1.4 million more than in 2015. Russia has also seen a clear pause in investment, which Andrei Illarionov metaphorically calls a "strike by entrepreneurs." The profit of enterprises in 2015 compared to 2014 grew by 76%, while investment in the economy in the same year decreased by 9%. In 2016, profits increased by another 23%, while investments in the economy fell by another 3%.[6] The transition to a new technological order is blocked by inadequate attitudes to the humanitarian component of economic development. "Optimization" of strategically important spheres for sustainable development (health, education, science) is carried out while state spending on them has dropped in absolute terms. Thus, there is a consistent decrease in the share of costs for education and health care in the state budget: 4.9% and 4.1% in 2015, 4.5% and 3.6% in 2016, 3.3% and 2.2% in 2017. All this is fraught with degradation of human capital and growth of social tensions. There is insufficient funding for basic research, which is approximately 0.15% of GDP (compared with 0.4–0.6% of GDP in developed countries). The new agenda and ideology of development are not formulated, the state loses its orientation to social well-being, the poor become poorer, and the rich become richer. Russians' material well-being in a stagnant economy and sharply reduced real incomes has not yet reached its pre-crisis level.[7]

While we could examine such inequality using typical political economy, Bourdieu's field framework allows us to take a different angle to reasons for instability of economic growth in modern Russia. Not only are global cyclical fluctuations that affect Russia's underdiversified economy, or structural and institutional factors, of decisive importance. Macrosocial structural imbalances, and quantitative and qualitative discrepancies in the distribution of capital, are also at the root of unstable economic dynamics, prolonged stagnation, low rates of economic growth, and negative synergetic effects that, in turn, reduce effectiveness of mechanisms for repairing macro-economic "equilibrium" and under-utilization of industrial and human potential. For example, consider one of the negative synergetic effects arising in the interaction of economic and social capital (and at all levels of management). The predisposition to speculative games in the currency market, underinvestment in the real sector, and rent-oriented behavior express definite dispositions of economic agents and their inherent reproduction strategy, which is very far from "industrial behavior" but that fully falls within the definition of "status rivalry" (to use Thorstein Veblen's term)—a strategy of enrichment and over-consumption that hurts economic development. The low level of trust between economic agents (unformed social capital), the high level of uncertainty and risks in business prospects, and the lack of clearly defined developmental benchmarks make it difficult for stable, long-term contractual relations, further undermining long-term planning and investment.[8] A strategic vision has been lost and poorly thought-out policy (with decision-making behind closed doors) dominates. These factors add up to block institutional changes and freeze social structures, prolonging ineffective economic practices and contributing to institutionalization of opportunistic behavior by a narrow layer of beneficiaries.

For several decades, the scholarly literature has discussed the problem of relations between inflation and economic growth, and discussions about causes and institutional mechanisms of Russian inflation are not small, either. It is important to emphasize that inflation in Russia is generated not only by technological and sectoral structures or by depreciation of fixed assets and the high level of monopoly that generates cost inflation. No less important are structures and behavioral models that fix and "preserve" factors identified as institutional "traps." In Bourdieu's logic, for a long time the redistributive strategy of reproduction, conditioned by dispositions of key agents in fields of economy and bureaucracy and oriented to rent-seeking and enriching monopolists and transaction agents, has long dominated in Russia. Simultaneously, there was a swelling of the bureaucratic superstructure resulting in an administrative corruption "tax," and administrative transactions suppressed market transactions, resulting in the growth of transaction costs. The economy of the bureaucracy succeeds even in conditions of stagnation and stagflation, and maintaining sustainable economic growth rates is beyond the scope of Central Bank policy. Apparently, anti-inflationary policy will remain its imitation without a clear target for achieving sustainable growth rates, which is impossible without the transformation of social structures and behavior patterns in the economy—a socio-economic renovation that would create a competitive environment and overcome rent-seeking. For this to happen, more than market competition is needed—the system of dispositions and positions of key actors in the economy must change.

Fields, the State, and the Nature of Power

The realization that Russia's economy is burdened by such anomalies and pathologies is necessary for turning to a full-fledged study of Russia's realities. Such anomalies and pathologies are connected not only with the "unfinished market economy" but with the intrinsic nature of transactions (including market transactions) in Russia's concrete network of power relations. Power is an inalienable attribute of all social relations (spheres of economy, politics, culture, and in everyday life). Economic and political forms of power characteristic of Russia hinder the establishment of effective mechanisms to ensure identification and dynamic balance of various interests. Society's interests have been replaced by narrow groups and hurt by redistribution of wealth in favor of those in power via extractive institutions. In this regard, an institutional analysis cannot ignore one key process of the 1990s and 2000s: the reorganization of the Russian state elite and subsequent creation of large personal fortunes with the help of resources of power. The bourgeoisie (in its broad sense) appears as a class that uses capitalism for enrichment. A "bureaucratic bourgeoisie" is inextricably linked with partial replacement of its public functions (as personification of the public sector) with hidden, semi-shadow, informal activity aimed at appropriating administrative and political rents. Parasitism on market and non-market relations is diverse and results in huge gaps in levels and quality of consumption of different social groups, as well as price exploitation of end users whose welfare in the system of corruption and kickbacks decreases. Note that the bourgeois nature of Russia's state elite is conditioned not by traditional ownership of the means of production, but by the specific role this management class plays in organizing various transactions. This makes it a "transactional-managerial bourgeoisie" that extracts rent from owning a state organization and an administrative resource. In the private sector, elites use patronage for protection from competitors and law enforcement bodies, while providing representatives of bureaucratic bourgeoisie with a constant flow of corruption income (Akinin and Shevelev 2012). In contemporary Russia, the tendency of the ruling class to waste leadership and renewal potential has prevailed, and this is directly related to deformations and partial degradation of its social functions and the inability to provide expanded reproduction of quality public goods and their equitable distribution. As a result, favorable conditions are not created for full reproduction of national human capital, which is key to any modern economy.

Structural distortions in the economy and the inability to mobilize scarce resources and productive businesses to realize development goals are due not only to the distribution of national income and wealth but also to behavioral characteristics and dispositions of the ruling elite. Practices of deferred gratification were once at the heart of economically rational behavior, and the process of accumulation was associated with tabooing consumer desires, to which Protestant austerity once contributed (Panarin 2003a: 169) The lack of self-restraint by state elites, manifested in the growth of officials' wages in comparison with other groups, is especially visible for top managers of state corporations. This narrow layer of the elite has accumulated an impressive amount of wealth—and spent that wealth in conspicuous consumption of expensive yachts or foreign sports clubs (and foreign citizenship). At the same time, growing inequality, underdeveloped credit mechanisms that could stimulate import substitution and investment (and the gap between savings and investment amounted to nearly 7% of GDP), low trust between business and government—all these are links in one chain that make up the complex pathology of the socio-economic system. Its basic characteristics are the transformation of power into dominant varieties of business and the prevalence of using the country's resources for personal enrichment rather than investment in expanded reproduction and innovation. "By the end of the 2000s, the main basic principle of the new Russian reality was formed: free conversion of power into money and property, and back. The elite has become consolidated and united. This is the elite of the authorities, perceiving their activities not as a service to society, but as a kind of business" (Inozemtsev 2013: 593). As a result, 71% of all personal assets of Russians belong to 1% of the population. These contradictions can be resolved only through a complex process of organizing collective action, and a synergy of civic initiatives, entrepreneurial activity, and state participation. However, results of various surveys consistently indicate that Russia consistently ranks low in almost all parameters of entrepreneurial activity.

At the same time, one result of the system of power in Russia is largely a stalemate, the characteristics of which are the inability to create and integrate drivers of systemic development and the imperative of a transition to a new technological order as the fourth industrial revolution unfolds, amorphousness of the bourgeoisie as a class and dominance of an ineffective state bourgeoisie oriented to preserving the *status quo* and that does not want to give up privilege of rents in exchange for loyalty or to renounce benefiting from corruption, the absence of conditions and incentives for a circulation of the elite because of the low competitiveness of the economic and political systems, and incompleteness of structuring society. All this has resulted in a suspended state between a decorative democratic system and authoritarianism, the lack of creative goal-setting, and the inability of the ruling class to implement policies and promises.

In practice, there is no movement to create a more modern and flexible party system (that would create a real civil society). Attempts to build a corporatist state have foundered because situational management cannot replace strategic vision, just as derivatives of social movements are powerless to compensate for the lack of full-fledged institutions and a structured and competitive political system.

An optimal situation would have an "open access" social order, characterized by: competitive economic and political systems; innovation rents rather than monopoly rents; universal rather than exclusive property rights, with legal protections; and open access to organization (self-organization) of public initiative. The last aspect is crucial. Consolidation of social and political forces, minimization of risks associated with civil initiatives and movements, and controlled deployment of democratic principles cannot be achieved through administrative restrictions and arbitrariness of power or the dismantling of competitive mechanisms in the political sphere. One cannot disagree with the assertion that Russia is a vivid example of a "natural state" (social order of limited access) and in recent years has undergone involution. "This happened as a result of building a more rigid power vertical, establishing control over the media, weakening institutions of civil society ... The entire system is becoming more fragile and prone to external shocks and self-destruction. For the elite and groups of elites—in this case, those who rent-seek—there is no guarantee that the institution of personal power will remain in the long term and have any permanence, which may present serious dangers in the near future ... The long-term outlook in the country for many reasons does not create reliable commitments" (Raskov 2011: 28–29). According to some experts, the period 2014–2016 became the next stage in the consolidation of an authoritarian regime in Russia, or "reverse transit from forms of competitive, soft authoritarianism, characteristic of the political system of the second half of the 2000s, to a more rigid form that relies not on co-optation mechanisms but on repressive strategies and expansion of the state's zone of control" (Rogov 2016: 51).

The growth of social inequality in the United States over the past decades, what Paul Krugman called the "Great Stratification," was largely a result of deregulation and the growth of the financial and speculative sector, but it was based on a dynamic economy with great potential.

As Acemoglu (2009: 11) noted, the profit motive can be an engine of innovation and growth, but without institutional regulators, the profit motive can degenerate into rent-seeking behavior, corruption, and crime. This is happening in modern Russia as a consequence of the dominant social structure and corresponding extractive institutions created for redistributing income and wealth in favor of one part of society at the expense of others. At the same time, it would be reckless to pedal the topic of "struggle" with excessive social and economic inequality in Russia, since overcoming its causes is of prime importance. In this regard, there is no substitute for a meaningful development strategy, clearly articulated priorities, new industrialization, quality and affordable education, elimination of decline in health, creating a full-fledged innovation environment to bring younger people into creative work and nurturing entrepreneurial talents. These are obvious goals, but for now they are unachievable tasks. No less important is the formation of institutional mechanisms to minimize or block monopolies, unfair competition, and clan capitalism, and to create that full-fledged competitive environment absent in Russia. The technological revolution (robotization, internet, artificial intelligence) will inevitably transform the labor force, shifting demand to highly skilled workers and depriving others of employment opportunities. Without an active employment policy, all this is fraught with increased inequality and social consequences. "Precisely because illnesses are not treated with the elimination of symptoms, the reduction of income differentiation cannot be an end in itself: the efforts of society should be focused on solving underlying problems that may be behind it and generate it" (Kapeliushnikov 2017: 135). Not as self-evident as the relationship between growth and inequality is the subjective perception of "justice." People attach greater importance not to the scale of inequality, but to its sources. In this regard, it is important to rank factors that determine Russia's economic inequality, to understand which of these can be managed, to create an institutional environment to ensure the growth of incomes from labor and entrepreneurship rather than from speculation and rent-seeking. The urgent task is to overcome excessive wage inequality driven by the state that has emerged recently but not brought improved quality and effectiveness of bureaucratic management. The highest growth rates of nominal wages, even under stagnation, are in spheres of financial activity and real estate; the lowest are in education and public health. Average wage figures mask unacceptable gaps between managers and producers of social services, and targets are achieved by reducing employment in social sectors and various manipulations with reporting.

Ways Forward

A constructive solution could be systemic modernization of the state: for example, a rationalized and professional bureaucracy, division between business and power, a competitive political process, separation of powers, mechanisms for effective feedback from society, promotion of civic initiatives, and support for socially oriented non-profit organizations. In the pre-crisis years that were favorable to structural reforms (ultimately not implemented), the ratio of expenditures in the federal budget between traditional functions of government and those more modern, connected with investment in human capital, was 3:1, versus a global norm of 1:3 (Grinberg 2010: Chap. 5). An upgrade of basic foundations to transform the state and create new political drivers of development is sorely needed. The most preferred option is managed evolution, excluding unacceptable risk and devastating consequences of political cataclysms. Yet mere imitation of reforms that preserves the status quo ("conservative modernization") leads to pathological situations and lost legitimacy and does not help avoid eventual systemic crises. To remake economic fields and structures of social positions, a critical mass of two sub-elite groups is fundamentally important: a modernizing sub-elite capable of developing and leading industrial modernization and a creative sub-elite for gradual cultivation of the postindustrial sector. If the first sub-elite can shape thinking in the upper reaches of the state, the second will wield an innovative consciousness and entrepreneurial predisposition to risk and success, the ability to produce new ideas and technologies outside the framework of state paternalism. We cannot do without a new political sub-elite, for which state service will prevail over "feeding" that currently exists. It is necessary to combine strategies of new industrialization and postindustrial modernization while simultaneously building a strong state with clearly defined powers, as well as tasks of socio-economic policy.

Turning to complex, contradictory, and ambiguously interpreted processes in the sphere of national culture, I argue that the field of culture in Russia has consistently degenerated into game interactions of imitations and simulations, creating a loss of authenticity and self-worth. Sociocultural disintegration was expressed as "erosion" of the ethos of the nation or, as Gustav von Schmoller defined it, a common spiritual and moral consciousness crystallized in custom and law. Over years of patterned manipulation, elementary respect for national history has ceased to be a cultural norm—and this against the backdrop of degradation of national cultural capital. Not only economic fields but also cultural fields in Russia are permeated with relations of power, asymmetric interactions of dominant subjects and subordinate agents often involving using markets and peculiar combinations of interests. Winners are received by all parties, and those dominated are forced to agree to minimize missed opportunities and not to maximize benefits (Oleinik 2011). The maximizer-beneficiary is the dominant entity, which is often a symbiosis of a parasitic state bourgeoisie and loyal big business. The transformation of the nation's cultural capital into a combination of economic assets, whose use is inextricably linked with commercialization, privatization, and corruption, became a reality in current power-market relations. In accordance with Habermas' (2005) concept of communicative action, the genesis of many social pathologies is associated with the intrusion of the economic system (money) and the administrative system (state bureaucracy) into lifeworlds in which interactions are mediated by communicative rationality, oriented to mutual understanding, and providing a triune process of transferring cultural traditions, social integration, and socialization of youth (Furs 2000).

It is obvious that no long-term projects of modernization can be realized without the integrity of the national community, formed not only by economic and political factors but also by collective values. However, the intangible substance of the latter is constantly thinned, replaced by manipulative technologies and aggressive ersatz ideologies. This destructive process is reinforced by negative trends of recent years, such as the substitution of the humanitarian component in educational services with a functional or applied component designed to ensure reproduction of the "mass person." The phenomenon of the "posteconomic self-fulfilling person"

(Panarin 2003b), corresponding to the national tradition of education, turns into intellectual exotics. A "stratified" education has emerged, which not only does not guarantee equal opportunity to acquire symbolic capital and access to prestigious and "rent-intensive" professions and positions. In comparison with the Soviet educational system, this means rejecting a high average level of mass education; in combination with additional fees (e.g. the market of tutoring) that increase the chance to study at a prestigious university, this means accepting degradation and inequality in education. In fact, this is a new mechanism for social stratification and stagnation. It is difficult not to notice that a complex of institutional mechanisms has developed that blocks Russia's dynamic development and condemns it to prolonged stagnation (Shevelev 2015).

A paradox is that Russia has considerable cultural capital and capacity for its reproduction and renewal; has a cultural, historical, and social memory of the nation, in addition to tragic failures; and retains considerable motivating potential for development. At the same time, the quality of public administration, current economic policy, socially unjust distribution of national income, and blocked social mobility call into question the possibility of sustainable, dynamic development. Mistrust of the bureaucracy and alienation of a significant part of Russians from the state (and vice versa) give rise to a persistent rejection of ways of forming and imposing symbolic matrices on society. A common symbolic field cannot form under a policy of double standards and total domination of social hypocrisy. Subordination to authority arises from the alignment of cognitive structures of collective history (phylogeny) and individual history (ontogeny) embodied in bodies and objective structures of the world (Bourdieu 2015). The ineffectiveness of state regulations and spontaneous rejection of practices of imposing false worldviews on social agents express a general discrepancy between the dominant social order and cognitive structures that have arisen recently. This is by no means a failure in the ideological work of the state, but rather is about the impact of powerful information and communication processes that emerged and continuously transform the worldview of its citizens. In the context of globalization, the initiative and dominance in the formation of cognitive structures passed from the state, which by inertia remained adherent to excessively violent, protective, and prohibitive practices affecting its citizens, to other producers of meanings disseminated through social networks.

Network culture is not only poorly accounted for and used in current policy; its role is clearly underestimated, by theorists and practitioners, in the formation of modern institutional structures, power, and social spaces. In this regard, Castells (2016: 466–467) emphasized that we must detect a specific network configuration of actors, interests, and values that participate in creating power through the connection of their networks and mass communication to construct meanings in the public consciousness. Integrative power in a networked society forms in relatively stable (but also dynamic) structures of public consciousness, which are the deep sources of genesis and formation of institutions (gestalts). Institutions appear as the result of an interactive mental construction of intersecting social practices. In the networked society, a mutual intersection of fields of economy, power, and culture forms a complex symbolic universe (Shevelev 2016). At the same time, national cultural self-consciousness is formed not only by state policy or universities or Twitter but also by the creative reproduction of cultural memory and its translation from generation to generation through ordinary families and schools, by a thoughtful comprehension of history and humanitarian culture.[9] A one-sided bet on the import of institutions and cultural patterns, along with a commitment to symbols of cultural tradition and historical memory, the destructive reformation of education,[10] partial denationalization of the elite, and the fallout of the current government from national culture would not facilitate the formation of a universally valid field of culture.[11] Therefore, the long-term goal is the formation of an ethos of a new Russian nation, turning to the future rather than to the past, capable of providing creative synthesis of cultural and historical traditions and innovative potential of the newest subcultures, and in the remote historical perspective moving beyond mere market and state fields and institutions (perhaps away from *homo institutius*) to renew relations between atomized individuals.

Summing up, the nature of power, and behavioral characteristics of the ruling elite, explain part of the emergence of structural problems and pathologies in today's Russia. The rise of a bourgeois bureaucracy has bred structures in which there is currently no strategically oriented, consolidated, and influential "development coalition." Structural distortions characterize the existing system of power-market relations, for which subordination of the business community to the power vertical is characteristic. Rent-oriented behavior of economic and administrative elite groups predominates, and these two are intertwined and do not accept competition, but quite successfully create different kinds of economic, administrative, and political monopolies, redistributing income and wealth to their advantage. Negative effects result in the invasion of falsely understood criteria of economic and bureaucratic rationality into spheres that ensure the transfer of cultural tradition and the growth of national cultural capital. Key social institutions are under pressure of the state bureaucracy, which is given carte blanche for sequestering social expenses. It is important to note that there is no other acceptable way forward than full-fledged democratic institutions (albeit with inevitable Russian specifics), competitive political and economic systems, and changes in the prevailing model of power. The state makes important contributions to production and reproduction of tools for building social reality. As an organizing structure, the state constantly carries out practices that shape dispositions through various kinds of coercion, as well as mental and physical discipline (Bourdieu 2014). It is clear that mental and corporeal discipline in the legal field should be applicable to representatives of the ruling class. At the same time, without a fundamental transformation of the state and social order and the formation of new tools for building social reality, neither sustainable development nor maintenance of long-term social stability is possible.

Notes

1. GDP growth in Russia for 2009–2016 was 2.7%, during which the global economy grew by 31.6% and economies of energy exporting countries by 23.8% (Illarionov 2017).
2. According to data from The Worldwide Governance Indicators, for the period from 1996 to 2015, Russia moved into the 50th percentile two times. The quality of state regulations improved between 2000 and 2004 (from the 28th to the 50th percentile), but afterwards dropped in ratings.

By various indicators, Russia is inferior to anywhere from 52% to 87% of other states (Mzokov 2016).
3. Cf. https://news.mail.ru/society/30202977 (accessed October 1, 2017).
4. This is similar to market distortions: if the key to market efficiency is prices, then anything that distorts signaling distorts perceptions of the market and ensuing preferences and strategies.
5. Bourdieu had an outstanding gift for thinking in terms of complexity and for discovering new dimensions in social space. Theoretical deployment and substantive content of his concepts (and their application to Russia's realities) requires a transdisciplinary approach and convergent methodology (hybridization of discourses) not yet developed in social science (cf. Shevelev 2017).
6. "Golodets: Zarplaty v RF ne sootvetstvuiut urovniu kvalifikatsii rabotnikov," *Gen eralnyi director*, 2017 (accessed October 1, 2017 at http://www.gd.ru/news/7070-qqn-17-m2-27-02-2017-golodets-zarplaty-v-rf-ne-sootvetstvuyut-urovnyu-kvalifikatsii-rabotnikov).
7. More than 80% of Russians consider one "poor" if he does not have enough means for food and clothing. In May 2017, according to a VTsIOM study, the "poor" might account for 39% of the population (with 54% of pensioners in poverty). In one survey, 10% of respondents had difficulty obtaining food, and 29% admitted barely having enough money for clothing. The percentage of respondents who can buy food and clothing, but who have difficulty buying furniture or large household appliances, was 40% in a similar May 2016 survey and 41% in May 2017. VTsIOM press release, June 29, 2017 (https://wciom.ru/index.php?id=236&uid=116289).
8. Between September 1998 and July 2012, the investment in the Russian economy grew annually by on average. In contrast, that average from July 2012 to April 2017 was −4.3% (Illarionov 2017).
9. "Power is the ability to subordinate people to one's will, despite their resistance. No matter how politics opposed those forces that come from science, religion, art, language, or the family, it often turns out to be helpless before these sovereign structures, although they do not have at their disposal such tools as the army, the police, prison, administration of all levels" (Epshtein 2016: 314–342).
10. In July 2017, another large-scale reform of education was announced: 44,000 schools not subordinate to the Ministry of Education and Science will be transferred from municipal to regional authorities. It is expected

the state will build public schools that, according to Minister of Education and Science O. Vasilieva, are now "outside state concern." The future will show what changes the restoration of the bureaucratic vertical will bring (https://news.mail.ru/politics/30300783/).

11. "Culture determines the reason for any state to exist. Any nation. Any people. This, it would seem, is more or less obvious, but not for our authorities … Culture is a thing extremely complex and delicate, you do not have to interfere with it, but you have to help it." So said author Daniil Granin (https://www.novayagazeta.ru/articles/2017/07/05/73021-daniil-granin-hamskoe-otnoshenie-k-kulture-prodolzhaetsya).

Works Cited

Acemoglu, Daron. 2009. Krizis 2008 goda: strukturnye uroki dlia ekonomiki. *Terra Economicus* 7 (2): 9–17.

Akinin, A., and A. Shevelev. 2012. K voprosu o prirode gosudarstva i praviashchego klassa v sovremennoi Rossii: otvet 'Ekspertu. *Filosofiia khoziaistva* 6: 61–77.

Bourdieu, Pierre. 2001. *Prakticheskii smysl.* St. Petersburg: Aletheia.

———. 2014. Formy kapitala. In *Klassika novoi ekonomicheskoi sotsiologii*, ed. V.V. Radaev and F.B. Iudin, 293–315. Moscow: Higher School of Economics.

———. 2015. Dukh gosudarstva: genezis i struktura biurokraticheskogo polia. In *Sotsiologiia sotsialnogo prostranstsva*, ed. N.A. Shmatko, 243–245. St. Petersburg: Aletheia.

Castells, Manuel. 2016. *Vlast' kommunikatsii.* Moscow: Higher School of Economics.

Epshtein, M. 2016. *Poeziia i sverkhpoeziia: O mnogoobrazii tvorcheskikh mirov.* St. Petersburg: Azbuka-Attikus.

Furs, V. 2000. *Filosofiia nezavershennogo moderna Iurgena Khabermasa.* Minsk: Ekonompres.

Grinberg, R.S. 2010. *Strategicheskie orientiry ekonomicheskogo razvitiia Rossii.* St. Petersburg: Aleteiia.

Habermas, Jurgen. 2005. *Politicheskie raboty.* Moscow: Praksis.

Illarionov, A. 2017. Lovushka nesvobody. *Vedomosti* (July 3).

Inozemtsev, V. 2013. *Poteriannoe desiatiletie.* Moscow: Moskovskaia shkola politicheskikh issledovanii.

Kapeliushnikov, R. 2017. Neravenstvo: kak ne primitivizirovat' problem. *Voprosy ekonomiki* 4: 117–139.

Kordonskii, S. 2008. *Soslovnaia postsovietskoi Rossii*. Moscow: Institut fonda "Obshchestvennoe mnenie".

———. 2012. Kak ustroena Rossiia. *Russkii reporter*, 8 (February 29): 237.

Mzokov, A. 2016. *Sushchestvuiushchie sposoby otsenki kachestva gosudarstvennogo upravleniia*. WGI и BTI//«Science. Thought: electronic periodic journal»—scientific e-journal #12: 127–128.

Oleinik, A. 2011. *Vlast' i rynok: sistema sotsialno-ekonomicheskogo gospodstva v Rossii "nulevykh" godov*. Moscow: Rospen.

Panarin, A. 2003a. *Iskushenie globalizmom*. Moscow: Eksmo.

———. 2003b. *Pravoslavnaia tsivilizatsiia v globalnom mire*. Moscow: Eksmo.

Raskov, D. 2011. Institutionalnye issledovaniia kak budushchee sotsialnykh nauk. In *Nasilie i sotsialnye poriadki. Kontseptsualnye ramki dlia interpretatsii pismennoi istorii*, ed. Douglass North, John Joseph Wallis, and Barry Weingast, 9–31. Moscow: Gaidar Institute.

Rogov, Kirill. 2016. Politekonomiia reaktsii: ekonomicheskaia dinamika i sotsialnye nastroeniia v 2014–2016 gg. In *Politicheskoe razvitie Rossii 2014–2016: instituty i praktiki avtoritarnoi konsolidatsii*, ed. Kirill Rogov, 76–98. Moscow: Fond "Liberalnaia missiia".

Shevelev, A. 2015. Evraziiskaia politicheskaia ekonomiia v kontekste tsivilizatsionnogo i formatsionnogo podkhodov. *Problemy sovremennoi ekonomiki* 3: 95–98.

———. 2016. Sotsialnye struktury rynkov i setei: novye institutsionalnye podkhody. *Problemy sovremennoi ekonomiki* 2: 64–65.

———. 2017. Institutsionalnyi analiz v epokhu chetvertoi promyshlennoi revoliutsii: stanovlenie konvergentnoi metodologii. *Problemy sovremennoi ekonomiki* 2: 40–43.

Shmatko, N.A. 2005. Sotsialnoe prostranstvo Piera Burde. In *Sotsialnoe prostranstvo: polia i praktika*, ed. N.A. Shmatko, 554–567. St. Petersburg: Aleteiia.

Zaostrovtsev, A. 2008. Neft, pogonia za rentoi i prava sobstvennosti (obzor kontseptsii). In *"Resursnoe prokliatie": Neft, gaz, modernizatsia obshchestva*, ed. N.A. Dobronravin and O.L. Marganiia, 3–30. St. Petersburg: Ekonomicheskaia shkola.

Index[1]

[1] Note: Page numbers followed by 'n' refer to notes.

J. K. Hass (ed.), *Re-Examining the History of the Russian Economy*,
https://doi.org/10.1007/978-3-319-75414-7